SEEKING TRUTH

The Honourable Roger North Esqr.
Ætatis cir. 30
P. Lely pinx 1680
Geo. Virtue sculp 1740

SEEKING TRUTH

Roger North's Notes on Newton and Correspondence with Samuel Clarke c.1704–1713

Jamie C. Kassler

Routledge
Taylor & Francis Group

LONDON AND NEW YORK

First published in paperback 2024

First published 2014 by Ashgate Publishing

Published 2016 by Routledge
4 Park Square, Milton Park, Abingdon, Oxon OX14 4RN

and by Routledge
605 Third Avenue, New York, NY 10158

Routledge is an imprint of the Taylor & Francis Group, an informa business

British Library Cataloguing in Publication Data

A catalogue record for this book is available from the British Library.

The Library of Congress has cataloged the printed edition as follows:

Kassler, Jamie Croy.
  Seeking Truth: Roger North's Notes on Newton and Correspondence with
  Samuel Clarke c.1704–1713 / by Jamie C. Kassler
  pages cm
  Includes bibliographical references and index.
  1. North, Roger, 1651–1734 – Philosophy. 2. Newton, Isaac, 1642–1727 –
  Influence. 3. North, Roger, 1651–1734 – Correspondence. 4. Clarke, Samuel,
  1675–1729 – Correspondence. I. Title.
  DA437.N87K38 2014
  941.06092–dc23   2014002736

ISBN: 978-1-4094-4921-8 (hbk)
ISBN: 978-1-03-292073-3 (pbk)
ISBN: 978-1-315-60819-8 (ebk)

DOI: 10.4324/9781315608198

CONTENTS

ILLUSTRATIONS

ABBREVIATIONS

In citing manuscripts, abbreviations are used for their repositories, the country being given first and then the library, as follows

Australia (A)

Nla	National Library of Australia, Canberra, Australian Capital Territory

United Kingdom (UK)

Ccc	Corpus Christi College, Cambridge
Iro	Ipswich Record Office, Ipswich
Lbl	British Library, London
Mch	Chetham's Library, Manchester
Nr	Private collection, Rougham, Norfolk
Nro	Norfolk Record Office, Norwich
Ob	Bodleian Library, Oxford

One manuscript consists of a catalogue of books mostly from Roger North's own library. This catalogue, which is described *infra* References (section 1), is cited by the following short title: RN Books (1).

With the exception of the abbreviation for the *Oxford English Dictionary* (OED), a number of other published works also have been assigned abbreviations for (1) the *North Papers* series; (2) recent editions of North's biographical writings; and (3) books frequently cited. See *infra* References (section 2), where the abbreviation is given prior to the full title. Note also that all footnote citations to North himself are given as RN.

ACKNOWLEDGMENTS

Shortly after completing his edition of North's *Notes of Me* and with little time left to live, Peter Millard sent me his copy of the first edition of William Whiston's *Historical Memoires of Samuel Clarke*. This generous gift was received as a wonderful memento from a fellow labourer in the field of North studies; but I did not then understand why he chose a topic so far removed from my scholarly interests at that time. On hindsight, however, I now believe that this was his way of planting a seed that lay dormant for some years until it began to germinate and then, in the following study, to bear fruit. I hope he would have been pleased.

Mordechai Feingold not only agreed to read a draft of the study, but also provided valuable information, comments and continuing encouragement when it seemed completion of the final version would not be possible. Alan E. Shapiro and John Gascoigne also read and commented on some parts of the study relating to their particular areas of expertise. In addition, David Fairservice translated North's (sometimes faulty) Latin phrases; Douglas Leedy provided information about the meaning and source of one of North's Greek terms; Graham Pont transcribed one of North's letters in the British Library; and Mary Chan supplied details regarding a manuscript of North's still in private hands.

Jane Muskett, archivist of Chetham's Library, not only expedited my request for copies of Clarke's letters (discovered only after Chapter 3 had been drafted), but also granted permission to reproduce as Figure 1 the scheme in one of his letters. As in the past, my husband, Michael, though busy with his own projects, assisted in solving computer-related problems, including those that arise from producing camera-ready copy; but he retained his wonderful sense of humour even when more recent events necessitated his taking on increased household responsibilities. I am grateful also to Rachel Lynch, who has been my editor over the last decade, and her colleagues, Kirsten Weissenberg, who looked after the book's production, and Jenny Hayes, who copy-edited its text. Each of them has ensured that the final process has been mostly trouble free.

Northbridge, N.S.W.
February 2014

SOME PRELIMINARIES

OUTLINE OF THE NARRATIVE

The story told in this book dates from the early 1690s, when Roger North was preparing to remove from London to Rougham, Norfolk. Although he had a number of reasons for this move, he may not have fully grasped how country living would deprive him of a milieu that could share his intellectual interests.[1] This is not to say that he had nothing in common with the Norfolk gentry, for he, like some of them,[2] attempted to implement improvements in agriculture that were to make Norfolk one of the most progressive counties in England. But in his writings on estate management, he hints that the gentry indulged in certain ways of living with which he had little affinity.[3]

From these hints it may be inferred that he was referring to country pastimes then practised, such as those briefly described by another Norfolk resident, Nicholas Le Strange, fourth baronet.[4] In a

---

[1] See Kassler, *HRN*, Ch. 1, pp. 17–75.

[2] E.g., the Coke and Walpole families; see Wilson, *England's Apprenticeship*, pp. 246–9. For RN's involvement with the latter family, see Korsten, *Roger North*, pp. 22, 266 (nn. 220, 221), and Appendix II (Correspondence), pp. 238–9, 240, 250.

[3] See Kassler, *HRN*, pp. 383, 387.

[4] The surname is variously spelled. Sir Nicholas, whose Norfolk seat was Hunstanton Hall and farm, was a godparent of RN's second son, Montagu. RN himself stood trustee to the marriage settlement of Thomas, the eldest surviving son of Sir Nicholas; and the latter's youngest son, Henry, 6th baronet, married RN's second daughter, Mary; see Korsten, *Roger North*, pp. 21–2 and 271 n.321; Cherry, 'Sir Nicholas L'Estrange', p. 319, and Kassler, *HRN*, p. xiv. The connections between the North and the Le Strange families date from RN's grandfather and father; see Ashbee, *The Harmonious Musick*.

short autobiographical memorandum of advice for his children,[5] he
stated:

> I think necessary here to observe that though wee alwayes kept
> up a fayr correspondence, and made and returned visitts as our
> neighbours did, we still avoyded goeing into the method used
> by them of supping at the same places they dined, which with
> cards after among the ladys and a glass of wine for the
> gentlemen, usually prolonged the visitts till one or two,
> sometimes till four or five in the morning. Which custom
> being attended with so great irregularityes and inconveniences
> both to the family that entertained, as well as that which must
> avayle and expect the returne of their master and lady home,
> we could not comply with; and so it fell off in our mutuall
> visitts, and by degrees among the neighborhood also.[6]

Although he kept 'his hounds and sometimes hunted in my own
feilds when leisure best served', Sir Nicholas grew increasingly
occupied with 'country buisiness, haveing my great farm at Heacham
and another at Sedgford'. In any case, he never cared for keeping set
days for hunting, which not only brought in 'all sorts of company'
but also pandered to 'other peoples diversion more then your owne'.[7]

During his residence in London, North cultivated, and later
wrote about the diversions that gave him most pleasure;[8] but he
especially valued the search for truth,[9] which for him meant
knowledge of nature, including human nature. Since neither his
neighbours nor his wife Mary, whom he married on 26 May 1697,[10]

---

[5] About this memorandum, see Cherry, 'Sir Nicholas L'Estrange', p. 320; for an
edition of it, see pp. 320–8.

[6] Ibid., pp. 326–7.

[7] Ibid., p. 327.

[8] For some of these, see RN, *Notes of Me*, pp. 99–160; see also Kassler, *HRN*,
pp. 21–4, 49–55.

[9] Regarding this search, see ibid., pp. 99–110.

[10] See ibid., p. 32. According to Gayer, *Memoirs*, pp. 27–8, she was a daughter of
Sir Robert Gayer by his second wife, Lady Elizabeth Annesley (daughter of James,

shared this passion, it is not surprising that North's writings contain a number of remarks expressing his feelings of intellectual isolation. In one text, for example, he writes that although he 'long had an intent to forme a body of natural philosofy', he discovered such a systematic design was 'too operose[11] for me, being in the circumstance of a retiredment from all proper conversation, and assistance, and not from but into a world of affairs, such as belong to country management and a large family'.[12]

To alleviate his isolation North found solace in reading ('study') leading out to a succession of conversations with himself in the form of 'second thoughts', that is to say, reflections on his experience of reading his own or others' writings. As a mirror of his own thinking, these reflections served as a means of self-criticism and reassessment and, hence, as a method of testing his thoughts. Indeed, this method of critical reflection, which was his constant practice, accounts for the problematic state of most of his manuscripts on the philosophy of nature, many of which were intended chiefly for his own eyes and rarely for the perusal of others.[13] But he also recognised that critical reflections were insufficient as tests of his thought, for, according to him, without contradiction or opposition 'we are captivated with the seeming justness of our thoughts, and prosecute consequences upon them into a mizmase[14] of error'.[15] Consequently, he supplemented his method

---

2d earl of Anglesey). Little is known about her except from her letters, for which see (UK:Lbl) Add MS 32501: ff. 55, 91, 93, 119, 213. Note that the first of these folios has a postscript by RN, but the whole letter has been misattributed to him in RN, 'Letters', p. 258, as it is in Korsten, *Roger North*, Appendix II, p. 234.

[11] I.e., too laborious.

[12] See f. 1v of (UK:Lbl) Add MS 32548: in Notebook 1, ff. 1–26v (undatable); for another example, see *infra* 1.1.2. (f. 3v).

[13] For some reasons why RN published so little in his lifetime, as well as for what he did publish, see Kassler, *HRN*, pp. 58–60, 363–84.

[14] I.e., mizmaze; see OED 1, obs.: labyrinth.

[15] RN, *Notes of Me*, p. 166. For details of RN's methods, see Kassler, *HRN*, pp. 55–60, *et passim*.

of critical reflection with a method of contradiction modelled on the adversarial method he had learned in London as a student and later practitioner of the law.

Although this book provides an illustration of North's use of these two tests, its underlying theme is broader, for it concerns his responses to new ways of philosophising that were introduced from the end of the seventeenth century. To develop this theme, the narrative is structured chronologically in five chapters. But the structure of each chapter allows some deviation from strict chronology, because the first section orients readers to the second section, in which an edition is presented of one or more manuscripts, whereas the remaining sections function as a commentary that moves the narrative forward. The only exception is Chapter 4, which has no editions, because it serves as an orientation to the editions in Chapter 5.

Chapter 1, which introduces readers to North himself, includes an edition of an autobiographical fragment intended as a preface to a collection of his thoughts in the form of essays, some of which had been written, others of which were in course of writing. Since this plan was interrupted by three subsequent events that form the core of the narrative that follows, the remaining sections focus on features of North's philosophy that are central to understanding his responses to these events.

The narrative proper begins with the first event, namely, North's reading of Newton's books on rational (i.e., quantitative) mechanics and on optics. The evidence for this consists of three short notes edited in Chapter 2, the last two of which were written shortly after publication in 1706 of Samuel Clarke's Latin translation of the book on optics. Although in these notes North raises a number of issues, two in particular recur afterwards in his critical reflections. One relates to Newton's use of abstract terms in philosophising; the other relates to a suspicion that Newton was initiating a trend to replace one presupposition as to the ideal of natural knowledge with a different presupposition and, hence, a different ideal of knowledge.

Nevertheless, at the conclusion of his last note, North expresses some uncertainty as to whether he had fully understood Newton's philosophical aims; and it may have been this uncertainty that led to the second event, namely, his 1706 exchange of letters with Clarke. I say 'may', because what now remains of their fragmentary and incomplete correspondence does not include any initiating letter. Nevertheless, from the editions in Chapter 3, there is sufficient evidence to suggest that Clarke, a dogmatic theologian, was well suited as an adversary with whom North could test his thoughts by the method of contradiction. As a consequence of these tests, he realised that Clarke was promoting Newton's dogmata, as well as his philosophical aims. And from subsequent critical reflections, North began to grasp some of the implications of these aims, which extended beyond the philosophy of nature to theology.

Since this latter topic was to be the focus of a further exchange of letters between North and Clarke, Chapter 4 provides a context for understanding this third event. The exchange itself started in 1712, when North tells us that he received a letter from Clarke requesting him to read his newly published book on the doctrine of the Trinity. Initially, he declined this request, because he disliked theological dogmatics. But after receiving a second request, this time through an intermediary, he felt compelled to comply. So he penned several drafts before sending his answer in 1713, after which a reply from Clarke and an answer from North terminated the correspondence.

From the 1713 correspondence and related material edited in Chapter 5, it is clear that North fully understood, first, that Clarke was introducing a new way of philosophising in theology and, second, that in so doing, he was seeking to replace one presupposition as to the nature of the Godhead with a different presupposit ion. At the time of his 1713 letters, however, North did not know that Clarke was making public a philosophy of the Godhead that Newton also held in private.[16] But whether this was still the case after 1713 remains to be discovered.

From this brief outline it will be clear that the narrative thread concerns North and his responses to the three events summarised above. These events not only challenged his own way of

---

[16] See, e.g., Wiles, *Archetypal Heresy*, pp. 92–3.

philosophising but also the values inherent in his philosophy of nature. Both, in turn, were guided by his sceptical epistemology that included a theory of information processing, which, in brief, is that perception works by inference from limited sensory processing. Neither Newton nor Clarke developed a philosophy of knowledge, although their writings contain a number of hints as to its nature. In Newton's case, for example, one hint was published in his 1704 book on optics, where he pointed out that within the human body there is a locality in the brain called the 'sensorium' to which are conveyed the images or visible 'Species' of external objects.[17]

If vision is a window by which humans passively (not inferentially) know reality, then the fundamental assumption underpinning such a theory of perception is the metaphysical doctrine of externalism[18]—that phenomena are the only objects of knowledge or the only realities. Since, for North, perception works as an inferential process, it is not surprising that he sometimes comments on instances in which this metaphysical doctrine was applied not only by Newton but also by 'Newtonians'.[19] But it seems that other issues (some related) posed a greater challenge to his values, at least during the period of my narrative. Nevertheless, as I hope to demonstrate, his responses to that challenge are those of a reasonable man, whose search for truth did not cease in 1713 but continued until nearly the end of his life.[20]

As a consequence, the manuscripts related to this search represent a process of thinking or what I have called his critical reflections. Most of these are either interspersed in revisions of

---

[17] See Newton, *Opticks*, pp. 136–7. In future I hope to clarify Newton's use of the term 'sensorium'.

[18] In modern terms, phenomenalism.

[19] See, e.g., *infra* 2.2.2: No. 1 (f. 180v) and 3.3.5. Note, however, that RN was unaware of this problem prior to reading Newton's book on optics; see, e.g., *infra* 2.2.1.

[20] See *infra* 5.5.2.

earlier writings[21] or expressed in exploratory drafts, incomplete essays and sets of fragments on what today we would describe as philosophical aspects of science.[22] In the 1720s, however, he began to put into more finished form his reflections on a much wider range of subjects. And shortly before his death, he entrusted some of these writings to his son, Montagu,[23] thus at last producing the collection of his thoughts first mooted in the autobiographical fragment edited in Chapter 1.[24]

EDITORIAL GUIDELINES

The guidelines, itemised below, which concern North's copy texts only,[25] are based on principles first worked out in previous editions of his writings on sound and music and afterwards adapted to suit editions of his writings on other subjects. These guidelines are applied not only to the copy texts but also to North's other manuscripts quoted in this study. Some of the copy texts edited here present special problems, especially when they are drafts, discards or notes made in haste. In these cases, the problems and their editorial solutions are discussed under 'Physical features' in the descriptions of the manuscripts that precede the editions in Chapters 1–3 and 5.

---

[21] See, e.g., *infra* 2.2.3. See also RN, *Cursory Notes of Musicke*, which was completed *c*.1703. But it was soon followed by the revisions and rewritings described in *NP 5*, which led to the three-part 'Musicall Recollections' (see also *NP 6*); and then after more revisions and rewritings, the process concluded in 1728 with a treatise in two parts: an unpublished theory of sound, and *The Musicall Grammarian 1728*.

[22] See, e.g., the miscellany of RN's texts, (UK:Lbl) Add MS 32545: ff. 1–346v, which includes explorations of hypotheses and experiments, as well as philosophical principles such as body, space, 'absolutes and relatives'.

[23] See Kassler, *HRN*, pp. 1–4.

[24] See *infra* 1.1.2. (f. 5v).

[25] For Clarke's copy texts, see *infra* 3.3.2.

In this place, however, I should like to draw attention to a consistent feature of North's style, namely, his use of conditional sentences—'if', 'then'—although he often omits 'then', which leads to problems in reading very long sentences. I have supplied this word, when missing, in square brackets, an editorial practice that may seem somewhat pedantic. But my decision to adopt this practice is based on the belief that conditionals are a sign of North's probabilistic way of philosophising and, hence, an important feature of his pattern of thought.

Foliation/pagination. At the appropriate places within the texts, the foliation of a manuscript is indicated in bold type enclosed in brackets. This foliation is used for all references to the editions not only in the foregoing study but also in footnotes to the edited texts.

Capitalisation, italicisation and paragraphing. All sentences begin with a capital letter regardless of North's practice. Capital letters mid-sentence, apart from proper nouns, are changed to lower-case. Italics are reserved for foreign words and expressions (e.g., '*instar omnium*') or for book titles, whereas roman is retained for 'Bible', 'Gospel' and 'Scripture'. North sometimes writes in long paragraphs that extend over a number of pages, especially in copy texts written hastily or as unpolished drafts. When sense requires it, therefore, paragraphs are added.

Spelling. Although North's spelling is often consistently irregular, his spelling is retained, including 'then' for 'than', 'porpose' for 'purpose', 'latter' for 'later'. When necessary for sound or sense, a letter is added in square brackets, e.g., thro[w], chang[e], strang[e], yo[u]ng. The use of apostrophe 's' for possession is standardised; but the use of apostrophe 't' for past tense or past participles is retained, e.g., publish't. When the apostrophe 't' is not used (e.g., profest, adapt), it is added only when sense requires it, e.g., 'past' becomes 'pas't'. The use of apostrophe 's' for possession also is standardised; but no additional 's' is added after the apostrophe in cases such as 'Cartes', 'Cartesius' or 'Descartes'.

North sometimes incorrectly spells foreign terms or phrases; but these spellings, which may include or exclude accents of various kinds, are retained in the edited text and corrected in the Glossary.

Contractions. Abbreviations still in common use are retained (e.g., Mr.); but 'Dr.' and 'St.' are retained only when used as a prefix to a name, otherwise they are expanded (e.g., 'the Dr.' or 'the St.' becomes 'the Doctor' or 'the Saint'). All other abbreviations and contractions are expanded: e.g., North's symbol 'ℋ' becomes 'pounds'; 'ye' becomes 'the'; 'qu' becomes '*quære*'; '8vo' becomes octavo; '&' becomes 'and' except in Latin phrases, where the abbreviation is retained); '&c.' becomes etc.; 'vizt.' becomes viz.; '1stly' becomes 'firstly'; and '2. points' becomes 'two points'.

Punctuation. North consistently uses an equal sign '=' for words which are broken at the end of a line in the manuscript; and in these cases, the word broken is treated as one word. For hyphenated words he uses an en-dash '–', and this is retained, even where modern usage might not add a hyphen. Sometimes, especially when writing cursorily, he omits a hyphen; and this is added without comment in cases where the prefix of a word is either 'anti' or 'non' (e.g., 'nonChristian' becomes 'non-Christian'). For lists, 'nay', 'viz.' or 'that is', a comma is also added without comment. Where punctuation is silently altered to conform with modern practice, the principal guide has been clarity of meaning, especially when breaking a long sentence at a colon or semi-colon; when replacing a colon with a full-stop, a semi-colon or a comma; or when removing commas that seem to function not as a grammatical mark but as a pause in North's thinking or writing. All other editorial punctuation is given in square brackets.

   As for quotation marks, North's use of these marks is highly irregular. For example, when quoting or paraphrasing a biblical text, he may begin with a double quotation mark but not end with it. In these cases, all quotations marks are dropped. Where he uses indirect speech but without quotation marks, I do not add such marks; and where he uses double quotation marks at the beginning and end of a long passage quoted from another text, these are retained. However, when instead of quotation marks he (more or less) consistently uses an em dash on one or both sides of the material quoted or paraphrased, I retain his symbol.

Errors. Obvious slips of the pen are emended without comment, including confusion of homophones such as there/their, to/too, of/off, 'falls/fals'. Sometimes North omits a word either inadvertently or because of hasty writing. When this occurs, an appropriate word is inserted and enclosed in square brackets, e.g., '[which]'. In cases where a letter or letters in words at the end of a sentence have been obscured by tight binding, I treat the words in the same way as errors; but if only the first letter remains, the rest of the word being obscured, the additional letter or letters are inserted in square brackets.

Footnotes and glosses. Footnotes are used for a number of purposes, including cross-references within a text or to other writings of North; interpretation of difficult or hastily written passages; comments on the manuscript itself, including North's comment (e.g., those few occasions when he adds a marginal note 'qu', i.e., *'quære'*); identification of persons, books, quotations and proverbial expressions in English.

Terms to be glossed are identified in the text by an asterisk and glossed in the Glossary. When a word glossed occurs several times in one folio, it is given an asterisk at its first occurrence but not at subsequent occurrences in that folio. When North uses the terms 'phenomenon', 'phænomenum' or 'phainomenum', I do not gloss the last two, which are his spellings, respectively, of the Latin and Greek terms for appearance.

Like many lawyers of the period, North has a tendency to interlard his texts with Latin phrases. But his Latin can be problematic, since he tends to reproduce phrases by writing from sound or by quoting from memory. Moreover, when a phrase is from a classical source, he rarely identifies his source. Because of these and other problems, I have greatly benefited from the assistance of the Latinist named in the Acknowledgments. For glossed foreign language terms in languages other than Latin, the relevant dictionaries were consulted.

For glossed terms in English, the principal source is *The Oxford English Dictionary* (OED). Particular attention has been given to terms or phrases that are not in current usage (i.e., they are archaic, obsolete or rare) or whose usage is different from that current or where North's usage is either earlier or nearly contemporary with the first use of the word or where his usage is in advance of the first use of a word. (The same attention is also given to quotations from his writings in the main text, where words are glossed in footnotes.) When words or phrases are not found in the OED or when the OED gives North as the only source, these are identified by the term, 'Northism', even though North's usage may reflect contemporary, colloquial or local practice.

A NOTE ON THE MANUSCRIPTS

The editions of North's manuscripts are referred to in footnotes by giving chapter and section in which they are edited, e.g., 2.2.2. When there are several editions in a chapter, these numbers are followed by the relevant number in the sequence plus the relevant folio in round brackets, e.g., No. 1 (f. 177), If the edition in question is a letter, further information is given in the citation, e.g., RN→[SC 1706]. However, the following narrative is also based on a number of North's unpublished manuscripts that are cited in footnotes to each chapter. Consequently, the information below concerns, first, the style in which these manuscripts are cited; second, the source of the copy texts for the editions; and, third, information concerning dating the manuscripts.

Regarding the first matter, it should be noted that North's manuscripts may or may not have a title. In the case of titled manuscripts, the title is given within single quotation marks, the first letter of the first word only is capitalised and the abbreviation '&' is expanded. When a manuscript has no overall title, it is cited without single quotation marks either as Untitled or as Mostly untitled, the latter indicating that there may be a title or running head at some point within the manuscript. One particular manuscript has a long overall title that includes the words: 'With a dissertation of the new and moderne (new) philosofye inserted'. Since this inserted dissertation is found in North's life of his brother John,[26] I have not cited the full title but only that part of it concerning the dissertation by omitting its first and last words as follows: '...a dissertation of the new and moderne (new) philosofye....'.

The title (in all the above variants) is then followed by the abbreviation used for the repository that holds a manuscript, e.g., '(UK:Lbl)',[27] the manuscript sigla, if there is one, the relevant foliation and then, inserted between round brackets, a date (see further below). In cases where North has entered information in one of a set of notebooks, only the folios of the specific entry in question are given, preceded by the words 'in Notebook 2'. In cases where the entry fills the entire notebook, the word 'in' is omitted, e.g., 'Notebook 2'.

The copy texts for the editions of North's writings are found in six British Library Additional Manuscripts: 32545, 32546, 32547, 32549, 32550 and 32551. Because the two last volumes are uniform in subject matter, I have described their contents in Chapter 5. But since the remaining volumes are not uniform in subject matter, I have given a brief description of them below as a way of drawing attention not only to the state of a large portion of the manuscripts but also to the problem of dating.

(UK:Lbl) Add MS 32545: ff. 1–346v. Most of the texts are exploratory drafts, incomplete essays or sets of fragments relating to philosophical aspects of physics. The exceptions are a discourse 'in series' relating to the barometer and meteorology; a text on sensory perception in which other topics are developed; and an

---

[26] See also *infra* Epilogue.

[27] See *supra* Abbreviations.

autobiographical fragment (edited in Chapter 1). None of these texts is dated.

(UK:Lbl) Add MS 32546: ff. 1–321v. In addition to the subject matter related to physics and its philosophy, there are texts on mechanics and sense perception; fragments relating to North's 1706 correspondence with Clarke (edited in Chapter 3); notes on reading Newton (edited in Chapter 2) and on harness-making; legal documents and letters; and sketches of St. Paul's Cathedral. Of the longer texts, some are drafts, whereas others are incomplete essays. Of the shorter texts, most are fragments. Dates appear only in the notes on harness-making, some letters and legal documents, the last of which includes a certificate dated 20 January 1730 and signed by Ambrose Pimlowe,[28] attesting that North's son Montagu had received the Sacrament of the Eucharist and, hence, confirmation in the Church of England.[29]

(UK:Lbl) Add MS 32547: ff. 1–430v. With the exception of two fragments—one on continuity and fluidity, the other relating to his 1706 correspondence with Clarke (edited in Chapter 3)—the majority of texts in this volume are devoted to mechanics, as North understands the term. Unfortunately, none of the contents are dated.

(UK:Lbl) Add MS 32549: ff. 1–121. This volume is made up of a set of five quarto notebooks, each with marbled covers, some of which have a label.[30] In Notebook 1 there are two incomplete moral essays, the remáinder being filled by entries relating to music.[31] Notebook 2 contains 'Solutiones' to various phenomena. These are carried over to Notebook 3, which concludes with a fragment of North's 1706 correspondence with Clarke (edited in Chapter 3) and some related material. The entries in Notebook 4 consist chiefly of

---

[28] Following his ordination (Ely 1708), Pimlowe served three parishes in Norfolk as vicar of Castle Acre (1709–50) and Rougham (1710–23), as well as rector of Dunham Magna (1721–50). From time to time he also assisted RN as an amanuensis.

[29] For RN as a non-juror, see Kassler, *HRN*, pp. 46–8.

[30] See Chan, 'Dating the Paper', pp. 90–91.

[31] For details, see *NP 6*, pp. 43–4, *et passim*, and *NP 2*, pp. 28–35.

critical reflections on aspects of theology and the philosophy of physics, whereas Notebook 5, which has the label, 'Experiments of natural things', also includes a small number of entries on mechanics.

Since notebooks consist of paper that has been bound prior to use, the paper itself cannot provide adequate clues for dating the contents,[32] a problem that is compounded when taking into account two features of the notebooks: one, North's use of them and, the other, his style of entry. Regarding the first, the notebooks record experiences, as well as thoughts of importance to him. Regarding the second, in recording such information, he often used the technique of commonplace entry.[33] As a consequence of these two features, many (but not all) entries are abbreviated versions from a previous, more complete text that may or may not be extant; and this creates a problem for determining the date in which an entry was made in any particular notebook, as well as the date when the original, if no longer extant, might have been written.

From this brief description of the sources of the copy texts, it should be clear that the lack of dates in so many of his manuscripts makes it difficult to trace developments in North's thought. In attempting to address the problem of dating, I have adopted principles established in the series of *North Papers* for assigning approximate dates to his writings on sound and music.[34] Because some of these writings were dated, it was possible to identify three periods—early, middle and late—and to establish approximate dates for each period. However, the principles were established only after the manuscripts in question had been transcribed and a digest of each produced that, for the purposes of dating, included (but was not restricted to) information concerning paper, binding and pagination; citations of names and/or titles (e.g., music, books) within a

---

[32] Only one other volume in the British Library (Additional Manuscript 32548) is constituted of a set of notebooks.

[33] Regarding this style of entry, see Kassler, *HRN*, pp. 60–6.

[34] See *infra* References.

manuscript; and North's references to his own work, written or projected.

When making subsequent transcriptions of North's unpublished writings, I have continued to follow the foregoing practice. However, as a result of my ongoing study of the manuscripts, modifications were required as more information came to light. In my previous book on North,[35] these modifications were as follows: early period (*c.*1690–*c.*1705), middle period (*c.*1706–*c.*1720) and late period (*c.*1721–*c.*1732). In addition, four other modifications were then introduced: (1) when evidence indicates a long overlap between periods, I add a question mark, e.g. '(middle period?)'; (2) when a manuscript has material inserted after the initial period of writing, I add 'mostly', e.g., '(mostly early period)'; (3) when there is insufficient evidence for dating or when more information needs to be sought, I use the phrase '(date uncertain)'; and (4) when there is no evidence for assigning a period, this is indicated by the word '(undatable)'.

In the course of this present study, the date 1706 turned out to be crucial for reasons that will appear later on. As a result, two further modifications were required as follows: manuscripts written between *c.*1705 and the first few years of the middle period are cited as '(early period, late)', whereas manuscripts written during the early years of the middle period are cited as '(middle period, early)'. As more of North's unpublished writings are studied, systematised and edited, it is probable that further modifications or refinements will be necessary until, eventually, it may be possible to determine more accurately the dates for the three periods.

---

[35] I.e., Kassler, *HRN*.

1

INTRODUCING ROGER NORTH

1.1. DISCOVERING NEW PHILOSOPHY

Between *c*.1693 and *c*.1698 North wrote most of his literary gem, *Notes of Me*,[1] an account of himself from childhood to about the age of forty.[2] Unfortunately, the manuscript is incomplete, possibly because it was work in progress or possibly because some parts were lost during the dispersal of his manuscripts.[3] Fortunately, North's lives of three of his brothers may be regarded as extensions of his autobiography, since he appears in each of them in a number of different roles.[4] These extensions also include writings that on first consideration do not seem autobiographical.[5] But neither *Notes of Me* nor the autobiographical extensions include details of the period after North removed from London to Rougham, Norfolk, a period that has

---

[1] According to Millard in RN, *Notes of Me*, pp. 57–9, 61–2 work on the autobiography was in hand after 23 November 1693 to sometime after 1698.

[2] See Kassler, *HRN*, pp. 77–89.

[3] See ibid., pp. 1–4, for information about the dispersal of RN's manuscripts.

[4] See ibid., pp. 69–70, for some of his roles, and pp. 4, 393–8, for the lives themselves. These were published posthumously in an 'edition' by RN's son Montagu, whose sequence was the reverse of what RN intended; see RN, *General Preface*, p. 82. As the basis of later re-issues, this edition has been frequently cited, though without any awareness that it is a patchwork of bits and pieces collected from RN's several different manuscript versions. This problem has been alleviated by the modern editors of RN, *Life/JN* and *Life/FN*; but *Life/DN* has still to find its modern editor.

[5] See Kassler, *HRN*, pp. 86, 89–90, *et passim*.

yet to be thoroughly investigated.[6] Nevertheless, a glimpse of his way of country living may be found here and there among his manuscripts, including an autobiographical fragment that takes the form of a preface.

This fragment, which is edited in the next section, begins with a brief sketch of North's early education and his year of study at Cambridge University between 1667 and 1668. Then follows a valuable account of some of his practices as a writer[7] and a short advertisement for his intended 'wares', namely, the collection of his thoughts to be presented in the form of philosophical essays. But the fragment is of additional value because, first, it includes some information that does not appear in his other life writings mentioned above; second, it hints at some of North's philosophical aims at the time of writing; and, third, it exemplifies, even as a draft, some elements of his literary style, which in his day would have been described as 'familiar',[8] with its use of the first person pronoun, idiomatic expressions,[9] quirky sense of humour and irony.

Since the autobiographical 'Preface' is incomplete, a few details require filling in here, two of which regard North's year of study at university, a year that was cut short by a 'fen-ague', so that he was sent home to recover and to study logic until the family's removal to London in 1669.[10] The first detail concerns North's student status, about which, in *Notes of Me*, he remarks: 'nothing

---

[6] See ibid. for some details of RN's life in Norfolk, see also Korsten, *Roger North*, who consulted some of RN's correspondence and provided a useful, though incomplete catalogue of it.

[7] See Kassler, *HRN*, pp. 49–75, *et passim*.

[8] (OED 6c, obs.) plain and, hence, easily understood.

[9] In the fragment, these expressions (perhaps Norfolk dialect) are 'em and durst. But RN's interest in dialects developed at the time he accompanied his brother Francis around the various Assize Circuits, two of which in 1677 included Norfolk; see RN, *Notes of Me*, p. 191, and Ashbee, *The Harmonious Musick*, p. 100. Regarding idiomatic expressions in general, see RN, 'Of Etimology', pp. 246–52, *et passim*.

[10] RN, *Notes of Me*, p. 94.

extraordinary happned to me, unless it were, that I was forct to live in the quality of a nobleman, with a very strait allowance'.[11] Since he was the youngest son of a baron, he had entered Jesus College, Cambridge, on 30 October 1667 as a gentleman commoner, so it might be thought that the North family was wealthy.[12] But this was not the case, for his grandfather had dissipated the family fortune; and despite years of careful management, his father 'never shook off financial problems' passed down to him.[13] And this situation accounts for North's statement that

> I did most extreamly envy the comon scollers for the joy they had at foot ball, and lament my owne condition, that was tyed up by quality from mixing with them, and enjoying the freedomes of rambling which they had. And not having either mony or assurance to mix with my equalls, who were wild and extravant enough, was oblidged to walk with grave seniors, and to know no other diversion.[14]

As a consequence, and 'also by reason of my caracter',[15] he was 'very recluse'.[16]

Second, because of his penury North resided in the same room as his brother John, who had chambers in Jesus College and who acted as his nominal tutor.[17] And it probably was in these chambers that North heard 'a sort of sly discours' about Descartes's philosophy, which he alludes to in the autobiographical fragment.

---

[11] Ibid., p. 90.

[12] This assumption was made by Westfall, *Never at Rest*, p. 337, when he described RN's brother John as 'Scion of a wealthy family'.

[13] Randall, *Gentle Flame*, p. 75.

[14] RN, *Notes of Me*, pp. 91–2. For an indication of the straitened circumstances of the North family, see RN, *Life/JN*, pp. 103–4; and for the family's 'frugality and sobriety', see the letter quoted in Korsten, *Roger North*, p. 259 n.106.

[15] Here, meaning his reserve.

[16] Ibid., p. 154.

[17] Ibid., pp. 91, 259; see also RN, *Life/JN*, p. 130.

Moreover, it is probable that such 'discours' occurred in conversations between John North and his friend, James Duport, who in 1668 became master of Magdalene College.[18] Although the latter has been stereotyped as an arch conservative, Mordechai Feingold's more nuanced portrait indicates clearly that Duport's irritation with the anti-Aristotelian, pro-Cartesian temper at the university was triggered by 'his perception of what he feared would be the consequence for religion and belles-lettres if such world views were allowed to take root'.[19]

And what those consequences might have been is suggested in *Notes of Me*, when North expounds further on the 'sly discours', namely, that the Cartesian 'New Philosofy was [regarded as] a sort of heresie, and my brother [John] cared not to encourage me much in it'. He then continues:

> But I found such a stirr about Descartes, some railing at him, and forbidding the reading him as if he had impugned the very gospell, and yet there was a generall inclination, especially of the brisk part of the university, to use him, which made me conclude there was somewhat extraordinary in him, which I

---

[18] John North entered Jesus College in 1661 as a fellow commoner; he was made fellow by royal mandate in 1664 and graduated B.A. (1664) and M.A. (1666). Duport's *Musæ Sibsecovæ, sue poetica stromata* (Cambridge, 1676) includes a 'flattering' epigram on his friend; see RN, 'Of Etimology', pp. 212–13, and RN, *Life/JN*, pp. 121–3. Note that Duport had been tutor to Isaac Barrow, so that he may have been an intermediary in the friendship that developed between John North and Barrow.

[19] See Feingold, 'Isaac Barrow', pp. 10–15, p. 13. For some other Cantabrigians who held that Cartesianism was a threat to religion, see *infra* 2.2.2: No. 3 (f. 186 n.138). Note that John North himself believed that Descartes, perhaps unintentionally, had been responsible for the rise of atheism at this time; see f. 176v in RN's copy of his brother's notebook, 'Notes of Dr. North' (UK:Lbl) Add MS 32514: ff. 167–227v (undatable). For some details of the original notebook, see Kassler, *HRN*, pp. 54–5.

was resolved to find out, and at length did so, wherein the nititur in vetitum,[20] had no small share....'[21]

From the context, North's term, 'New Philosofy', denotes the Cartesian way of philosophising about matter and motion, a way in which God's veracity and causal powers are required only to set the general background for any particular interactions among bodies.[22] And thus supernaturalist and teleological explanations of natural events were superseded by a way of philosophising that promised 'a comprehensive heuristic for the explanation of all corporeal phenomena', [23] for its central concepts were a theory of matter and its properties and a theory of motion and its laws, the latter of which were based on an analogy with the principles of simple machines such as the balance, lever, pulley, wedge and screw. Accordingly, synonyms for 'new' in new philosophy were 'mechanical' or 'corpuscularian', the latter word referring to parts of matter called 'corpuscles' and their contact action.[24] But it is quite misleading to add 'the' to a philosophy variously described as new, mechanical or corpuscular, since a number of different versions were developed not only contemporary with Descartes but also afterwards.

Of course, there also were critics of one or another version, including North himself. For after his admission to the Middle Temple on 21 April 1669, where initially he resided in chambers

---

[20] I.e., *nimitur in vetium semper cupimusque negata* (Lat., Ovid, *Am.* 3.4.17) we ever strive for what is forbidden, and ever covet what is denied.

[21] RN, *Notes of Me*, p. 92. According to Gascoigne, 'Isaac Barrow's Academic Milieu', p. 277, Cartesianism was particularly suspect to some Calvinist and some Anglican defenders of theological orthodoxy, so that in 1668 an edict of the vice-chancellor forbade 'sophists and B.A.s from keeping acts based on Descartes's work, and ... stipulated that Aristotle's writings should be used instead'.

[22] For the functional role of God in Descartes's philosophical physics, see Collins, *God in Modern Philosophy*, pp. 65–9; see also *infra* 1.1.4 and n.117.

[23] Anstey, *Robert Boyle*, p. 4.

[24] According to Anstey, ibid., p. 12 n.2, the term 'Mechanick Philosophy' was first used in 1659 by Henry More, whereas others, e.g., Shapiro, 'Experimental Philosophy', p. 207, have claimed that the term 'mechanical philosophy' was introduced in 1661 by Robert Boyle.

with his brother Francis, he began to learn some practices that would eventually develop into the two tests mentioned previously, namely, the methods of critical reflection and contradiction.[25] For example, during his seven-year probation, he was able to hone his adversarial method (contradiction) with other contending lawyer advocates, because he was required to participate in moots or mock actions and arguments upon legal points.[26] Moreover, in private he and Francis practised putting cases by debating not only legal issues but also philosophical ones, including, for example, the 'reason' of the barometer, gravity and harmony.[27]

As a consequence of this training and later experience in the common law, he sought to improve, not to overturn new philosophy, an approach that contributes to his eclecticism. But he also applied this approach to other domains of knowledge where he opposes doctrinaire change and advises amending cautiously on the basis of existing precedents. As the last part of this sentence indicates, North's approach, about which I will have more to say later on,[28] is indebted to his training in the common law; and while it provides him with a methodological constraint on innovation, he does have an innovative streak,[29] which also contributes to his eclecticism.

Finally, North's autobiographical fragment was written at a time when he had attained a relative tranquillity of spirit, whereas prior to his removal to Norfolk, he had experienced a number of

---

[25] See *supra* Some Preliminaries.

[26] Regarding his student days in the law, see Kassler, *HRN*, pp. 19–21.

[27] See RN, *Notes of Me*, pp. 159–60, 201; RN, *Life/FN*, pp. 261–2 (including a paper by a young RN, who does not identify himself as the author, since it was not 'worthy of memory' except as a context for his brother's criticisms); Kassler, *The Beginnings*, pp. 26–7; and Kassler, *HRN*, pp. 52–4.

[28] See *infra* 1.1.3.

[29] See, e.g., ff. 9–10 in RN, 'The Theory of Sounds....' (UK:Lbl) Add MS 32545: ff. 1–82v (middle period), which incorporates RN's understanding of Newton, *Principia mathematica*, Bk II, Prop. 47. For an assessment, see Chenette, *Music Theory*, pp. 106–9; for some other innovations in acoustics and music, see Kassler, 'Roger North'.

turning points or 'crises' (his term).[30] The major crisis occurred during the period when the three brothers who were his closest friends died—John in 1683, Francis in 1685 and Dudley in 1691. In 1686 he removed from his Middle Temple chambers to a house in Covent Garden;[31] in 1689 he retired from most of his public appointments; and 1692 he contemplated taking his own life.[32] As a lawyer he knew the legal prohibition, so he gave consideration only to the religious prohibition. And this consideration led to two conclusions, one general and one applied to his own case: first, that self murder is 'lawfull, and done in a right mind, without conscience of fault in so doing, is no sin';[33] and second, that

> I have this advantage [over those who prohibit suicide], that I have eas and repose in my mind, so as if ever I am overwhelmed in irremediable calamity, and paines, I may, if I find I cannot bear them, put a period to free my self. And in this thought I have great comfort, and have had and shall continue to have all my life long.[34]

As a consequence of this resolution, North was even more determined to implement plans already made to start a new life away from London. These plans had been laid in 1690, when he signed a title deed for Rougham Manor,[35] an ancient estate situated near the village of Rougham, Norfolk.[36] During 1693 he devoted time and

---

[30] RN, *Notes of Me*, pp. 94, 96 ('time of tryall'), 110, 161, 240.

[31] RN, who was one of the executors of the painter Peter Lely, leased the house from the Lely estate. This house, now occupied by the Tavistock Hotel, was located in the Great Piazza of Covent Garden.

[32] See Kassler, *HRN*, pp. 34–49, 80–86.

[33] RN, *Notes of Me*, p. 207.

[34] Ibid., p. 210.

[35] See Kassler, *HRN*, pp. 31–3.

[36] For a history of Rougham (sometimes spelled Ruhham, Rucham, or Ruffam), see the account of the Hundred of Launditch in [Armstrong], *History and Antiquities of the County of Norfolk*, vol. 8.

thought to the kinds of work required to improve the estate; and then for the next few years he commuted either by horse or by coach between London and Rougham to organise and oversee the improvements.[37] Then, in March 1696, except for occasional visits to London, he settled permanently in the newly renovated Rougham Hall, taking up his role as lord of the manor.[38] In May of that year he married; and not long afterwards the first of seven children was born. During this transitional period, which dates from *c*.1693 to *c*.1705, he wrote a number of long and short essays, many with the word 'Notes' in the title.[39] But, as will become apparent, sometime towards the end of the transitional period and before the end of 1713, a train of events would lead to a further crisis in North's life, this time an intellectual one.

1.2. A FRAGMENT OF AUTOBIOGRAPHY: DESCRIPTION AND EDITION OF THE MANUSCRIPT

(UK:Lbl) Add MS 32545: ff. 2–6 [f. 6v is blank].

Title. There is no title; but the running head is 'Pre' (i.e., Preface), which appears only on ff. 3–6.

Physical features. The volume containing the copy text is tightly bound, and this has affected some letters or words near the margin. Folio 1 is blank except for the title, recto, 'Notes of my Father', which is in the hand of North's son, Montagu. The text itself is incomplete both at the beginning, which is missing a leaf, and at

---

[37] For improvements to Rougham Hall, see RN, 'Cursory Notes of Building'. For improvements to the grounds, see RN, [notebook on planting, etc.] (UK:Lbl) Add MS 32505: ff. 1–24 (some notes are dated 1691–1709); see also Williamson, 'Roger North at Rougham'.

[38] RN's manor included a type of tenants called copyholders, whose lands, though substantially their own property, were nominally part of the his estate; hence, the term 'manor' meant the jurisdiction exercised by the lord, i.e., judge, in his court, the law of which was the custom of the manor. On the subject of court keeping, see RN, *Notes of Me*, pp. 173–5.

[39] For some of these, see *infra* 2.2.3.

the end, where the final punctuation is a comma. To mark the sequence of pages from f. 2v, North has used alphabetic letters, which I have not included in the edition. The sequence is as follows: B. (f. 2v), C. (f. 3), D. (f. 3v), E. (f. 4), F. (f. 4v), H [=G] (f. 5), H. (f. 5v), I. (f. 6). There are a number of emendations (additions and insertions) and slips of the pen, as well as very long paragraphs and, in one instance, a space left for the later insertion of text (f. 2v). In general, the text indicates a lack of attention to detail, probably resulting from cursory writing, which suggests that the fragment is a draft.

Dating. The paper, which has no visible watermarks, is quite distinctive, very thin and used by North only in octavo; hence, the name assigned to this paper, 'Thin 8vo'. However, the suggested date for this paper of '1710s' cannot be relied on, because no dated documents were found using it.[40]

Internal evidence is somewhat more helpful, since it provides three clues. One is North's reference to himself as 'grim sir in a country farm' (f. 3v), so that we may infer that he was writing the text in Rougham Hall, his Norfolk residence between 1696 and 1734. Second is his statement that he sits in his closet, where 'I ruminate my no[tion] of the generall perfidy but more inexpressible folly' of mankind (f. 3v). These ruminations resulted in a number of moral essays, most if not all of which were written prior to 1701.[41] And third is his declaration that 'I doe not pretend to compile an intire body of phisicks' (f. 5), which is reaffirmed in *Notes of Me*,[42] written sometime between the dates assigned to North's early period (*c*.1690–*c*.1705).[43] Tentatively, therefore, I would date the writing of this draft to the period between *c*.1697 and *c*.1700.

---

[40] Chan, 'Dating the Paper', pp. 64, 85.

[41] Kassler, *HRN*, pp. 67–8.

[42] RN, *Notes of Me*, p. 142.

[43] For problems in dating this manuscript, see Millard in RN, *Notes of Me*, pp. 57–60.

A FRAGMENT OF AUTOBIOGRAPHY
(UK:Lbl) Add MS 32545: ff. 2–6

[f. 2]

Preface

[...] Thus farr I thinck wee are sure of them, for few see a book without being kind readers of the titlepage, and a line or two of the preface. But how long wee shall so fairely jogg on together, or when part, is past my skill to devine. I know if any thing hold them [then] it is curiosity, and telling of tales. Therefore take notice I am about to give some account of my self, and my wares here.

My very good parents[44] made me learne Latine at a free scool in the country,[45] where I also buisied my self with squibbs, crackers, [*]starr-kites,[46] melting mettals, turning joynery, and such like exercises, as the severity of our orders afforded time, and the place opportunity. After being returned home, I conceived a designe of making an organ out of my owne invention, a senceless but daring project,[47] which with other like designes, were interrupted by an

---

[44] Dudley, 4th baron North (1667), married Anne Montagu in 1632. Of their nine children RN was the youngest; see Randall, *Gentle Flame*, pp. 11–27. For comments on his parents, see RN, *Notes of Me*, pp. 79–82; for what he learned from them, see Kassler, *HRN*, pp. 86–9.

[45] I.e., the free grammar school, Bury St. Edmunds in Suffolk. In 1661 RN's father became governor of the school, where at least three other of his male offspring were educated; see Hervey, *Biographical List of Boys educated at ... Free Grammar School*. RN was then removed to the free grammar school, Thetford; see RN, *Notes of Me*, pp. 86–8. Note that this sequence in his schooling corrects an error in Kassler, *HRN*, p. 18.

[46] For a vivid description of these, see RN, *Life/DN*, Appendix I, pp. 215–16.

[47] For RN's curiosity about organs and how they work, see Kassler, *HRN*, pp. 23–4, and Kassler, *The Beginnings*, pp. 113–17.

importune reverend freind[48] who would needs read Seatore's logick[49] to me. But a dispatch to the university put an end to all my delights. **[f. 2v]** I was to stay a year there, being thought enough to initiate my braines in order[,] as others doe[,] to sell their brilliant [*]escume. There I was preferred to Burgersdicius' logick[,][50] where I learnt [*sic*].[51]

And likewise had a touch at metaphisicks[,][52] where I found out that *ens est quod habet entitatem.*[53] I had a Furnier on the six books of Euclid,[54] which I read by my self, and wondred I could not find out why such stuff was so celebrated and recomended. He that was called my tutor[55] told me if I was not delighted with it, [then] it was certein I did not understand it. So I laboured on 'till I was

---

[48] I.e., the rector of Ashley cum Silverley (1639–82), Ezekial [Exechiel] Catchpole, who taught RN from the age of five and, hence, prior to the latter's education at the free grammar schools (see n.45 above). From 1544 the advowson of Ashley, along with the Kirtling estate, had descended in the North family; and Catchpole weathered the Interregnum through their friendship; see British History on-line (Ashley cum Silverley).

[49] I.e., the Cambridge writer, John Seton, whose *Dialectica* (1545), published in numerous editions, seems to have been the last major work of scholastic logic published in England; see RN Books (1), which lists two copies, both octavo, one undated, the other dated 1611.

[50] Franco Petry Burgersdijk, *Institutionum logicarum synopsis, sive rudimenta logica*; see RN Books (1), which lists an octavo edition dated 1637.

[51] MS has a space for several sentences here, so RN probably intended to add some Latin phrases from Burgersdijk's book, as he does (ironically) in the next paragraph by showing, not stating, some nonsense in the books mentioned.

[52] Possibly, Christopher Scheibler, *Metaphysica duobus libris unwesum hujus scientiæ systema comprehendens....* (Oxford, 1665); see RN Books (1).

[53] (Lat.) a being is something which has beingness.

[54] I.e., George Fournier, *Euclidis sex primi elementorum geometricorum libri demonstrati* (Paris, 1644). This book, perhaps in the Cambridge edition of 1665, was procured for RN by his brother John; see RN, *Life/JN*, pp. 130 and 186. But RN, *Notes of Me*, p. 93, describes Barrow's book as 'much the best', meaning either *Euclidis Elementorum libri XV breviter demonstrati* (Cambridge, 1655) or its 1660 English translation.

[55] John North.

pleased, and having once [got] hold, could never since let goe such engaging truths, which are more detach't from the wickedness of the world then justice or religion both, for they suffer corruption under it. But all my penchant was after naturall philosofy, [which] they called phisicks[,] for out of that I conceived hopes of understanding the reasons of all things in earth and heaven too. A freind[56] lent me Sennertus' physicks[,][57] which I read all over, and found many dainty **[f. 3]** determinations, as *iris est reflectio solis in nube concava;*[58] *lumen est actus corporis diafani;*[59] *motus est actus entis in potentia quaternus in potentia,*[60] and such like which I could not controvert or contradict, but continually suspected my owne understanding rather then the seeming sterility of my author.

I [also] observed a sort of sly discours about Des Cartes' philosofy as being new and extraordinary. And found no repose till I had made a purchas of his works in octavo[61] and read em' over and over without any considerable light, but [with] infinite [*]wonder and pleasure from the novelty and variety I found there. I was inquisitive as farr as I durst[62] to know the meaning of his maine designe, as well as particular applycations, but never mett [any] other encouragement then frownes, and —why did I medle with such authors?

---

[56] Presumably, John North.

[57] I.e., Daniel Sennert, *Epitome naturalis scientiæ*, published in 1618 and afterwards in a number of editions. According to RN, *Notes of Me*, p. 92, he read most of this book on 'old physics' but 'without content'.

[58] (Lat.) a rainbow is the sun's reflection in a concave cloud.

[59] (Lat.) transparency is the act of a diaphanous body.

[60] (Lat.) motion is the act of the potentiality of a being while it is in a state of potentiality.

[61] Possibly, an error, because between 1645 and 1670 numerous quarto editions of Descartes's *Opera* were published; and one is listed in RN Books (1).

[62] Dialectal for dare.

In short, I had not many books and no lectures att all. A freind[63] lent his name, as tutor, to enable my admission [to Jesus College;] but I read to myself. So a litle phisicks, mathematicks, the peaceable side of logick, **[f. 3v]** some musick, and no conversation were my cursus at the university.[64] And from thence I was sent to shift in the world,[65] in which I had as others my frisks and turnes at sea and land and [*]proved all that was good and bad in the bustle of a very active life. But however imployed or retired I had alwais a book of some sort or other within my reach.

And now I am a grim sir in a country farm[66] and spend my time partly moving to and fro, partly reposing in a hole by the chimny corner with a round window called a closet[,] where like an old bird in an hollow tree I ruminate my no[tion] of the generall perfidy but more inexpressible folly of all the feather'd kind in yon' forrest. And whither I walk or sitt[,] no sort of thinking is so agreable to me as the speculation of naturall things and [the] ordinary but [*]wonderfull phenomena of light, sound, fire, explosions, etc. with the interest they have in our facultys.[67] And conceiving the power of them and desiring to register as well as test my thoughts, I dress them in words and sentences whereby I c[an] see how they look, and as I like[,] reject or reserve them for future diversion.

This practice I have bin the more **[f. 4]** inclin'd to, becaus I have ever found I did not well know my owne thoughts, till I have wrote and reviewed them; and then for the most part, mists fell away and [*]fondness and failings appeared in a clear light. It is no small assurance even of knowledge it self to have the perpetuall checks of

---

[63] John North.

[64] See RN, *Notes of Me*, pp. 91–4, 154.

[65] For details of his making 'shift' in London, see Kassler, *HRN*, pp. 19–34. By 'turnes at sea' RN alludes to his love of sailing, both as applied mathematics and as the act itself; see RN, *Notes of Me*, pp. 101–2, 105–10, and RN, *Life/DN*, p. 103.

[66] I.e., the estate in Rougham, Norfolk.

[67] Some of his writings are preserved on all these topics, including what RN here calls 'facultys', i.e., the sensory-motor 'capacity'; see Kassler, *HRN*, p. 89.

ingenious freinds,[68] but it is almost necessary[69] for[70] what is to appear abroad. For men know no more how their sentiments, as they happen to express them, will be accepted, then they can guess at their readers' complexions. And it is my unhappyness to want such helps. Therefore our forct use [is] the onely succedaneum, my owne reiterated self[,] expecting thence perhaps half a freind, such as it is.

I have bin often admonisht by the [*]*d[e]uterai phrontides*, so this my half freind is styled[,] to forbear scribling as not my talent, that even knowledge such as I court brings no profit, wast[e]s time, diverts buissness, and mends no man's condition of life. And after all[,] wrighting without an extraordinary felicity of style and plausibility of subject[,] such as [*]flatters, powers, factions or follys, is much wors. [f. 4v] But all in vain, for to say truth I cannot baulk the humor or rather variety of wrighting and the pleasure that it affords. I doe not excuse my self more then men use for other vanitys less reasonable tho more indulged. I know well a fop-wigg,[71] fine equippage, idle company keeping, gaming, drinking, swearing, [*]side-boxing, [*]faction, [*]politicks and many other such fashionable injoyments pass muster, and doe not sully their declared lovers. But if I should appear, as they, barefac't in this my dull, solitary dress, [then] I must expect contempt and scorne. Therefore since I mean not to compare my diversion with theirs, I shall crave leav to wear my mask and so hold my counte[nance] indemnifyed, which concludes the account I intended to give of myself. Next for my wares.

The circumstances of my education and cours of life will priveledg me in that supine sort of wrighting called essays, being

---

[68] I.e., three of his brothers—Francis, Dudley and John—whose 'checks' were in the form of opposition and contradiction.

[69] MS has the numbers '(1)' over the word 'knowledge' and '(2)' over the word 'perpetuall', as well as between the words 'almost necessary'. In other manuscripts RN uses this means to indicate a transposition of words or phrases, although what he intends here is unclear.

[70] MS has 'as'.

[71] I.e., one who is vain in appearance.

much of late in use, especially when gentlemen scrible, who care not to be confined to such strickt order and method, as compleat tracts pretend to. They have thoughts which must be vented, and they will not take that pains as is necessary in composing well. When the worme works, they write, and ceas when the flash is gone. And such is my case. For I have ever bin inclined to thinck on the subjects I here touch upon,[72] and so willingly[,] as never to fail at a pinch to divert me, they make even want of sleep[,] which [is] most hate[ful] to me[,] a pleasure. **[f. 5]** Nor could I ever shake off the [*]fond opinion most have, that thincking they understand, they think also they can express it more clear[ly] and intelligibly then others have done[;] and then the next step is to doctor it. But my vanity goes farther. I cannot but thinck in some things I have thought farther then others and here and there added some new discoverys, which seem to fill divers blanks and lacunas extant in philosofy[;] and [I have] also observed some failings in argument and application of experiments worthy to be noted and applyed. But what if all this be[,] as I suspect[,] fond, vaine, and groundless? I am not yet convinc't of it, and till then, I may be allowd my foible on account of good will.

I doe not pretend to compile an intire body of phisicks; there are so many of them. They are fulsome, and are but wrighting one after another, with some variety, as the authors in some particular things happen to differ from one and other; to find out which[,] one must have the fatigue of passing over all that vulgar stuff they are filled with. It were well if many of them had spared the press all but their owne singularitys, and proposed them onely, with the subjects concerned. This I have designed to doe; that is, treat no subject but such wherein I thinck I vary from or add to what hath gone before somewhat considerable. And I doe the rather decline the part of a compiler, because I know it betrays men into shallow determinations, least they doe not comprehend **[f. 5v]** every doubdt; and often rather then leave a gap they insert the best account they can find or make, altho they are well satisfied of none. It seems to me that nothing hath

---

[72] I.e., in the philosophical essays projected to follow this 'Preface'; see *infra* ff. 5–5v.

betrayed great witts into more failing then the designe of resolving every thing. If they would hold to what they bring new and clear to their owne minds, as their [*]nostrums in philosofy, [then] books would be less voluminous, the improvements and advances greater and every one's [*]propriety in them preserved. The world is very buysy in collecting experiments. I thinck a collection of thoughts would be very considerable. And that is what I here propose, adding to [the] work of others onely what I thinck may make my owne understood.

It may be sayd I take upon me too much, pretending to add to the arguments and discoverys of such great spirits as I referr to. Which granted I have not much to answer for; [*]*facile est inventis addere.* And if the authors were ali[ve] and spoke with, [then] they would encourage rather then blame such undertakings, and none bu[t] the impertinence of critiscisme can blame it.

It will be also charged upon me that I use too many words and that a philosofer should be concise and sententious.[73] I grant the shorter if u[n]derstood [is] the better, and if length obteins then the excess is not a mortall fault. The [*]contracting is after work, when notions have bin ventilated, and adjusted among the vertuosi, [for] then they may, by such as have a dexterous [f. 6] pen, be collected and contracted[74] for the benefitt of some, who care not to dwell long upon any thing. Of that sort is Le Clerck's phisicks[,][75] which had he compiled without pricking in his owne [*]conceipts, that are the worst of the pack, the designe was well enough executed. But such as

---

[73] The remarks in this paragraph contain an implicit defence of what RN called 'the *copia* of expression and order' against the perspicuity advocated by the language purists of his day. For an explicit defence, see RN, 'Of Etimology', pp. 229–69; see also Kassler, *HRN*, pp. 1, 129–78.

[74] I.e., abridged.

[75] I.e., Jean Leclerc, *Physica, sive de rebus corporeis libri quinque*.... (Amsterdam, 1696); see RN Books (1). After describing the problematic nature of this work, Thorndike, *History of Magic*, vol. 8, pp. 614–17, concluded that it provided 'a not unfair reflection and retrospect of the ups and downs, the change and confusion, the back currents and the stream of progress, in the scientific thought of the seventeenth century'.

invent[,] especially in matters of abstruse thincking, must be copious or obscure; and surely the latter is the worst fault. And to say truth, in such subjects language is defective, words are not to be had to note the various images in the mind, so recours is had to sc[h]emes, alusions, perifrases,[76] and figures in order to make others apprehend what we thinck, [...]

1.3. *LEARNING HOW TO PHILOSOPHISE*

In the autobiographical fragment North tells us that he purchased Descartes's collected works during his year of study at Cambridge. These works would have included the *Principia philosophiæ* and *Discours de la méthode* with its appended essays on dioptics, geometry and meteorology.[77] The latter book was especially suitable for the untutored but intelligent layman, because its rhetorical strategy was pedagogical and not 'theoretical discovery or proof'.[78] Indeed, in *Notes of Me* North further indicates that 'Nothing more gained on my judgment, as to his peice De Methodo', especially 'the rule of not building upon doubdts, but first to find out what is most clear, and thence as from a foundation proceed to other matters as farr as you can walk, with like clearness'.[79] But it was *Principia philosophiæ* that introduced North to the full extent of Descartes's new philosophy, concerning which he singled out for special praise

> ...the shaking off [occult] qualitys, which term confesseth ignorance[,] and reducing all things to longum, latum and

---

[76] I.e., circumlocutions.

[77] In a fragment of a draft preface, RN, 'Praefanda' (UK:Lbl) 32526: f. 2 (early period), RN also praises Descartes not only for publishing his correspondence, but also for sending forth his *Meditationes de prima philosophia* (1641) to 'all the sects in Europe' and then publishing it together with 'the essays', i.e., the objections, sets of which were contributed by Pierre Gassendi and Thomas Hobbes, among others.

[78] Eastwood, 'Descartes on Refraction', p. 498.

[79] RN, *Notes of Me*, p. 92–3.

profundum.[80] His notion of motion, as nothing in it self, but as
there is regard to the stations and positions of other body[s].
And then his laws of motion; all which are improvements
introduct, I may say invented in philosophy by that great
man.[81]

In order to improve on this, as well as other versions of new
philosophy, which he declares as his stated aim in the
autobiographical fragment,[82] he would have to learn about the theory
of mechanics, as well as acquire competency in philosophising. The
former process began shortly after his admission to the Middle
Temple, when he read a book called *Mathematical Magic*, in which
its author, John Wilkins,[83] drew from at least two important sources
on mechanics: *Liber mechanicorum* (1577) of Guidobaldo Del
Monte and *Cogitata physico-mathematica* (1644) of Marin
Mersenne.[84] Written in the vernacular for the unlettered artisan, as
well as for gentlemen in the improvement of their estates,[85] the book
provided an introduction to the principles of simple machines. And it
was this introduction that increased North's desire, briefly glimpsed

---

[80] I.e., length, breadth and depth.

[81] RN, *Notes of Me*, p. 93.

[82] See *supra* 1.1.2. (f. 5).

[83] Published in 1648 and re-issued a number of times, Wilkins' *Mathematical Magic* consisted of two parts,'Archimedes; or, Mechanical Powers' and 'Daedelus; or, Mechanical Motions', the latter of which included self-movers or automata. That this witty book stimulated RN's imagination is clear not only from his overt citations of it but also from his covert uses of it. For instances of the latter, see RN, 'Cursory Notes of Building', p. 8 (an imaginary self-sweeping broom as a labour-saving device), and *NP 3*, pp. 74–5 (an 'occular' model adapted from Wilkins to show pulses in a full accord).

[84] For the first author, see Bertoloni Meli, *Thinking in Objects*, pp. 18–39; and for the second, see pp. 105, 125–6, 129–30, 265. Unfortunately, Bertoloni Meli seems unaware that Mersenne's book contains expositions of some of Hobbes's ideas; see Kassler, *Music, Science, Philosophy*, p. 86 n.13.

[85] For Wilkins' educational aim, see Aarsleff, 'John Wilkins', pp. 362–3; for the book itself, see pp. 365–8.

in his autobiographical fragment,[86] to understand how things work. Indeed, as he himself recounts in *Notes of Me*, Wilkins' book gave him

> ...full insight into the theory of mechanicall powers [i.e., the simple machines], which I drove as farr as I was capa[ble] of doing, resolving not to owne any position of naturall philosophy, which did not quadrate with the generall laws of mechanisme. And I reviewd all that philosophy I had, and considered it with peculiar regard to them. For I concluded that whatever laws governed body, prevailed thro-out, and if any observation seemed to thwart them, it was not rightly understood.[87]

As is clear in statements following the passage just quoted, North conceived mechanics as a study of motion against a resistance, whereby motions, when opposed, will order themselves so as to balance or equalise the distribution of force. For he writes:

> ...the same body in a swifter motion hath more force, or in other termes, ... time is equivalent to quantity in the measure of powers. Whereby it is found that if it can be contrived that a less quantity shall work in opposition to a greater, in an over-proportion of celerity; it shall, tho of it self, the weaker, being so circumstanct, prevaile. This appearing to be the measure of all opposed powers, which is the subject of mechanicks, ... I conceived to take place in all cases of collision of body, and so satisfied my self of the reason of the laws of impulses, which have of late, from the hints of Mr D[es] Cartes bin improved by ... others to a refined height, and I thinck I have carried the

---

[86] E.g., the organ he made as a youth, see *supra* 1.1.2. (f. 2). Later on, he provided the first modern explanation of the initiation of vibrations in the flue pipe used in organs; see Kassler, *The Beginnings*, pp. 113–15.

[87] RN, *Notes of Me*, p. 100. For his early 'dabbling' in the mechanics of perpetual motion (a topic treated by Wilkins), see ibid., pp. 100–1; see also RN, *Life/FN*, p. 266, where the 'young artist' is RN himself.

doctrine farther then they, I mean to cases of all irregular bodys,[88] and their unaccountable collisions, shewing that those are governed by as steddy laws, as the most regular, and by this have explained the windmill, ship-sailing to windward, and other exotick forces....[89]

This kind of mechanics, the lineage for which is traceable to Galileo, is modelled on mechanical oscillators such as the pendulum, musical string and helical spring, which are objects that work always towards equilibrium when left to their own devices.[90] And this suggests that during the period he was writing *Notes of Me*, North's understanding of mechanics had advanced beyond what he had learned from reading Wilkins' book.[91]

In the passage quoted above, North alludes to his own work in mechanics; and, indeed, in *Notes of Me* he makes specific reference to 'my Mechanicks',[92] as well as to other writings, including 'the systeme of nature, as I have elsewhere proposed it, which still magnifies the artifice, and Artificer of nature',[93] and the 'notes I have made of my thoughts upon most physical heads'.[94] These physico-mechanical assays would have been drafted or produced during the period *prior to* his leaving London for Norfolk in 1696. Although a few manuscripts from that period have been preserved,[95] many more now extant in the British Library and elsewhere represent his

---

[88] I.e., irregularly-shaped bodies.

[89] RN, *Notes of Me*, p. 100.

[90] See Kassler, *The Beginnings*, pp. 121–33. For the linked developments between objects and seventeenth-century mechanics, see Bertoloni Meli, *Thinking in Objects*.

[91] For evidence that supports this statement, see RN, *Cursory Notes of Musicke*.

[92] RN, *Notes of Me*, p. 100.

[93] Ibid., p. 140; see also pp. 96, 141.

[94] Ibid. p. 142.

[95] E.g., RN, 'An Essay concerning the reason and use of the baroscope', (UK:Lbl) Add MS 32543: ff. 23–33v (*c*.1678–*c*.1682), written in a legal hand (not RN's); see Kassler, *HRN*, pp. 52–4.

processes of thinking *after* leaving London for Norfolk. Nevertheless, information about his earlier thoughts may be gleaned from manuscripts written in Norfolk, especially if they show signs of revision made under the duress of certain events to be described later on.[96]

As for the way in which North acquired competency in philosophising, it will be useful first to comment briefly on Descartes's ways, since, whether correct or not, many, including North, considered him as the 'inventor' of new philosophy. To bring his philosophy to fruition, Descartes identified two principal methods. One, the preferred way, was to study causes and the art of demonstration, seeking epistemological certainty in mathematics. Warning against mixing conjectures into the judgments we make about the truth of things, Descartes recommended the study of mathematics as a way to learn how to argue demonstratively rather than probably. Nevertheless, he conceded that, where truth was yet to be established, we might use a second method by making any hypothesis about the matters at hand, with the proviso that all the consequences of our assumptions agree with experience.[97] And he had illustrated this way of philosophising in some of the essays published together with his *Discours*.

In the essay on dioptrics, for example, Descartes did not rely on mathematical demonstration, because he did not seek the true nature of light. Instead, he aimed to summarise data and suggest properties that are to be assumed—suppositions—even though the properties are unobservable.[98] For this purpose, he relied on a method that consisted of a number of formal analogies—for example, a cane, wine in a perforated vat, a tennis ball and racket—as 'building blocks' for representations of direct, reflected, and

---

[96] See *infra* 2.2.3 and 2.2.4.

[97] See, e.g., Descartes, *Principia philosophiæ*, Pt III, Arts 44–47.

[98] I.e., inscrutable by sight, hearing or any other sensory mode.

refracted light.[99] Although Descartes presented his explanations as suppositions to be verified by their experimental consequences, he believed that suppositions themselves are deducible from certain *a priori* principles. And it was this procedure that was to lead Newton to understand, and later to reject, hypotheses in the Cartesian sense of unverified deductions from first principles.[100]

During his student years in the Middle Temple, North learned a different way of philosophising, one that takes the form of a 'put-case'—a legal case argued from an hypothesis derived from experience,[101] not from *a priori* principles. Elsewhere, I summarised this procedure with reference to North's writings concerning the physical phenomena related to the production, transmission and reception of sound or what we now call acoustics.[102] I shall provide the same illustration here, since it provides a pattern for understanding some of his other writings that also exhibit this way of philosophising. In the writings on sound, his procedure is to begin with a hypothesis or general principle (e.g., elasticity) derived from experiment and, hence, assumed to be true. He then proves the truth of other independent propositions by two principal means: repetition of instances (statistical means) and analogical reasoning (rational means, i.e., deduction from previous knowledge).

However, not all independent propositions are known to be true, because North's examined instances, in which a general

---

[99] See Eastwood, 'Descartes on Refraction', who also pointed out that the 'descriptive, illustrative analogies' were an essential tool in Descartes's 'pedagogical rhetoric'.

[100] See Hanson, 'Hypotheses Fingo'.

[101] I.e., such experience as rational adjudication upon the evidence of witnesses, documents, physical objects, etc. For a brief discussion of RN's use of this experience in some of his literary writings, see Kassler, *HRN*, pp. 69–75, *et passim*.

[102] See ibid., pp. 99–110. RN's writings on this subject commenced after his removal to Norfolk and continued close to the time of his death; but the increasing demands of estate management, as well as personal and professional duties would limit the time he would have for writing. As a consequence, he tended to write on the run—cursorily—so that rigor is not to be expected in any of his (quasi) systematic writings.

proposition is verified, are not known to be representative samples of the entire class to which they belong. The problem—afterwards known as the problem of induction—is to determine to what extent the samples are fair. In this determination analogy also plays a role, for if North's examined instances (e.g., water, air, sound) are known to be analogous in certain important respects with other phenomena (e.g., pendulums, musical strings, helical springs) that previously had been better explored, a generalisation which is known to hold for the latter may be adopted for the former with practically no changes. Consequently, 'wee have onely to take care, that wee hold to strickt analogy with knowne things, and not thwart experience. And then wee are in the best way, that is possible to be contrived for natural knowledg'.[103]

North, therefore, combines hypothetico-deductive and inductive procedures (*a*) by assuming a general principle from which all sorts of consequences can be deduced with the aid of logic and other independent propositions, (*b*) by proving some of these independent propositions by multiple instances using one or the other sampling technique (statistical, rational), (*c*) by presenting his demonstrations as suppositions imposed by hypothesis, and (*d*) by confirming his suppositions by experiment or experiences that serve to validate the hypothesis, promoting it to the rank of theory. His inductive argument does not demonstrate a general principle in the strict mathematical sense, but it may prove it to be sufficiently evidenced, so that it can be concluded with probability that the principle is applicable to the case at hand.

Having learned this way of philosophising as a law student, he continued to rely on it as a practising lawyer, especially during his years in London.[104] But he also applied the procedure to other domains of knowledge from the example of two particular writers. One was his brother Francis, who used the put-case method in a tract, published anonymously as *A Philosophical Essay of Musick*

---

[103] RN, *Cursory Notes of Musicke*, p. 8.

[104] For some aspects of his legal life, which has not yet been the subject of detailed investigation, see Kassler, *HRN*.

(1677).[105] The other writer was Christiaan Huygens, whose *Traité de la lumière* (1690) begins with a succinct summary of a conjectural way of philosophising as follows:

> ...whereas the Geometers prove their Propositions by fixed and incontestable Principles, here the Principles are verified by the conclusions to be drawn from them; the nature of these things not allowing of this being done otherwise. It is always possible to attain thereby a degree of probability which very often is scarcely less than complete proof.[106]

In *Cosmotheoros* (1698), published posthumously, Huygens presented another instance of a probabilistic way of philosophising; and, as will appear later on, North will draw from this example a succinct statement of what, for him, constitutes sound reasoning.[107]

1.4. SEEKING KNOWLEDGE OF NATURE

Since the nineteenth century it has been common to apply the epithets, rationalism and empiricism, as a means of differentiating ways of philosophising. In the rationalist way, the sources of knowledge are intellectual intuition and deductive reasoning, so that concepts, which are clear and distinct, are arrived at independently of experience, all true statements regarding such concepts being either

---

[105] For this tract, see Kassler, *The Beginnings*, pp. 37–88, 131–70; for Newton's comment on it in letter to John North, dated 21 April 1677, see pp. 94–104, 175–8.

[106] Huygens, *Treatise on Light*, p. xi. See also Shapiro, 'Huygens' "Traité de la Lumiére"', pp. 225–7, 229. For RN's unacknowledged use of this work, see RN, *Cursory Notes of Musicke*.

[107] See *infra* 5.5.2: No. 1, RN→SC 1713 (ff. 19v, 24v). For some of his references to the book, see ff. 67v–68 in RN, 'Preface to a philosofick essay' (UK:Lbl) Add MS 32526: ff. 64v–68 (early period); f. 20v in RN, Mostly untitled (UK:Lbl) Add MS 32546: ff. 1–18v, *continued* ff. 25v–32v, *continued* ff. 19–24v (early period); f. 99 in RN, Untitled draft (UK:Lbl) Add MS 32545: ff. 92–151v (early period); and f. 108v in RN, 'Physica' (UK:Lbl) Add MS 32544: ff. 1–274v (late period).

self-evident or logically deducible from propositions that are self-evident. In the empiricist way, the sources of knowledge are observation (taken in the broad sense to include various sensory modalities) and inductive reasoning. Accordingly, concepts are grounded in experience, and all true statements regarding such concepts are those which have been tested by experiment directed to the natural world.[108]

In the case of North, however, there is a problem with relying on this means of differentiating ways of philosophising, for it fails to account for variations. And there also is a problem in the claim that most, 'if not all, seventeenth-century philosophers', who had recourse to the two methods described above, believed that the faculties of reason and sense led to 'the uncovering and recognition of truth *without recourse to any other authority, training, or tradition*'.[109] North did not share this belief, because his way of philosophising is indebted to legal reasoning, the dominant characteristic of which can be conceived as a means of reaching a probable decision.

To achieve this purpose, legal reasoning is constrained by certain authoritative sources used as standards, whose claim to be binding is that they were laid down in the past. But these standards—on the one hand, precedent (i.e., case law or prior court decisions) and, on the other hand, rules of legal procedure (e.g., 'to condemn none without hearing' or 'to hear both sides'[110])—are themselves constrained, because both precedent and procedural rules are to be evaluated in terms of procedural fairness ('like cases should be treated alike').

Indeed, procedural fairness is not exterior to, but inheres in the common law, for without this equitable spirit a law will eventually

---

[108] Perhaps to overcome this problem, Oldscamp in Descartes, *Discours de la méthode*, p. xxxiv, concluded that the rationalist and empiricist methods are merely different in degree than in kind.

[109] Leyden, *Seventeenth-Century Metaphysics*, p. 64 (italics mine).

[110] RN often cites or alludes to the latter maxim; see Kassler, *HRN*, pp. 119–20, and *infra* 5.5.2 n.200.

die, that is to say, become obsolete by disuse or void by legislation. Consequently, the early seventeenth-century jurist, Edward Coke, identified procedural fairness with the 'life' of the law and the law itself with 'the artificial perfection of reason'. As the latter phrase suggests, Coke's conception of reason is constructive. For law is made by, or results from artifice; it is brought about by skill and not spontaneously; it is skilful in result as well as in process; and it represents the historical experience and reasoning—common sense— of learned jurists. North shares this model of reason as trans-subjectivity and as the best way of reaching a probable decision not only in law but also in other domains of knowledge. Hence, it is an important feature in his pattern of thought.[111]

Although, as indicated in the previous section, Descartes conceded the need for hypotheses, he remained 'a persistent foe of probabilism', insisting instead that the ideal of an epistemic enterprise was metaphysical certainty.[112] For him, as for some other seventeenth-century philosophers, certainty was to be found in a conception of reason modelled on mathematical demonstration from axioms or propositions assumed to be true.[113] North allows that the mathematical arts have 'a great prerogative in being capable of demonstration', so that in that domain of knowledge complete certainty is attainable.

> And the reason is, because quantity of extension is certein, and so much as is made use of is agreed. So there can be no deceipt[114] in the principles, and the operation is but an application of one thing to another by way of measuring,

---

[111] It is important to note that in the common law tradition, Coke's model of reason as trans-subjectivity differed from the models developed in two versions of the natural law tradition—one, classical (reformulated from Thomas Aquinas by Richard Hooker), the other, modern (reformulated from Hugo Grotius by, e.g., Hobbes and John Locke); see Kassler, *HRN*, pp. 110–28, *et passim*.

[112] See Voss, 'Scientific and Practical Certainty in Descartes', pp. 574–5.

[113] See further *infra* 4.4.2.

[114] See *infra* Glossary.

which produceth demonstration. And this certeinty is not found in any other science.[115]

In the passage quoted above there is an implied distinction between demonstrative and non-demonstrative science. But North also makes a further distinction between mathematics and the sciences concerning 'naturall agencys'.[116] In the former case, the foundation is identity to which all 'mathematicall axiomes' are reducible. But in the latter case, the foundation is an almighty power that can create something out of nothing, as well as annihilate, since the 'continuance' of natural agencies requires 'a perpetuall creation, which is referred wholly to the omnipotent God'.[117] And from this latter foundation North derives axioms that respect 'our capacitys'; for example, that sensation is true, though inferences from sense may be false; that nevertheless, a true judgment of real entities may be made if (among other requirements) 'all tryalls, and consequences agree'; and that such consequences may be made intelligible when terms are clearly and distinctly understood.

Now, if mathematics alone is a demonstrable science, then it cannot be a standard for any other science. And regarding those other sciences, North is of 'the opinion that there is no true knowledg in the world but of things naturall'. Religion, for example, is 'built upon knowledg for miracles that require faith', and even faith in miracles could not be effected, 'if wee did not know so much of nature as to

---

[115] See f. 93v in RN, 'Mechanick notes' (UK:Lbl) Add MS 32540: ff. 81–118v, *untitled continuation*, ff. 119–137v (early period, late).

[116] See RN, 'Of axiomes, and the foundations of them' (UK:Lbl) Add MS 32548: in Notebook 3, ff. 75v–77 (undatable). Note that RN's philosophy of nature includes human nature and, hence, morality; see Kassler, *HRN*, pp. 89, 104.

[117] RN singles out God's causal power as the foundation for natural science, whereas Descartes required both divine veracity (God is no deceiver) and causal power as the foundation; see Collins, *God in Modern Philosophy*, p. 68. Note, however, that since not only Descartes but also most other philosophers of the period assumed that matter is inert, some kind of doctrine of continuous creation was required; see, e.g., Gaukroger, *Descartes*, pp. 150–2, 343–4, 350–1.

determine what may, and what may not be by naturall means'.[118]
Yet, even when suspense or doubt results from insufficient evidence,
there still is a rational path to knowledge founded on degrees of
probability, whereby 'very strong' arguments may persuade, for
example, that 'the sun hath ever rose in the east, therefore it will doe
so still'.[119] In short, North's rational path is the probablistic
reasoning, a kind of reasoning that he learned in the law and
afterwards applied in his search for knowledge of nature, where 'the
question is of truth', that is to say, reality.[120]

But he acknowledges that the search for truth has many
difficulties, since there is a discrepancy between reality and
appearance. And the reason he gives for this discrepancy is that
nature has limited our sensori-motor capacity for information
processing to what we can mentally or physically number or count.[121]
Consequently, 'all our failings in comon judgment of things' are due
to this natural limitation, for it is certain that

> ...our sensations are received thro the interposition of bodys,
> and particularly that machine wee claime the governement of
> and call our owne. The constitution of that is such, (at present
> no matter how or why,) that wee can move our members, but
> in a certein manner; and with a determined celerity. ... So men
> that count any thing passing, either by nodds or motion of the
> hand keep company with them, and if they pass faster then
> those nodds will distinguish, the account is lost. ... The
> consequence of these defects (if I may so call our corporeal
> powers, tho regulated by the laws of body, which are perfect)
> is, that if motions have returnes swifter then any motion of our
> bodys can keep pace with, all distinction ceaseth, and the idea

---

[118] See ff. 108v–109 in RN, 'The infelicitys of natural philosophy' (UK:Lbl) Add
MS 32549: in Notebook 5, ff. 103–109v (middle period). (The notebook cover has
the label, 'Experiments of natural things'.)

[119] See f. 109 in RN, 'Mechanick notes' (UK:Lbl) Add MS 32540: ff. 81–118v,
untitled continuation, ff. 119–137v (early period, late).

[120] See ibid., f. 84v.

[121] See Kassler, *HRN*, pp. 99–110.

[i.e., appearance] is confused, and the sensation is as of a thing continued. So wee distinguish the movement of the musitian's hands, but not the vibrations of his strings, and the like.[122]

Our beliefs, therefore, fall short of reality, so that we must continually live with the risk of being in error. Although it is for the judgment to know the evidences or 'signes' of reality and to distinguish them,[123] yet for this we require 'needful discoverys called experiments'. And since there is a dearth of such discoveries, 'wee doe not comand a clear knowledge of the universe and to say truth but of very few things in it'. Consequently, 'if truth be the point', 'wee must for the most part confess our ignorance and say to many things wee neither doe [n]or can know them'.[124] For what knowledge we do have consists 'in different degrees of probability, some more and some less, and scarce any, in absolute certeinty but our immediate sensations and what is included in them'.[125]

North's sceptical account of the weakness of human capacity derives from his belief that the essential mark of wisdom is recognition of one's own limitations. But this does not mean that his scepticism is negative or destructive; rather, for him, it is a positive, enabling intellectual activity, for his belief concerning wisdom derives from mainstream Christian thought, in which reason is employed 'to the full extent of its natural usefulness without overextending it and giving it authority beyond the bounds of its nature'.[126] In seeking truth, therefore, North is motivated by the belief that natural knowledge is possible, because 'nature is governed

---

[122] See ff. 253v–254 in RN, Untitled (UK:Lbl) Add MS 32546: ff. 247–262v (early period), where at ff. 254–256 he treats mathematics as a means of overcoming nature's limitation.

[123] See f. 84v in RN, 'Mechanick notes' (UK:Lbl) Add MS 32540: ff. 81–118v, *untitled continuation*, ff. 119–137v (early period, late).

[124] Ibid., f. 81v.

[125] See f. 252v in RN, Untitled (UK:Lbl) Add MS 32546: ff. 247–262v (early period).

[126] Frye, 'Reason and Grace', p. 404.

by a rule, that rule is uniforme and universall. Nay[,] the rule is
reasonable, and according to our understandings necessary'.[127]

Of course, the belief that the natural world exists and is law-
governed is not provable beyond doubt. Accordingly, North's belief
is not only an act of faith, but also a sign of is his realism. Elsewhere,
I described his realism as a critical realism, a kind of realism that
occurs at the level of reflection on ordinary discourse.[128] Here,
however, it is necessary to be more specific about the nature of his
realism, which is defined by an independence thesis, namely, an
ontological commitment to entities that exist independently of
knowledge or experience of them.[129] But, as will become apparent
later on, this thesis can be held in different ways.

In the case of North, his ontological commitment is
epistemological, because knowledge is, for him, always and
necessarily a reflection of the structure of understanding we as agents
bring to our practices in the world. Accordingly, the study of nature
includes 'the principles of which all things are composed, the
varietys in the working of these principles [and] the severall
phenomenas they produce', as well as 'the consideration of our
natures and manner of perception, the errors it is subject to, [and] the
disadvantages of it'.[130] Indeed, his commitment to a sceptical
epistemology accounts for the frequent reference in his writings to
learned ignorance, as, for example, in his autobiographical
fragment[131] or in his treatise on mechanics, in the latter of which he
writes:

---

[127] See f. 81 in RN, 'Mechanick notes' (UK:Lbl) Add MS 32540: ff. 81–118v,
untitled continuation ff. 119–137v (early period, late).

[128] See Kassler, *HRN*, pp. 101–13.

[129] For a modern treatment of the nature of realism, see Grayling, *Truth, Meaning
and Realism*, to whom I am indebted.

[130] See f. 84v in RN, 'Mechanick notes' (UK:Lbl) Add MS 32540: ff. 81–118v,
untitled continuation, ff. 119–137v (early period, late).

[131] See *supra* 1.1.2 (ff. 4v–5).

> ...while I am in an opinion that I can lay together the severall parts of philosophy in a better coherence then hath yet bin done, which opinion I cannot shake off[,] it is my infirmity, I have no peace with my self without venting the fancy by writing....

Although he hopes that this tendency may be cured 'by a conviction of vanity', yet until the time of that conviction, he begs to be excused, because 'there cannot be too much caution in foundation work'; and 'when I come in to contingent and uncertain matter, I hope my profession of ignorance may pass instead of dogmaticall witt, upon the account of safety and care not to err'.[132]

North's use of the defining thesis of realism is strategic, for he assumes that it has practical value in making sense of experience (including experiment), either by inference *from* observables *to* unobservables[133] or by perception of the effects of observables. In the case of nature, both related forms of access assist his numerous attempts to give an account of causal relations in the world to which his theories assign properties that transcend human investigative powers. But he does not make the mistake of believing that his strategic assumption is thereby shown to be true, for he had sufficient knowledge of logic to understand that nothing derived from an assumption employed as a premiss in that derivation can, without circularity, prove the truth of the assumption itself.

If, for North, the mark of wisdom is awareness of his own fallibility and if the search for truth is a process that precludes demonstrative certainty, then his sceptical epistemology and probabilistic way of philosophising are closely related. It might be supposed, therefore, that for these two patterns of his thought, he may have been indebted to a group of English philosophers and

---

[132] See f. 88v in RN, 'Mechanick notes' (UK:Lbl) Add MS 32540: ff. 81–118v, *untitled continuation*, ff. 119–137v (early period, late).

[133] E.g., see ibid., f. 96v, concerning RN's probable conclusion that two bodies may be similar but never identical, since 'as wee say[,] no two eggs or peas are alike'; nevertheless, we have 'reason enough to conclude there is the same œconomy in insensible things'.

Church of England theologians, who sought to solve the philosophical problem of certainty. According to Harry van Leeuwen, Richard Popkin and Barbara Shapiro, members of the group in question, whom they class as latitudinarian, include (in chronological order) William Chillingworth, John Wilkins, Robert Boyle, John Tillotson, John Locke, Edward Stillingfleet and Joseph Glanvill.[134] By stressing the difficulties faced in the search for truth and the need for caution based on an awareness of human fallibility, these men arrived at a theory of limited certainty that emphasised a probabilistic way of philosophising, an approach now called 'mitigated' or 'constructive scepticism'; and, according to the three commentators, it was this approach that contributed to the emergence of modern empiricism, as well as to a modern scientific outlook.[135]

North once owned a number of works by so-called latitudinarian writers;[136] and in his published and unpublished writings, he also cites various works of some members of the latitudinarian group mentioned above. But as I have suggested in my previous study devoted to North's practical philosophy—his 'wisdom' about life, morality, law and tradition—a principal source of his scepticism seems to be the *Essais* of Michel Montaigne.[137] What is more, he also shares Montaigne's fideism,[138] a component of

---

[134] See Leeuwen, *The Problem of Certainty*; Popkin, *The History of Scepticism*, pp. 65–6, Chapters 7 and 13; and Shapiro, *Probability and Certainty*, Chapter 2. The term 'latitudinarian' is insufficiently analysed, one of these commentators describing it as 'moderate' Anglicanism, another, as low church. But the term itself undergoes a change in meaning, particularly after the turn of the century, when party theology began to emerge in the Church of England; see *infra* 4.4.2.

[135] In casting his net beyond England, Popkin, *The History of Scepticism*, pp. 92, 94, also claimed that Gassendi's 'tentative Epicureanism' was a form of 'mitigated scepticism' that also 'comes close to the empiricism of modern British philosophy'.

[136] E.g., Simon Patrick, Edward Stillingfleet and John Tillotson; see RN Books (1).

[137] See Kassler, *HRN*, pp. 12, 77–82, 182, 185, *et passim*.

[138] Of the three commentators, only Popkin, *The History of Scepticism*, devoted attention to fideism, including that of Montaigne, pp. 47–9, 52, 54–6, and his

North's thought I did not then explore in any depth. But his reasons for holding it seemed to demand further investigation; and it was to satisfy this demand that I returned to North's writings, which provided evidence for a more complex and quite different story that is told in the following pages.

Nevertheless, it is possible to discern the nature of North's fideism from his separation of appearance from reality, noted above, for this implies a distinction between knowledge of natural phenomena (appearances) and knowledge of reality (truth). In seeking the first kind of knowledge, North proceeds on the assumption that the natural world is accessible to human understanding, because there is a rational path to such knowledge (probabilistic reasoning). Recall, however, that he grounds this assumption on two interlinked beliefs about reality. First, the foundation of natural knowledge is an omnipotent God who creates and perpetually maintains the natural world. Second, and following from this, nature exists and is rule-governed, not irrational. The implication of these beliefs is that knowledge of reality is a kind of knowledge that may be rendered plausible or probable from the evidences of natural knowledge. But, as North will indicate later on, those evidences can never be adequate as demonstrations of truth, because belief in reality (including the supreme reality) is an act of faith, not of understanding.[139] In short, North's fideism is an extension of his sceptical epistemology to theology.

---

humanistic disciple, Pierre Charron. Note that RN owned English translations of Montaigne's *Essais* and Charron's *La Sagesse*; see RN Books (1).

[139] See *infra* 4.4.5 and 5.5.2: No. 1, RN→SC 1713 (e.g., ff. 1v, 2v).

2

CHANGES IN PHILOSOPHICAL METHOD

2.1. DISCOVERING NEWTON'S CHALLENGE TO NEW PHILOSOPHY

Before the seventeenth century the terrestrial world was regarded as an imperfect reflection of the perfect mechanisms of the heavenly world. Since perfect knowledge was denied to humankind inasmuch as earthly existence was imperfect, certainty was precluded in studies of the terrestrial world. As a consequence, the natural sciences were qualitative, not quantitative, a conception inherited by the seventeenth century, along with other parts of Aristotle's philosophy. In alluding to this old philosophy, North points out that 'it was ever a failing to ascribe exactness to the heavens and disorder to the earth'.[1] By the time he put this thought in writing, Aristotle's philosophy had been gradually superseded by a philosophy that unified the terrestrial and celestial worlds, namely, new philosophy not only in its Cartesian form but also in the variants that were produced during Descartes's lifetime, as well as those that appeared after his death.[2]

New philosophy taught three things: first, that the cosmic entities ('particles', 'corpuscles') are ultimate and unobservable; second, that those entities and their various configurations cause everything that happens in the world; and third, that since ultimate

---

[1] See ff. 24–24v in RN, Mostly untitled (UK:Lbl) Add MS 32546: ff. 1–18v, *continued* ff. 25v–32v, *continued* ff. 19–24v (early period). This remark may reflect his belief that irregularities are aspects of the infinite cosmos, which is perfect, and that uniformity belongs not to nature but to its laws which alone are 'regular'; see *infra* 3.3.2: No. 7, RN→[SC 1706] (ff. 282, 287v).

[2] For a brief statement concerning Descartes's 'ambitious plans' to provide 'an account of the system of the world aimed at replacing scholasticism'; see, e.g., Bertoloni Meli, *Thinking in Objects*, p. 141.

entities are unobservable, we may understand them only by what we can observe.[3] In short, the observable became the key to understanding the unobservable, a key that was variously interpreted. In North's case it meant that unobservable entities such as air particles can be supposed to move in simple harmonic motion only after we possess all the signifying analogies of bodies which we can see and put to the test of experiment—bodies such as pendulums, springs and musical strings.

From its inception, however, new philosophy reconstructed ultimate entities out of the concepts of number, extension and motion, so that the ground was laid for mathematics to become an universal instrument of analysis. Initially, this ground was developed not only by the use of objects as models but also by geometric representations. But in the second half of the seventeenth century, a number of philosophers began to move away from objects towards mathematical principles and representations.[4] And this move, in turn, was to lead to a new way of philosophising about nature which depended on a method of inquiry that some termed analysis and synthesis.[5] In his 1687 *Principia mathematica* Newton exemplified this method in which mechanical problems are analytically resolved and synthetically constructed.[6] And it was this twofold procedure

---

[3] See Kassler, *Music, Science, Philosophy*, pp. 44–8.

[4] For details, see Bertoloni Meli, *Thinking in Objects*; see also Guicciardini, *Isaac Newton*, p. 27.

[5] Note that this method of inquiry differed from that of the scholastics, in which analysis and synthesis (or resolution and composition) were defined as demonstration *a posteriori* and demonstration *a priori*. For this latter method, see *infra* 3.3.4.

[6] Very roughly, the method consisted in discovering analytically the structure and operations of nature, reducing these discoveries to mathematical principles, establishing the principles by experimental test, and thence deducing synthetically (by geometrical constructions) the causes and effects of things. For his first public pronouncement on the use of this method in natural philosophy, see Newton, *Optice*, Qu. 23, pp. 347–8. For his changing conceptions of analysis and the reasons for his privileging geometrical synthesis, see Guicciardini, *Isaac Newton*.

which constituted a new, mathematical way of philosophising,[7] the use of which, he hoped, would refute probability and foster certainty in natural philosophy.[8]

Before 1687, the mutual action of collision was the basic phenomenon of seventeenth-century mechanics. However, there were different ways of treating such action, some philosophers concerning themselves solely with the forces involved upon impact, others with the problem of the communication of motion from one body to another. Moreover, such treatments depended on conceptions of matter—on the one hand, whether bodies were conceived as observable in ordinary experience or as unobservable and, hence, hypothetical; and on the other hand, whether such physical properties as inelastic hardness or flexibility were assigned to one or both kinds of body. In 1687 Newton introduced a new concept of matter as inertial mass, its parts acted on by attractive and repulsive forces, thereby challenging the previous concepts. And from this challenge arose one of the most important questions of seventeenth-century cosmology: what holds the universe together?

For Newton, the answer was attraction—that all parts of matter attract each other reciprocally as the square of the distance—a concept also known as universal gravitation.[9] But Descartes had accounted for gravity on the theory that celestial bodies are carried by vortices whirling around a central body—the sun in our system or other stars elsewhere in the universe. According to him, their curvilinear motions result from the combination of the pressure of the vortex toward its centre and the body's tendency to move away along the tangent. Consequently, in his account, there is no real

---

[7] According to Guerlac, 'Where The Statue Stood', p. 324, what was important about Newton's method is 'the degree of emphasis to be placed on experiment and induction', and what was new is 'the language in which the operations are exclusively to be expressed', namely, the language of mathematics.

[8] See Guicciardini, *Isaac Newton*, pp. 321–7.

[9] Bertoloni Meli, *Thinking in Objects*, p. 258.

attraction; rather, what appears to be attraction actually results from the pressure of the vortex or what he called its '*conatus ad motum*'.[10]

Even though Newton made use of some of Descartes's concepts, a number of commentators have noticed that he disliked the French philosopher's solution;[11] but more recently, two commentators have made more specific claims. One put forward the suggestion that 'the more one studies Newton's *Principia*, the more one discerns a far-reaching plan aimed at banishing Cartesian notions'.[12] The other provided the first extensive evidence for this agenda, from which the author concluded that Newton was 'opposed to the anticlassical stance' he seemed to detect in Descartes's essay on geometry, so that he 'portrayed himself' as indebted to the ancient Alexandrian geometers, whose lost analytical method he was attempting to restore.[13] But a much earlier critic had already suspected that Newton's aim was to devalue Descartes and to overthrow new philosophy. Since this critic was Roger North, we might ask: what works of Newton did he read?

Although some lists of his book purchases are extant[14] and although a catalogue is preserved of books once owned by him,[15] these represent only a small sample of what once constituted his very extensive library. However, since neither the lists nor the catalogue

---

[10] (Lat.) tendency to motion. For his conceptual framework, which relied on the rotation of a stone in a sling, see Descartes, *Principia philosophiæ*, Pt I, Arts 57–59.

[11] See, e.g., Cohen, "'Quantum in Se Est'".

[12] Ibid., p. 256.

[13] Guicciardini, *Isaac Newton*. In addition to the founder of the Alexandrian school, Euclid, who lived in the time of the first Ptolemy, the other ancient geometers were Apollonius of Perga, who was educated at Alexandria under the successors of Euclid, and Pappus, who worked at Alexandria.

[14] Lists of book purchases (or books desired to be purchased) are now preserved in the Cobbold MSS Box 17 (UK:Iro) HA 49/C2/10/1 (two fragments, undated) and HA 49/C2/10/1 (1719). Note that RN's name also appears on some subscription lists.

[15] See *infra* References (section 1).

include any works by Newton, the chief source for answering the question posed above is North's own writings, even when no specific work by Newton is cited.[16] For example, in *Notes of Me* he offers a few remarks about 'the theory of light, which leads to opticks, and the other mathematicall sciences derived from it'.[17] He states that he had 'much pleasure' in this theory; and in going on to detail some of that pleasure, it is possible to discover the path that led to his reading an early paper by Newton.

Recall that in 1667 he purchased Descartes's collected works as a student at Cambridge,[18] a purchase that included the *Discours de la méthode* with its appended essays. In the essay on dioptrics the French philosopher explained the reflection and refraction of light by conceiving bodies as inelastic and treating collision on the analogy of the motion of a tennis ball after impact, an analogy that during the seventeenth century was used to account not only for light but also for sound, including their deviation in reflection and refraction from a rectilinear path.[19] And according to North, it was this very essay that showed him 'the nature of reflection, and refraction, and the use, as well as the (mathematicall) caus of curve glasses'.[20]

But the same essay also awakened his curiosity about the principles of light, and in satisfying that curiosity, he was led to criticise Descartes's reason 'for the equiangularity of reflection, which is built on a supposition that the body moved, is affected with a compound direction'.[21] North describes this supposition as

---

[16] Korsten, 'Roger North', provided a checklist of names cited by RN in his writings on 'science'; but the checklist, while useful, contains a number of errors and many omissions (e.g., none of RN's writings on sound were consulted).

[17] RN, *Notes of Me*, p. 140.

[18] See *supra* 1.1.1 and 1.1.2 (f. 3).

[19] Shapiro, 'Kinematic Optics', pp. 136–7, 140; see also pp. 189–91 for Descartes's theory of colour in the essay on meteors that concludes the *Discours de la méthode*.

[20] RN, *Notes of Me*, p. 140.

[21] Ibid., pp. 140–1.

'precarious' (and also provides a reason for his claim), after which he states:

> As to the reason of light, I have ever admired Mr. Newton's hypothesis, as new, and most exquisitely thought. Which is in short that light generally speaking is a blended mixture of all colours; and that these colours are the different effects of certein rays.... And these coloured rays, are not refrangible to the same angle, but some to a greater and some less, which makes the distinction of colours, in all refractions. And they are not accidentally created by that action, but onely separated, existing all.[22]

During the period North was writing *Notes of Me*, Newton's *Opticks* had yet to be published. Hence, the remarks quoted above suggest that he had read the first of several papers on light and colours that Newton contributed in the 1670s to the Royal Society's *Philosophical Transactions*, namely: 'A Letter of Mr. Isaac Newton, Professor of the Mathematicks in the University of Cambridge; containing his New Theory about Light and Colors: sent by the Author to the Publisher from Cambridge, Febr. .6. 1671/72; in order to be communicated to the R. Society'.[23] In North's opinion, Newton's 'hypothesis' lacked 'a physicall solution', 'without which, however plausible, it will not be admitted'.[24] He therefore sought, and thought he had found such a solution; but he did not positively determine whether 'it will square with this hypothesis'.[25]

---

[22] Ibid., p. 141.

[23] *Philosophical Transactions* (1672), No. 80: 3075–87; reprinted in Cohen (ed.), *Isaac Newton's Papers*, pp. 47–59. In the paper Newton addressed a geometrical problem of refraction that had been ignored by Descartes and others, namely, the problem of 'explaining the spectrum *for a given position of the prism*, and this on the supposition that the rays from the sun were all equally refrangible'; see Sabra, *Theories of Light*, p. 240.

[24] RN, *Notes of Me*, p. 141.

[25] Ibid.

By a physical solution North means a natural history in the form of a description of events.[26] And in *Notes of Me* he provides a brief summary of his solution,[27] which begins as follows:

> ...when a force falls upon a fluid, there is 2 sorts of action comunicated; one progressive as waves on the surface of water, which produceth sound. The other in an instant, by reason of the perpetuall contiguity of matter[.] This is light.

Insofar as his statement concerns light, North is indebted to Descartes's *Principia philosophiæ*, for this book included a continuum theory of light, in which the transmission from the luminous source to the organ of vision is instantaneous.[28] But that theory neither provided an account of reflection and refraction nor included any elements of a wave theory. Hence, according to Alan Shapiro, Descartes and 'those who rigidly adhered to the Cartesian system' did not contribute to the development of the latter kind of theory, even though the French philosopher

> ...had considerable indirect influence on the seventeenth-century continuum theory of light. This, however, arose from his standing as the foremost mechanical philosopher of the seventeenth century, and not from his theory of light.[29]

Insofar as North's statement concerns sound, it is clear that he did not adhere to a Cartesian type of continuum theory. Indeed, his first physical solutions for sound were written during the same period that he was writing *Notes of Me*, that is, sometime before 1693 and

---

[26] Since the 'phisicall part' of any subject is the same as the 'philosophicall', the aim of a description, or natural history, is 'no more then to say what, not why'; see f. 160v in RN, 'Of light and colours' (UK:Lbl) Add MS 32546: ff. 143–168v (early period, late).

[27] Ibid., pp. 141–2.

[28] For details of Descartes's 'celestial optics', see Shapiro, 'Light, Pressure, and Rectilinear Propagation', pp. 243–66.

[29] Shapiro, 'Kinematic Optics', p. 140.

sometime after 1698 (but before *c.*1706). And in the more finished of these solutions,[30] he relies on an analogy between sound and surface waves on water as a means of constructing an extended metaphor, that is, a continuum theory in which sound is produced by a 'comprest wave' in the air, this wave is propagated 'by vertue of its spring', and 'falling upon our organ of hearing', it conveys to our minds 'the sense of sound'.[31]

From the disparity between the two statements concerning sound and light, therefore, we may infer that during the time he was writing his autobiography, North had yet to criticise Descartes's continuum theory of light, possibly because he had not read Newton's 1672 paper carefully enough. But, as we shall soon discover, not long after April 1706 he had the opportunity to re-acquaint himself with Newton's theory; and, as a consequence, 'second thoughts' were soon to follow.[32] It also should be noted that at some point, North also seems to have read one or more of the papers on Newton's reflecting telescope published in the *Philosophical Transactions* (1672), No. 81: 4004–10 + plate; (1672), No. 82: 4032–5; and (1673), No. 83: 4056–9.[33] For in an unpublished set of essays concerning the sensori-motor capacity, North resembles

---

[30] For the earliest preserved text, see RN, 'Some Notes upon An Essay of Musick'. For the later treatise, see n.31 below.

[31] See RN, 'The caus of sounds', *Cursory Notes of Musicke*, p. 71, prior to which the subjects treated are: Of springs, An hypothesis of comon springs, Experiments of springs, Propertys of springs, Of Pendulums, Of the monochord, Of procession [of springs], Watery surfaces, The spring of the air. Note also the subtitle to the whole, p. 1: 'Essay of musick, being a philosophicall introduction, tending to unfold the mechanisme of the whole art, and practice of it'.

[32] See *infra* 2.2.2: Nos 2 and 3, and 2.2.4.

[33] For reprints, see Cohen (ed.), *Isaac Newton's Papers*, pp. 60–75. According to Shapiro, 'The Gradual Acceptance of Newton's Theory of Light and Color', pp. 64–5, Newton's reflecting telescope, 'the first ever constructed', was delivered to the Royal Society at the end of 1672, where it was received with acclaim, thus leading to his immediate nomination for membership.

the 'seat' of thought to 'the object convex mettall in Mr. Newton's reflecting telescope'.[34]

By far the most important of Newton's publications, however, were two books, the first of which was on rational mechanics, namely,

Philosophiæ naturalis principia mathematica. London: J. Streater, 1687.

Philosophiæ naturalis principia mathematica. Editio Secunda Auctior et Emendatior [edited with a preface by Roger Cotes]. Cambridge: Cornelius Crownfield, 1713.

North owned copies of both editions, the second of which was published in June 1713. But he is not systematic in citing either of them. For example, he may refer to the book either as 'Principia' or as a garbled version of its title (e.g., 'Principia naturalis philosophiæ mathematica'.[35]) Or he may refer to Cotes either by name or only as 'editor'. More often, however, he provides no citation at all.

As for Newton's second book, two versions appeared in print during the period covered by this narrative:

Opticks: or, A Treatise of the Reflexions, Refractions, Inflexions and Colours of Light, also Two Treatises of the Species and Magnitude of Curvilinear Figures. London: for S. Smith and B. Walford, 1704.

Published in quarto by 16 February 1704,[36] this first edition of the *Opticks* has separate pagination for Book I (pp. [ii], 1–144 + plate) and Books II and III (pp. 1–137 + plate), the last of which concludes

---

[34] See f. 14v in RN, 'Some essays, concerning our manner of sense, or perception of things' (UK:Lbl) Add MS 32526: ff. 8v–34, *continued as* 'Of humane capacity, ff. 34v–47v (mostly early period).

[35] See, e.g., f. 30v in RN, 'Of distinction and indistinction' (UK:Lbl) Add MS 32548: in Notebook 2, ff. 28–31v (middle period).

[36] See Shapiro, 'Beyond the Dating Game', p. 222.

with Queries 1–16 (pp. 132–7). The two mathematical treatises, which are in Latin, follow: *Enumeratio lineis tertii ordinis* (pp. [i], 139–64) and *Tractatus de quadratura curvarum* (pp. [i], 165–211).[37] To the best of my knowledge, North does not cite these mathematical treatises in any of his writings, although it should not be ruled out that he might have made use of them. However, when the title 'Opticks' appears in North's manuscripts, it may not designate this or any of its subsequent editions (see further below).[38]

Optice: sive de Reflexionibus, Refractionibus, Inflexionibus & Coloribus Lucis Libri Tres, Authore Isaaco Newton, Equite Aurato. Latine reddidit Samuel Clarke, A.M. Reverendo admodum Patri ac D^{no} Joanni Moore Episcopo Norvicensi a Sacris Domesticis. Accedunt Tractatus duo eiusdem Authoris de Apeciebus & Magnitudine Figurarum Curvilinearum, Latine scripti. London: for S. Smith and B. Walford, 1706.

Published in quarto on 2 April 1706,[39] this Latin translation of the 1704 *Opticks* was undertaken (at Newton's request) by Samuel Clarke with the assistance of Abraham de Moivre.[40] It included an additional set of queries (Queries 17–23, pp. 293–348), a number of

---

[37] For details of these Latin treatises, see Guicciardini, *Isaac Newton*. Note that extracts from the second treatise were included in John Wallis, *A Treatise of Algebra* (1685), a work that RN cites; see, e.g., *infra* 2.2.5 and n.207.

[38] During RN's lifetime, the 1704 edition was re-issued, with some additions, in 1717, 1721 and 1730.

[39] See the advertisements in *The Post-Man* (30 March to 2 April 1706), no. 1594, p. 2. The 1706 edition was re-issued in a second, revised version in 1719, when Clarke also re-issued revised versions of some of his own publications.

[40] Shapiro, 'The Spectre', pp. 189–90, who also pointed out that in his translation Clarke completely eliminated the word spectrum, replacing it with the Latin word *imago*. Little notice seems to have been taken of the dedicatee, John Moore (Westfall, *Never at Rest*, does not mention him), so it should be noted that he was a patron of Clarke; see *infra* Chapter 3.

revisions to the main text itself[41] and a separately paginated section for the two mathematical treatises (pp. 1–43).[42] Sometimes, North refers to the Latin translation as 'Optica', sometimes, as 'Opticks', although here too he may not provide any citation at all.

Finally, one other book by Newton should be mentioned, namely, the book published anonymously under the editorship of William Whiston.[43]

> *Arithmetica Universalis; sive de compositione et resolutione arithmetica liber. ... In Usum Juventutis Academicæ*. Cambridgiæ: Typis Academicis. Londini, Impensis Benj. Tooke, 1707.[44]

To date I have no definite evidence that North owned or read this book. Still, it is possible that he did, for on 22 December 1706 he wrote his nephew, Philip Foley, then at Peterhouse College, Cambridge, that

> Mr. Newton's algebra will be very abstruse, but extraordinarily good and new, for that is his masterpiece. And if he had kept to that and mechanics, and let dabbling in physics alone, he had done no harm to his fame. For he hath broached a sort of philosophy more occult than that of the peripatetic scool. And his aim, as I guess, is to sanctifye all vulgar and natural prejudices in a philosophick dress, and to

---

[41] For some changes in the Latin translation to the first set of queries, see Koyré, 'Les Queries'.

[42] According to Priestley, 'The Clarke-Leibniz Controversy', pp. 37–8, during the course of publication a revised version of an important passage in Qu. 20 was substituted for the cancelled original in the majority of extant copies, only five of which now have the unrevised version.

[43] I.e., a disciple of Newton; see *infra* 3.3.1.

[44] This book is discussed in Guicciardini, *Isaac Newton*, who, p. 63, detected a tension in it. On the one hand, in its conception of algebra as the tool for the analysis of problems, the book might be understood as a fulfillment of the program outlined by Descartes regarding the use of algebraic criteria in the construction of geometrical problems (see his essay on geometry in *Discours de la méthode*). On the other hand, the book ends with pointed criticisms of Descartes's program.

keep the world from looking further, [whereas] in geometry he
is boundless[ly] liberall.[45]

Since North's meaning will become clearer later on, I shall merely
draw attention here to his reference in the above letter to 'Mr.'
Newton, who had been knighted on 16 January 1705. Nevertheless, it
cannot be assumed that North was ignorant of this fact. Indeed, in
1706 one of his correspondents referred to Newton under the
denomination of both 'Mr.' and 'Sir' in each of two letters.[46]
Unfortunately, therefore, North's use of one or the other title cannot
provide evidence for dating his manuscripts, including those that are
the subject of the next sections.

2.2. NOTES ON READING NEWTON: DESCRIPTION AND EDITIONS OF THE MANUSCRIPTS

To my knowledge, there are only three of North's manuscripts that
record his experience of reading Newton's two main books. Written
in the form of notes, these three manuscripts provide the copy texts
for the editions included here. But they are notes in two respects.
First, two of the manuscripts have the term 'notes' in their title.
Second, each manuscript is written wholly or partially in the style of
commonplace entry, a style North elsewhere describes as abridging
('contracting') the meaning of a 'point', 'period' or legal case and
then writing down the meaning as a short note or memorandum.[47]

The first note is devoted to *Principia mathematica*, although
the title is not mentioned. The second, much shorter note concerns
the book on optics, although here too no title is mentioned. Unlike
these two notes, however, the third concerns both books, citing the

---

[45] RN→Philip Foley, 22 December 1706 (UK:Lbl) 32501: f. 54. Note that there
are a number of errors in the edited version; see RN, 'Letters', pp. 254–5.

[46] See *infra* 3.3.2: No. 4, SC→RN, 28 September 1706 (f. 1) and No. 8, SC→RN,
21 November 1706 (f. 1).

[47] See Kassler, *HRN*, p. 64.

first as Newton's 'astronomy' and the second as his 'Opticks'. Moreover, its style suggests a re-reading, not a first reading, which means that North is beginning to move toward a more critically reflexive stage in his thinking.

In the descriptions of the manuscripts and editions that follow, the notes are ordered according to a probable chronology based on inferences from stylistic matters just discussed, as well as from other internal evidence (see below *Dating*).

No. 1 (UK:Lbl) Add MS 32546: ff. 174–178v.

Title. When North began to make his notes, he used 'Newton' as a title; but afterwards he relied on running heads as follows: 'Newton' (ff. 174, 174v, 175, 175v); 'Newton. p. 382' (f. 176); 'The Mathematick Style' (f. 176v); 'Mathematick Method' (ff. 177, 177v); 'Of Absolute Motion' (f. 178v). I have retained the title 'Newton' but dropped the running heads for the reason given in the next paragraph.

Physical features. The handwriting indicates that the notes were written on the run—cursorily—for there are numerous slips of the pen, a feature more obvious from a transcription than from an edition. At f. 175 there are instructions for moving a text from ff. 175–175v to a different location in ff. 178–178v. Because of this, the running heads previously mentioned (see above *Title*) are omitted in the edition.

Dating. The paper on which the note is written falls within two classes: 'Unidentified' (ff. 174–174v), because lacking both watermark and countermark, and 'Shield unidentified', because lacking a countermark (ff. 176–178v).[48] Hence, the paper cannot be dated. As for internal evidence, there is reference to 'Mr.' Newton; but for reasons given in the foregoing section, this reference does not necessarily mean that the text was written before the date when Newton was knighted (16 January 1705). Fortunately, however, North refers to some of his own essays, written and projected, though without providing specific details (see ff. 176, 178). In the case of his written essays, the main clue is that they contain his 'reason' of

---

[48] See Chan, 'Dating the Paper', pp. 63, 64, 86–7.

gravity. If I have identified these writings correctly, then this note was written before *c.*1705, the end of North's early period, and almost certainly before he read the Latin *Optice* (see below *No. 2*).

No. 2 (UK:Lbl) Add MS 32546: ff. 180–180v.
    Title. 'Some Notes of Mr. Newton's Singular Insinuations'. These 'Notes' are unrelated to the three folios that follow ('Some hints of light & colours', ff. 181–183v), which are discards from a longer essay.
    Physical features. At f. 180 the old British Library foliation (f. 420) has been crossed out. Tight binding has affected f. 180v, so that some words are partially or wholly obscured. Although there are few emendations (deletions and insertions), there are a number of slips of the pen.
    Dating. Since the paper falls into the class 'Unidentified', it cannot be dated.[49] However, internal evidence indicates that North is referring to several of the new queries added to *Optice*.[50] Consequently, North wrote this text sometime after its publication on 2 April 1706.

No. 3 (UK:Lbl) Add MS 32546: ff. 184–187.
    Title. 'Notes on Reading Mr. Newton's Opticks'.
    Physical features. The verso of each folio is blank; and the old British Library foliation (ff. 424–427) has been crossed out.
    At the beginning of the fragment, North promises to mark out some of Newton's unusual opinions for judgment. Hence, he begins by numbering the two paragraphs that follow as '1.' and '2.', after which he drops the numbering altogether. I have retained the numbering, since it reflects the style of commonplace entry prior to his moving toward a more critically reflective style. There are surprisingly few emendations (deletions or insertions) or slips of the pen. Nevertheless, hasty writing is the source of a number of

---

[49] See op. cit., pp. 64, 87.

[50] See *supra* 2.2.1.

problems, including difficulties in deciphering some words and, especially, in determining to whom North refers when writing 'he' or 'him'.

Dating. The paper is classed as 'Unidentified' and, hence, is undatable.[51] Moreover, the reference to the 'Opticks' is misleading, because ff. 184 and 185v contain remarks that derive from one of the new queries added to the 1706 *Optice*. And this means that the reference to 'Mr.' Newton is also misleading. From stylistic features, however, it is probable that the text represents North's final note, written sometime before the latter half of 1706.[52]

No. 1 (UK:Lbl) Add MS 32546: ff. 174–178v
[f. 174]

Newton

1. fol.[53]   *Aer duplo densior in dupla spatio quadruplus est.*[54]—
[*]*Rectissime*.

2. [fol.]   It is supposed matter is not disturbed in any state without difficulty.[55]

I doe not like that way of expression, such as [*]*vis inertia*, and the like. For it is no more a shew of force for bodys that meet to part,

---

[51] See Chan, 'Dating the Paper', pp. 64, 87.

[52] The date is inferred from evidence *infra* 3.3.2.

[53] I.e., Definitio I. in Newton, *Principia mathematica*, p. 1: The quantity of matter is the measure of the same, arising from its density and bulk conjointly. In modern terms: mass (quantity of matter) as the product of specific gravity (density) and volume (bulk).

[54] Ibid., p. 1, comment on Def. I., which begins: Air of a double density in a double space is quadruple.

[55] Ibid., p. 2, comment on Def. III., which includes the statement that a body, from the inert nature of matter, is not without difficulty put out of its state of rest or motion.

then to goe on in a direct cours; but the fancy is from our conserving our owne force, and comparing with that.[56]

5. [fol.]   [*]*Tempus absolutum*, etc.,[57] that is, a true and mathematicall time, without relation to body, flows equally.[58]

This I deny [because] time, abstracting body, is nought. Body is measure, or demension, which is transferred to time. Take away all measure by body or its image (as wee conceive)[,] space and time must be gone.

It is not strang[e] that this [*]conceipt is so difficult, I might say absurd to our understanding, which cannot make an image or idea of time-less-ness: It is just the same as when wee conceiv any thing, it is under some measure. For all wee know of our selves is body, and that wee feel it. I wish philosofers would not [have a disposition] to argue the necessity of things, from the necessity of their conceptions.[59]

6. [fol.]   There are much paines taken to make men conceiv the relation of motion, so that in some respects things move, and in others stand, and thence grow distinctions of motion and of place, absolute and relative, and men walking in ships under sail, which considerations enterteined the vertuosi in the infancy of new philosofy; as may be seen in the disputes of Gassendi against the

---

[56] In op. cit., Def. III. Newton used the term *vis insita* (innate force). But in the comment he added that this *vis insita* may by a most significant name be called *vis inertiæ* (in modern terms, inertial mass).

[57] Ibid., Scholium (to the Defs), pp. 5–11, which includes the distinction between absolute and relative time, space, place and motion.

[58] Ibid., Scholium, p. 5: absolute, true, mathematical time, of itself and from its own nature, flows equably without relation to anything external, and by another name is called duration.

[59] A hint of RN's epistemological realism, reiterated *infra* f. 174v.

Tolomaik systeme,[60] and so here [in *Principia mathematica*].[61] Whereas one would thinck the world should have now a clear intuition of that whole matter which is but, in short, that as you goe to or suppose bodys continuing or changing place and distance with respect to each other, they are **[f. 174v]** sayd to move or not to move; for look at one thing you move[,] at another thing you rest. So that the result is that motion is but a word by which men mean they know not what; they would have it somewhat, if they could and yet they cannot be without it. And in truth motion and no motion, respecting such body as to it self[,] is all one, and nothing more can be said of a body in motion (as it is termed) then may be sayd of the same, going no farther, then if it (as they[62] say) rested. And if one would give a description, or caracter of motion, or of all that wee can find true in it, [then] it is chang[e] of position and distance, and nought els. So if you could look for motion, [then] examine distance and position, and you have it. What happens upon the [*]occurs of bodys, I may touch after.[63]

7. [fol.]   I desire to know what absolute time is? He says the perseverence of things in their being;[64] which returnes us to our old

---

[60] I.e., Gassendi's assessment of the astronomical systems of Claudius Ptolemaeus, Tycho Brahe and Nicolaus Copernicus in the first (of two) volumes of his *Syntagma philosophicum Epicuri* (posthumously published, *Opera omnia*, 6 vols, Lyons, 1658).

[61] I.e., Descartes and Newton have opposing views concerning a system of two bodies in relation to ships under sail. See Descartes, *Principia philosophiæ*, Pt II, Arts 13–15, 24–31. See also Newton, *Principia mathematica*, Scholium, pp. 6, 8 and Corol. V (to Axiomata sive Leges Motus), p. 19.

[62] I.e., philosophers.

[63] Missing in this text; but see ff. 108–118v, *et passim* in RN, 'Mechanick notes' (UK:Lbl) Add MS 32540: ff. 81–118v, *untitled continuation*, ff. 119–137v (early period, late).

[64] Newton, *Principia mathematica*, Scholium, p. 7: '*Eadem est duratio seu perseverantia existentiæ rerum....*' (The duration or perseverance of the existence of things remains the same....).

way of conceiving time, which is from comparison of movements,[65] when all that while they persevere, etc. But if time be any thing, [then] it consists in measure, els why say they[,][66] [*]*equabiliter fluit, et mutari nequit*?[67] What can that measure be, but demension under which wee conceiv body. Then take that away[,] the other vanisheth.

Then the paralell between time and space, [the] one, succession, the other[,] place, which is it els, but referring the [*]essence of things to our imaginations?[68] I would know if the Almighty can annihilate all space? [If] they will say, yes; then he might create it, and as much of it as pleased him; and it is (to him) possible, that there should be any or none without implying any contradiction. Why then, all these distinctions?

8. [fol.]   The gusto of disputing Cartesius' definition of motion, *qui sim[il]ia corporum qu[a]e tanquam quiscentia spectantur,*[69] shews[,] with other beating the bush[70] to define what every one knows[,] motion to be nothing **[f. 175]** except in a relative sense, as bodys mutually measure distance betwixt each other. For if you see no alteration of distance among bodys in their parts, which wee call position[,] [then] you know no motion, alltho you and all

---

[65] I.e., of the heavenly bodies.

[66] Note the change from 'he' to 'they', perhaps a clue that in addition to Newton, RN may have in mind one or another of his followers.

[67] RN here paraphrases Newton's statement that the flowing of absolute time is not liable to any change (a statement that precedes that quoted in n.64 above).

[68] For another who used this method and for the reason why RN rejects it, see *infra* 3.3.4.

[69] (Lat.) which are similar to bodies that are observed as if in a state of rest. RN here paraphrases Descartes, *Principia philosophiæ*, Pt II, Art. 25: '[*Motus est*] *translationem unius partis materiæ, sive unius corporis ex vicinia eorum corporum quæ illud immediate contingunt & tanquam quiescentia spectantur, in viciniam aliorum.*' ([tr. by Miller and Miller, p. 51] it [motion] is the transference of one part of matter, or of one body from the vicinity of those bodies immediately contiguous to it and regarded as at rest, into the vicinity of others.)

[70] 'beating the bush', i.e., expending labour of which the fruit is not gained by oneself.

you see, with others out of view[,] may chang[e]; but the chang[e] is not in one more then in another.

Therefore once for all let us lay aside these words, as to the ordinary sence of them, as giving, receiving, generating, indeavouring, partaking, joyning, determining, etc. of motion, and confine our selves to know[,] if wee can, what changes in the casuall [*]occurs of bodys may happen, and to that onely referr all our language.

A property of motion, that such as keep position, partakes of ye same motion.[71]

9. [fol.]   Distinction of true motion, and relative motion.[72] The former is not without impuls, tho [the] lat[t]er may be from other bodys moving. All this is built upon our fancy that force is as it were a spirit, for wee feel it, and know the paine and concussion of our flesh, and therefore attribute to strokes of inanimate things the like. Whereas upon every stroke there is as regular and determinate

---

[71] MS has a semi-colon, perhaps signifying RN's intention to add a comment, although little space is left for it. Regarding the sentence itself, see Newton, *Principia mathematica*, Scholium, p. 8: '*Motus proprietas est, quod partes quæ datas servant positiones ad tota, participant motus eorundem totorum.*' (It is a property of motion that the parts which retain given positions to their wholes, do partake of the motions of those wholes.)

[72] Newton's distinction between true and relative motion is detailed over a number of pages, ibid., pp. 6–11; but RN is responding to the first paragraph, p. 9, which begins: '*Causæ, quibus motus veri et relativi distinguuntur ab invicem, sunt vires in corpora impressæ ad motum generandum.*' (The causes by which true and relative motions are distinguished, one from the other, are the forces impressed upon bodies to generate motion.) The import of the paragraph is that to understand the relationship between true motion and impressed force, motion is to be understood only in terms of a body's relation to places within absolute space and not in terms of its relation to other bodies (i.e., Descartes's definition). In the paragraph that follows, Newton begins his description of the first of two experiments that, according to him, prove true motion, namely, his bucket experiment, concerning which see n.73 below and *infra* ff. 178–178v.

consequences, and as easy as when bodys move without any concussion.[73]

[f. 176]

[p. 382[74]]   The objection to the vortexes[75] is that[76] by the motions of the planets, the periods are timed as 1½ of their distance; and [that] the fluid must move (as before proved by a seeming experiment, and arguments[77]) [and] must be in the quadrate, or duplicate proportion of the distance. To which I say

[*]*Negatur*, that the vortexes must by parts turne swift in such proportion, [and thus] they may goe slower further off the centre.

He argues that the force beginning either from the center or circumference of a fluid inclosed (as [a] vortex) will never leav till it bring the fluid to an uniforme motion, such as a wheel the diametrall points of which moves in time in duplicate proportion with the distance from the centre.

True, but suppose the parts from the center of a fluid such as are concerned mainely (for the subtile matter whatever it is, tho somewhat, yet is not intirely concerned as passing between) doe become grosser and grosser, it cannot be say'd that ever, the subtiler from the center shall bring the grosser outwards to the same degree

---

[73] MS has a new paragraph after this one, which begins: 'The experiment of the pail of water, hung by a cord and turned, see another paper, to which add.' The instruction, 'see another paper', refers to the paragraph that follows, but which, according to RN's directions, I have transferred to ff. 178–178v below.

[74] MS has this page number as part of the running head (see the description of this copy text above).

[75] I.e., of Descartes; see Newton, *Principia mathematica*, Bk II, Prop. LIII, Theorem XL, pp. 382–3.

[76] MS has: 'The objection to the vortexes is that,) in view,)'. I do not understand the meaning of the parentheses.

[77] See Newton, *Principia mathematica*, Scholium to Bk II, Prop. LII, Theorem XXXIX, pp. 375–81, in the last page of which he made clear that the purpose of Prop. LII was to investigate the properties of vortices in order to find whether the celestial phenomena can be explained by them. Note also that in the concluding sentences of Bk II, p. 384 (mispaginated, p. 400), he warned that the hypothesis of vortices serves more to perplex than to explain the heavenly motions.

of swiftness (or uniforme as in [a] wheel) but the heavyer will hang back somewhat, as the greater being struck, the less shall rather reflect then excite the same degree of swiftness.

That this is so, the discours of the reason of gravity demonstrates.[78]

[f. 176v]

It [the mathematical style of demonstration] seems appropriate to the subject of precise quantity. Cartesius first introduc't it in philosophy, and others since have affected the same, particularly Newton. But comendable as it is, like other sciences [the style] grows rude by unskillful handling, and not a litle, by sheltring conjectures under an abstruce method, which all will not nicely examine, and then [*]*Q.E.D.* closeth the paragraff.

I thinck it wholly improper in phisicks, which deall not more in substance, but events. And the axioms that relate to events are more exposed to error then those that belong to pure quantity, as none can doubdt, but the whole is equall to all the parts, and the like. But when wee say, like caus hath like effects, [this] may be also clear, but there is more [difficulty in] the application wherein both caus and effect may be mistaken; as in that elench[us][79] [*]*non causa, pro causa.* But this is yet good method in the way of argument and ordinary reasoning provided wee doe not pretend to the same rigid demonstration as belongs to quantity subsisting.

To instance. Cartesius was mistaken in his laws of motion grossly,[80] as arguing that a less body would not move a greater at

---

[78] See ff. 31v–32v in RN, Mostly untitled (UK:Lbl) Add MS 32546: ff. 1–18v, *continued* ff. 25–32v, *continued* ff. 19–24v (early period). At f. 32v he states that gravitation deserves 'a particular discours', for which see RN, 'Gravitation' (UK:Lbl) Add MS 32546: ff. 91–111v (early period). In both texts RN assigns the cause ('reason') of gravity to the mechanics of the vortex; and in both (especially the latter) he answers objections of 'virtuosi' that were made previous to Newton's critique of Descartes's vortex mechanism.

[79] I.e., cross-examination or refutation.

[80] I.e., not Descartes's seven rules for calculating motion but his three 'laws of nature' concerning the transference of motion; see *Principia philosophiæ*, Pt II,

all,[81] and [he also was mistaken] in other instances of that sort which abound in his works [and which] he holds as demonstrated, because he applyed axioms to a wrong subject. So also [*]parties that set up for his correction, much wors. He [Newton] would have the least body move the greatest with all its swiftness, but it must be done [*]*in vacuo*; and both [men] **[f. 177]** attribute the variation in experiment to the medium[82] which they say alters the case from the strickt energy of the force [*]*movant*. All which is [*]*tromperie*. For if a body be struck [*]*in pleno*, [then] at the instant of the stroke it is in all circumstances as [*]*in vacuo*, onely so much as the circumjacent matter hinders the force, which is reduceable to a stated quantity. It is just as if the body *in vacuo* were so much bigger then *in pleno*; for the plenitude of quantity makes it [so]. And yet the

---

Arts 37–42, especially the third law and its two proofs, as the rest of the sentence implies.

[81] See the fourth rule for calculating motion, op. cit., Pt II, Art. 49. According to Hall, 'Mechanics and the Royal Society', pp. 25, 33–4, both Hooke and Huygens rejected this rule, the latter asserting that any [one] body 'however large' is moved by any other body 'however small', moving with any velocity. According to RN, in his *Discours sur le mouvement local* (Paris, 1670), Ignace Gaston Pardies had asserted that *in vacuo*, 'the least body shall give the greatest its full celerity' and that *in pleno* 'great bodies are hindered by the medium'; see f. 107 in RN, 'Mechanick notes' (UK:Lbl) Add MS 32540: ff. 81–118v, *untitled continuation* ff. 119–137v (early period, late). Against the latter, RN's 'maxim' is that *in pleno* the least body striking the greatest moves it somewhat; see, e.g., f. 147v in RN, 'Of light & colours' (UK:Lbl) Add MS 32546: ff. 143–168v (early period, late).

[82] For Descartes, the principal factor in determining resistance was the subtlety of the parts of the medium, whose motion he conceived as a series of tiny impacts deflecting or retarding a body. By 'experiment', RN probably means Descartes's illustration of impeded or resisted force by analogy with a stone rotated circularly in a sling (the force of the stone to recede from the center is restrained by the tension of the string); see *Principia philosophiæ*, Pt III, Arts 57–59. For Newton, the density of the medium, not the subtlety of its parts, is the principal factor in determining resistance; see especially *Principia mathematica*, Bk II, where the subject of bodies moving in resisting mediums is treated at length. By 'experiment', RN may mean the Scholium (to the Axiomata), pp. 20–5, in which Newton described an experiment with pendulums of equal and unequal bodies (bobs) and pointed out, p. 21, that to avoid '*errore sensibili*' concerning velocities, all such experiments should be made as if the resistance of the air were taken off, i.e., as if the pendulums were really placed *in vacuo*.

buissness of motion, becaus that is in all respects adjusted by quantity of matter in which it is considered, is the onely branch of phisicks that the mathematical method can be reasonably used with.

Experiment and not axiom is the ground of phisicks.[83] I grant multitude of experiments succeeding alike argue inductively, so as to make probable, but not demonstrate. As [for example] the great luminarys for many centurys have held a certein cours, it is morally safe to conclude[84] the same will continue; but not so proved as mathematicks require. If it be say'd, without miracle interposing [then] it must, for bodys in motion continue, etc., that is[,] all [bodies] wee know doe so. But it is yet easyer and fairer to reply that some caus of chang[e] may occurr that never appeared to us, then to deny mathematick axioms. In short, mathematicks properly demonstrates, that is, shews things as they actually exist by their parts,[85] or which is the same thing[,] supposed to be, and in truth are really so in all lumps of quantity, tho no knife can practise the devisions. But physicks deals wholly in events and reducing them to their primary[86] causes; but causes are so latent, and various in divers subjects, that events of similar causes may not alwais answer.

[f. 177v]

But the greatest and worst [*]inconvenience is loss of a world of art and labour, such as Mr. Newton's work, which is compleat in the mathematick way, but in phisicks[,] barbarous.[87] Nor doth his

---

[83] I.e., op. cit., p. 20, the sentence that begins the Scholium to the Axiomata, in which Newton stated that hitherto he had laid down the principles received by mathematicians (who are named) and confirmed by a multiplicity of experiments. The logical implication is that the mathematicians' principles ('axioms') are empirical generalisations from experiments, an implication RN misses here.

[84] I.e., safe to conclude from probablistic reasoning.

[85] RN provides here a hint of his conception of analysis, which he elsewhere likens to algebra, see Kassler, *HRN*, pp. 102–3.

[86] I.e., principal.

[87] The relationship between mathematical method and physical problems was not yet understood; indeed, many philosophers, including Huygens, presumed that there was a basic difference in kind between mathematical sciences and natural philosophy; see Dear, 'What is the History of Science the History *Of*?', p. 398 and

apology, of not asserting phisically anything,[88] absolve him. For in conclusion he doth plainely and stricktly assert an hypothesis more precarious then any of the peripatetick traine, viz., the mutuall attraction of matter, supposing all bodys to attract each other in proportion to their quantity,[89] whereby the planets, working reciprocally towards approach, regulated [*]*secundum majus & minus* by distance and quantity and opposed by the principle of receding from the center, ballanceth their motions according to their phainomena. What are [the peripatetic] qualitys, substantiall formes, intentionall species, other [than] this? For may wee not say, when one quality works counter to its opposite, the consequence is a mediate station[,] as Mahomet in his iron tombe between the loadstone and the earth.[90] And if you will suppose quality, [then] what may not be demonstrated? For you must suppose degrees of

---

n.32. Even the president of the Royal Society did not understand the relationship, according to Feingold, 'Mathematicians and Naturalists', p. 82. At the time he was reading Newton, RN also shared the presumption of Huygens and others; but subsequently, he sought to gain a clearer understanding of the relationship, as also to find a justification or rationale for physics; see, e.g., RN, 'Preface to a philosofick essay' (UK:Lbl) Add MS 32526: ff. 64v–68 (early period, late), and RN, 'Physicks' (UK:Lbl) Add MS 32545: ff. 7–14v, and its draft, ff. 15–20v (middle period, early). However, to understand the relationship, a distinction was required between mathematics in respect to formal correctness and in respect to use. In the former, the method of mathematics is a logical matter, whereas in the latter, 'the method cannot be divorced from the subject matter—the "operative" limitations connected with physical problems'; see Strong, *Procedures and Metaphysics*, p. 215.

[88] See Newton, *Principia mathematica*, Def. VIII, pp. 3–5, where, pp. 4–5, he pointed out that the terms attraction, impulse, or propensity of any sort towards a centre are used by him indifferently and promiscuously one for another, since he has considered those forces not physically but mathematically; and where he afterwards warned the reader that when speaking of centers attracting or as endued with attracting powers, he is not to be understood as attributing forces in a true and physical sense to certain centers, which are conceived only as mathematical points.

[89] Ibid., Bk III, Prop. VII. Theor. VII, including (in RN's comment) Cor. I and II, pp. 411–12. The proposition itself states, p. 411: '*Gravitatem in corpora universa fieri, eamque proportionalem esse quantitati materiæ in singulis.*'

[90] For the source of RN's irony, see Browne, *Pseudodoxia epidemica*, Pt III, Ch. 3, pp. 78–9; see also RN, 'Of Etimology', p. 248.

more and less, [and] then enters arithmetick, geometry, etc. But to what end, when the very principle is [*]supposititious. [*]*Labor actus in orbem*?

Now, I not onely deny, as an opponent, these notions of centripetall and centrifugall forces,[91] but as a philosofer sincerely declare, I thinck them fals. For I conclude there is no such attraction nor any thing like it so universall as is pretended. Nor is there, originally or otherwise, then as a result of compound motions[,] such a thing as recess from the center of motion which leaves[92] the discours of the subject there.

[f. 178]
9 [fol.]   Experiment.[93] A pail hung by a cord, and turned till twisted very hard; then filled with water and lett goe; as it turnes back, the water at first is flatt, at length riseth at the sides and becomes more concave, till conformed in its motion with the pail.[94]
[f. 175]   It [the water] rose not at first, for the sides of the pail twitcheth and so holdeth the contiguous matter, and gradually in time, that is, the next[,] then [the] next and so on to the center or axis, which parts striking one another, drives them forewards, with a force in tangents, which makes it swell about the sides, and dish in the midle till the water moves all [in conformity] with the vessell[,] then

---

[91] For centripetal force, see Newton, *Principia mathematica*, Defs V–VIII, pp. 3–5, the first definition of which states that centripetal force is that by which bodies tend to a point as to a center. (Later in the book Newton referred to this force as 'attraction'.) As for centrifugal force, since in the first edition Newton did not provide a definition, some commentators have failed to agree on its meaning. For a plausible interpretation of centrifugal force as a reaction to (the action of) centripetal force, see Bertoloni Meli, 'Inherent and Centrifugal Forces'.

[92] MS has 'lets'.

[93] I.e., referring to the very first experiment described in Newton, *Principia mathematica*, Scholium (to the Defs), pp. 9–10. According to Hesse, *Forces and Fields*, pp. 138–9, there is a logical circle in this and in the experiment directly following.

[94] After this sentence MS has the words, 'Which is'. These words have been repositioned below just after the conclusion of the inserted passage from ff. 175–175v, regarding which see *supra* f. 175 n.73.

it sinks flatt, and it must be so when there is no striking at all. For as to the pail or vessell[,] there is no more motion then **[f. 175v]** was at first; and as to the air, there is motion, but not to lay hold, or draw the parts as the walls of the vessell did,[95] **[f. 178 cont.]** which is to prove, that it is movement absolute (as he calls it) and not relative, that creates the recess from the center.[96] For while the pail went round and the water stood still, which is a relative chang[e] of position, no effect followed; but when [the water] moved round and kept place with the pail, the full effect was [observed].

And therefore there is somewhat absolute in motion; and [if] it consists not wholly, [then it] is relation[,] so as to say, it's no matter on which body a moving force falls, so they separate or approach.

Now I conceiv this experiment doth not prove an absolute nature in motion, so as to ascribe the force or [*]essence of it to one more then another body, whilst they approach or separate, but it doth indeed prove[97] the motion or recess of adjacent bodys to, from or by any fluid **[f. 178v]** or any aggregate combination of bodys so as to create a recess from any point, untill, by impelling them, they are put into a state of translation, relative to some fixt point, which wee call a center. For till the [rotating] pail by the snatching [of] the fluid hath brought it into its cours, [the water] is not to be concerned in the buissness, more then the passage of a body at any distance. But the water is as it was at rest with respect to all center.

---

[95] On f. 175v MS begins a new paragraph with the words, 'Hence the author makes a'. But these words should have been deleted there, because they do not make sense in the context of f. 175v (where they are left hanging) or in the new context (where there is no text that would attach to them).

[96] The bucket experiment was intended to prove that in the case of rotation, absolute motion not only exists but also is detectable. Hence, it is an implied critique of Descartes's principle of relativity of motion, in which motion is defined with respect to contiguous bodies. Note, however, that, Newton, *Principia mathematica*, p. 10, hints ('*Unde & in Systemate eorum qui Cælos...*') that his critique is more broadly aimed at the world system of Descartes and his followers, though they are not named.

[97] MS has 'prove that'.

And when it is put to move with the pail[,] tho it may be called absolute, being raised to that gradually by the convulsion of the sides upon the fluid till it conformes, yet with respect to the center or fixt point, it is still relative. And be the cours at unity with the sides of the pail, it is in constant variance with the center, supposed a fix point, or considering a cilinder with the axis, a fixt line.

But still[,] goe to the reason of the recess from the center, elswhere given,[98] and it will appear to have no distinction from the case of comon impulses, and the consequences. Therefore these words, absolute, relative, etc. are but confounding, and not informing, or fitt for demonstration.[99]

No. 2 (UK:Lbl) Add MS 32546: ff. 180–180v
[f. 180]
Some Notes of Mr. Newton's singular Insinuations

1. That light is corporeal,[100] and darted with incredible celerity from the luminary all about,[101] and accordingly hath powers[102] attractive and protrusive;[103] and [that] divers happen in refractions, from the formation of the sides of such rays, as the usual and

---

[98] I.e., RN's 'reason of gravity'; see *supra* f. 176 n.78.

[99] MS concludes with a flourish of the pen.

[100] See Newton, *Optice*, Qu. 21, p. 315 (light rays are composed of small bodies [*corpuscula*] emitted from shining substances) and Qu. 23, pp. 335–6 (rays of light seem to be hard, impenetrable bodies without pores).

[101] I.e., the propagation of light from luminous bodies is not instantaneous but in time, ibid., Bk I, Pt I, Definition II, p. 2 and Bk II, Pt III, Prop. XI, Prop. XI, pp. 236–7. Newton estimated that from the sun to the earth the time involved was about seven or eight minutes.

[102] I.e., small particles of bodies have certain virtues or forces, ibid., Qu. 23, p. 322; namely, attractive and repelling forces, ibid., pp. 338–9.

[103] I.e., having a power or tendency to thrust forward or onward other bodies; hence, having a repulsive power.

unusuall rays together refracted in talk,[104] in one piece one way, and in an other the contrary [way].[105]

Answer.

That light is performed by the means of body, and consequently, must exhibit appearances conformable to body, as reflections, etc. is most true, but that rays of light are a stream of body[s] flowing from the luminary is absolutely fals and impossible. For

If [according to Newton] light be body, [then] wherever light is seen, there is that body, and accordingly, the light of a candle fills a room, for no point is to be found free from light. But that is not all. Let the room be walled and sealed with looking glass; each reflection fills the room, againe and againe. And what is more [*]wonderfull, these corporeall rays, tho each fill the room, doe not so much as justle one and other, for no one light is the more confused for another's being present. And this holds not in two or three instances, but miriads of light in a cathedrall church are seen each by his peculiar right-lined emanations as if there were none other there. Now is not this a prodigy of an opinion, if our notion of body,[106] ever hard and impenitrable[,] be true.

---

[104] I.e., talc, a name commonly applied to various transparent, translucent or shining minerals such as talc proper (i.e., hydrated silicate of magnesium), selenite and mica or Muscovy glass. Note that RN's spelling is also found in Hooke, *Micrographia*, Obs. IX, p. 48, and in the 1717 and later editions of Newton's *Opticks* (see Qu. 25).

[105] For the first mention that rays have sides, see Newton, *Optice*, Qu. 17, pp. 302–3; for the usual (*usitatæ*) and unusual (*inusitatæ*) refractions of the Iceland spar, see all of Qu. 18, pp. 304–6; Qu. 20, pp. 308–9; Qu. 21, p. 318; and Qu. 23, pp. 334–5. RN would have been acquainted with the previous work of Huygens relating to this crystal, for which see Bell, *Christian Huygens*, pp. 176–92.

[106] I.e., 'body' as a singularity. For Newton's conception, see ibid., Qu. 23, p. 343, where he stated that at the beginning God formed the properties of (atomic) matter as '*solidæ essent, firmæ, duræ, impenetrabilies*[,] *innertes & mobiles*'. For RN's conception, see *infra* 2.2.3, 2.2.4 and 3.3.3.

[f. 180v]

As to his colours, appearing upon refractions, by mea[ns] of different refrangibility[107] that is inherent in metall which exhibite them,[108] it must be admitted, that degrees refracted doe constantly produce such colours. But it follows not that there are corporeall rays, distinct from the matter of the world, imbued with the faculty[109] of[110] those colours and of being refrangible. It is enough[111] that[112] refracted ordinary light allwais occasions in us, suc[h an] idea, as wee have of colours, and that colours onel[y] made by refraction are not altered but continue the[m][113] thro all future refractions. But as to the caus, it may be fro[m] the nature of our organs, or any thing rather then corporeall rays.[114]

---

[107] According to Newton's theory, colours are original and connate properties of the rays, just as their respective degrees of refrangibility are. First published in 1672, it was developed at length in the 1704 edition of *Opticks*, as well as in its 1706 Latin translation, where it is also briefly recapitulated; see Newton, *Optice*, Qu. 21, p. 317.

[108] For Newton's brief observation concerning colours that arise on heating metals, see *Optice*, Bk II, Pt I, Obs. 19, pp. 184–5.

[109] RN here refers to the nature of Newton's (atomic) rays, which are not themselves coloured but which have a 'property', 'disposition' or 'power' for colour-making that is immutable and connate, a property that differs in each of the sides of the rays. For the property, see, e.g., ibid., Bk II, Pt II, pp. 204–5; for the sides of the rays, Qu. 18, pp. 304–6; Qu. 19, p. 306; and in Qu. 21, pp. 318–19.

[110] MS has in right-hand margin what looks like 'of th', the latter word being cut off by tight binding.

[111] MS has in right-hand margin 'enough t', but the latter word is cut off by tight binding; perhaps 'enough t[ho]'.

[112] MS has what looks like 'c-' above the word 'that'.

[113] MS has in right-hand margin 'y' with the tail of the 'y' continued in an arch around it, apparently with some continuation of the bottom of the arch to its right, though what this means is obscured by the tight binding.

[114] For RN, colour is an appearance (not real), the physiological explanation for which is to be found in his theory of information processing; see *supra* 1.1.4 and *infra* 3.3.5.

No. 3 (UK:Lbl) Add MS 32546: ff. 184–187
[f. 184]

Notes on reading Mr. Newton's *Opticks*[115]

Here [in his book on optics] as in his [book on] astronomy,[116] he brings in his dogmata, which are extraordinary in the way of phisicks, and so [since they are] much counter to the received opinions in that learning[,] I cannot but mark them out for judgment.

1. In generall I find a continued designe of depretiating Cartesius. For his [book on] astronomy seems calculated to prove [that] the planets [*]movent [*]*in vacuo*,[117] and [that] the vorticall movement of the etheriall matter[118] [is] inept to direct their motions.[119] And in this [book on optics] he continues his notions of vacuum[120] and colours,[121] for which lat[t]er, there may be reason [to depreciate Descartes]. However, he is the solver of the rainbow,[122]

---

[115] I.e., the 1706 *Optice* (see nn.120, 124, 134 below).

[116] I.e., the entire *Principia mathematica*, not merely Bk III (De Mundi Systemate), because 'astronomy' is not necessarily restricted to our solar system.

[117] I.e., the immense celestial space void of resistance because (mostly) void of air, exhalations and vapours; see ibid., Scholium (to the Defs), p. 11 (*in vacuo quovis immenso*); for cognate terms, see also, e.g., Bk II, p. 400[=384] (*in spatiis liberis*) and Bk III, p. 401 (*spatia corporibus vacua*).

[118] I.e., first element matter in Descartes, *Principia philosophiæ*. Note, however, that RN does not follow the French philosopher's distinctions concerning sizes and shapes of material particles; see, e.g., *supra* 2.2.2: No. 1 (f. 176) and *infra* 2.2.3 and 2.2.5.

[119] Newton, *Principia mathematica*, Bk II, Sect. ix (on the circular motion of fluids), especially Prop. LII, Theor. XXXIX and its Scholium, pp. 375–84 [=400], where Newton demolished the Cartesian vortex theory by showing that it is irreconcilable with astronomical phenomena.

[120] I.e., continued in Newton, *Optice*, Qu. 20, pp. 310, 312–14, which also includes his 'final sally' at continuum theories of light; see Shapiro, 'Light, Pressure, and Rectilinear Propagation', p. 296.

[121] I.e. continued from Newton's papers published during the 1670s in the *Philosophical Transactions*; see *supra* 2.2.1.

[122] I.e., in the essay on meteors that concludes Descartes, *Discours de la méthode*.

tho in that he [Newton] makes him plagiary of Antonio de Dominis, a sorry tutor for such a philosofer.[123] So for his thought of continuity;[124] and indeed [Newton] batters all the systeme called the mechanicall, and in no place gives him any credit.

2. He seems to designe an overthrow of all naturall philosofy, and to reduce it from inquiry into the mechanicall texture and caus of things to some granted principles, as attraction, repulsion, and the like, upon which he may exercise his excellent skill in geometry and fluxions.[125] One would thinck that philosofy were bewitched, so that when on[c]e sifted and made clean[,] it must be mingled againe with its chaff and straw. Or rather that there is more vanity then sincerity in the great geniuses of all ages; and no one will be contented without being a sole inventer of all, which inclines them to a spirit of contradiction. And under the influence of that[,] so farr from carrying on the discoverys of others and **[f. 185]** building on them as on so much ground gained,[126] they study more to confute then add of their owne, altho the matters[127] are in many things most reasonable; and if

---

[123] For this charge, see Newton, *Optice*, Bk I, Pt II, Prop. IX, Prob. IV, p. 140. For the 'sorry tutor', see Antonio de Dominis, *De radiis visus et lucis* (Venice, 1611), where is showed that in the primary bow one reflection and two refractions were involved. But Descartes achieved much more, so that Newton's characterisation of him as a plagiarist is 'wide of the mark'; see Boyer, *The Rainbow*, pp. 187–92, and Buchwald, 'Descartes's Experimental Journey', p. 7.

[124] I.e., RN's term for cohesion, a subject treated by Newton, *Optice*, Qu. 23, pp. 335–40, where he mentioned and rejected three theories purporting to explain how the parts of homogeneal hard bodies stick together. Although he did not mention any names, the first theory, that bodies are held together by hooked atoms, is that of Gassendi; the second theory, that bodies are glued together by rest, is that of Descartes (whose name for inertia was 'force of rest'); and the third theory, that bodies unite by '*plane Nihilo*' (nothing intelligible), probably refers to that of the animists (see n.133 below).

[125] I.e., the method invented by Newton for determining the ratio between the velocities—fluxions—of two motions when an algebraic relation between the generated magnitudes was given.

[126] I.e., RN's way which builds on precedent, like the common law, but innovates by using hypotheses (the analogue of legal fictions).

[127] I.e., discoveries of others.

any guesses or expressions are open to cavill [then] be said to peck there.

If Mr. Newton had erected a better sc[h]eme of his owne then had gone before him, [then] wee had approved his designing to pull downe the former. But [he is] like a second Aristotle, [who] overturned the learning of the ancient naturalists,[128] and set up qualitys and quidditys in its room, [whereas] now the world allows [the learning of the former] to have bin in a better way then the learning of [Aristotle] himself.[129] And what doth Mr. Newton doe other, setting up his powers attractive, dispersive,[130] centripetall and centryfugall, and I know not what, with which he supposeth matter specifically and variously possest[131] and to work withall like living creatures, loving and hating as appetite moves.[132]

This I thinck is [either] to overturne all naturall knowledg, or to make it childish and contemptible.[133] As if you ask[,] why the needle

---

[128] I.e., the atomists; see *infra* 2.2.5.

[129] MS has 'But like a second Aristotle, he overturned the learning of the ancient naturalists, whom now the world allows to have bin in a better way then himself, and set up qualitys and quidditys in the room.'

[130] I.e., having the power of dispersing, diffusing or spreading. The technical meaning came later (1801); see OED *adj.* b. *Optics*, of a refractive medium: having the quality of causing the different-coloured rays of light to diverge.

[131] Since in *Principia mathematica*, Newton used the term *vis insita* (innate force) as a synonym for *vis inertia*, it is possible that RN here understands the other forces as innate; see *supra* No. 1 (f. 174 and n.56). But he was not alone in this, for in the 1690s Richard Bentley was another who understood forces in this way; see, e.g., Hesse, *Forces and Fields*, pp. 150–1, whose entire chapter on 'The Theory of Gravitation' provides considerable insight into the problems of interpretation for contemporaries of Newton.

[132] I.e., Newton's central forces, like Aristotle's 'qualitys and quidditys', are unintelligible or 'occult', as RN hinted *supra* 2.2.1, RN→Philip Foley, 22 December 1706, and 2.2.2: No. 1 (f. 177v)

[133] Note that in the following series of questions, the answers—that matter has soul-animated 'loves and 'hates'—imply those that would have been given by animists, including Johannes Kepler, who supposed that magnetic attraction resulted from a 'mutual affection'; see Chalmers, 'The Lodestone', and Bell, *Christian Huygens*, p. 105.

comes to the magnet? Answer, the magnet hath an attractive power. Why doe heavy things goe to[,] and light things from the center, etc.? Answer, becaus bodys draw one and other. Why doth air rarifie? Answer, becaus the parts have an aversion, and avoid one and other. Why doe chimicall spirits act and influence? Answer, becaus they love not one and other, and will not, as birds of a different feather, be kept in one cage, and the like, as may passingly be observed in his *Opticks*. Where by the way, he hath crammed in his dogmata,[134] as desiring to possess the world, but not to take the paines to forme a systeme, wisely knowing the method will not bear it.

[f. 186]

It is hard to guess at a man's mind, but I, with [*]reverence to Mr. Newton,[135] beleeve his [intent] is, in his attempts of this kind, pious. He must be allowed a prodigy of good sence and clearness of thought; but yet in many subjects, which fall not in the pale of geometry, to labour under some prejudices of education, and particularly in the case of Cartesius' philosofy. He entered into his studys, when that [philosophy] entered into the world.[136] And it is well knowne how the academys received it, scarce otherwise then as herisy or anti-Christ; and the men of severer mood, among whom Mr. Newton may be placed, battelled it with all their might, scarce permitting students to read in it.[137] I fancy at that time Mr. Newton embraced a porpose to confute Cartesius, and bent all his studys that

---

[134] I.e., in Qu. 23, *Optice*, pp. 322–48, which includes Newton's theory of atomism, his conception of the proper method for investigating natural phenomena, his teleological argument for God's existence and his conception of God. Note that it served as the basis of the more frequently noticed 'Scholium generale' that Newton added to *Principia mathematica* [1713].

[135] Despite his criticisms, RN's reverence to Newton does not end with this note.

[136] Descartes's *Principia philosophiæ* 'entered into the world' in 1644. Newton, however, entered Trinity College, Cambridge, in 1661; became scholar in 1664, B.A. 1665, minor fellow 1667 and major fellow 1668. Although RN was in residence at Jesus College during the latter two years, I have no evidence that the two men actually met at that particular time, although the rest of the paragraph suggests that RN is recalling his year at Cambridge.

[137] See *supra* 1.1.1.

way, and so, as many doe, studyed himself into opinions, which he would not have taken up in other termes. And if he[,] as others pointed at,[138] thought that Cartesius' philosofy was injurious to religion in generall, as tending to atheisme, [then] I do not blame his Christianity and morality in seeking to ore'turne it. And if his admirable analitick genius hath given him a credit in the world, as Aristotle had, sufficient to found a sect and beat downe what he [Newton] pleased by dent of such authority, [then] he [Newton could] not better imploy it then in such a caus. And I scarce thinck any man in Europe more competent then himself for compassing such a designe.

But the failing, if any, is in thincking at this time of day to impose on the world, who will search all channels of [f. 187] philosofy and be their owne judges.[139] What els means [*]*Nullius in verba*?[140] And nothing will chang[e] the inquisitive and censoriall disposition of men. Perhaps accident[,] as befell Aristotle's works upon the revivall of learning[,] may give, as there, a prepossession, with a corrupt [*]politick aiding; but the literati will in time early or late, break thro all prejudice what ever. And so happened to Aristotle's philosofy of dreams, till the age of those vigilant hero[e]s, Copernicus, Gassendus, Bacon, Galileo, and in the

---

[138] The Cambridge Platonist, Henry More, was one of those who 'pointed at' Descartes's atheism and subsequently, in his *Enchiridion metaphysicum* (London, 1671), concluded that the Cartesian philosophy was deterministic and, hence, atheistic; see Feingold, 'Isaac Barrow', pp. 25–9, and Gabbey, 'Henry More'.

[139] Expressions of the principle of free inquiry, as well as its related principle (freedom of utterance) appear in a number of RN's writings; see, e.g., RN, 'Præfanda' (UK:Lbl) Add MS 32526: f. 2 (early period) and *infra* Epilogue (n.23). However, Colligan, *The Arian Movement*, p. 78, described this principle as part of 'the permanent deposit in the Nonconformist tradition', defining the latter tradition in a narrow sense. But as a non-juror (for a period of time), RN may be described as a non-conformist, since he was concerned to maintain a separation between church and state; see Kassler, *HRN*, pp. 46–7, and ff. 84v–85 in RN, 'Of the Clergy of England' (UK:Lbl) Add MS 32526: ff. 79v–87v (early period).

[140] I.e., the motto of the Royal Society, from Horace, *Epistles* 1.1.14: *Nullius verba iuratus me per omnes philosophiæ magistros funderen.* (Having bound myself to swear in the words of no master.)

[*]'rere[,] Cartesius. Wherefore it is my opinion that, if there are methods of philosofy which ultimate intens thinking will establish, as I take the generall cast of the mechanicall philosofy to include, [then] it is better to expose and imploy it to pious and moral porposes. And I am of opinion that much more such use may be made of the hipothesis of plenitude,[141] then of Mr. Newton's vacuity; but of that elswhere.[142] And whither I have a right notion of this author or not must appear [afterwards] by the unfolding some of his opinions, or suggestions.[143]

2.3. BEGINNING THE PROCESS OF UNFOLDING NEWTON'S OPINIONS

In the three notes edited in the previous section, North identifies a range of issues arising from some of Newton's concepts[144] or what he elsewhere refers to as 'indefinites' or 'abstracts'. By these terms he means principles that concern 'the whole world', that is to say, they are universal.[145] But since such principles are 'notions purely of the mind, and to be weighed by reason', they are not reducible 'to any certein test of measure or experiment'; and, as a consequence, they 'ever were and will be obnoxious to various fancys and opinions'.[146] As will appear, an important (perhaps the important) example is the universal principle, matter, which Newton had reconceived in 1687.[147] Recall, therefore, that at the conclusion of his

---

[141] I.e., the universal plenum.

[142] See *infra* 2.2.3 and 2.2.4.

[143] I.e., in RN's subsequent critical reflections on Newton's two books.

[144] From a modern point of view, the better term might be constructs; see Margenau, *The Nature of Physical Reality*, pp. 70–1.

[145] See f. 50v in RN 'Abuses of words' (UK:Lbl) Add MS 32548: in Notebook 2, ff. 50–53 (undatable).

[146] See f. 231 in RN, 'Indefinites' (UK:Lbl) Add MS 32546: ff. 231–246 (early period); see also Kassler, *HRN*, pp. 101–3.

[147] See *supra* 2.2.1.

third and last note, North expresses uncertainty whether he has understood Newton's meaning, so that 'whither I have a right notion of this author or not must appear [afterwards] by the unfolding some of his opinions, or suggestions'.[148]

Now, since all knowledge presupposes a fidelity to unspoken values, North's unfolding will also make more explicit some of his own values. In the following two sections, therefore, I draw on three manuscripts that were written during or shortly after completing his notes in order to exemplify some aspects of this dual process, beginning in this section with a manuscript that consists of an introduction and three parts.[149] From the introduction, which presents a detailed five-part plan, we discover that the manuscript is incomplete, for the second half of the first part is missing, and the third part falls far short of what the introduction had promised.[150] Nevertheless, the plan itself seems to depict what North would have written out as part of the collection of thoughts mooted in his autobiographical fragment.[151] Moreover, the physical features of the incomplete first part indicate that he also is seeking to provide in a more finished form his philosophy of nature.[152] Although undated, the manuscript is transitional, for it was written partly before North read Newton's *Principia mathematica* and partly afterwards, as will become apparent below.

---

[148] See *supra* 2.2.2: No. 3 (f. 187).

[149] RN, Mostly untitled (UK:Lbl) Add MS 32546: ff. 1–18v, *continued* ff. 25–32, *continued* ff. 19–24v (early period). Note that after a preliminary but untitled introduction, the running title is 'System of the World' and in the third part of the manuscript, 'The World'.

[150] This manuscript bears a close relation to RN, Untitled (UK:Nr) Rougham MS. According to Chan, 'Dating the Paper', pp. 55–6, 99, the MS is bound in a large folio volume with '5 Quire: Ridd: common-P' written on the spine—that is, it is a commonplace book. It is probable, therefore, that the transitional manuscript, discussed above, is based on notes in this commonplace book.

[151] See *supra* 1.1.2.

[152] After his move to Rougham, this aim proved 'too operose'; see *supra* Some Preliminaries.

In the first, incomplete part, North sets out, and comments on four universal principles (ff. 4v–16v): perception (*percipio*, not *cogito, ergo sum*); 'body' or the nature and constitution of matter; motion or changes of its several parts; and time. Since the second principle includes different mathematical methods of treating body,[153] he indicates that it is an error to believe that where there is no formal demonstration, there is no knowledge, because 'many things are clear enough to us from dayly experience and observation, which cannot be proved but by discours relating to observation'. Moreover, since his 'propositions relate to futurity and are not so much of present existence', his principles will not be demonstrated.[154]

Next follows a section entitled 'Abstracts' (ff. 16v–18v), which, North indicates, has been included to provide brief reasons that determine him to 'one or other opinion' about infinity (of extent as well as smallness), eternity of time, vacuity and elasticity. Concerning the first abstract term, North tells his reader that philosophers differ on infinity of smallness even to the point of demonstration, some holding there are, others there are no minima or indivisible parts. He identifies the first group only—they are atomists. But as will become apparent, the second group are Aristotle's successors, the scholastics. North claims that his definition of body (taken as a singularity) includes both positions and, hence, maintains a mean between the two extremes. For, on the one hand, in the section devoted to principles he had postulated that matter has actual parts of 'infinite or perfect hardness', 'without pore' and 'capable of no chang[e] att all'; that is, the parts 'cannot be actually broken, or made less', so that as it was at 'the dauning of the world' so it is 'at this day and will remaine for ever' (f. 11). And, on the other hand, in the same place he had also postulated that matter fills space in three dimensions and, hence, is geometrically extended

---

[153] I.e., arithmetic, geometric and algebraic techniques.

[154] Unfortunately, RN does not elaborate on this suggestive statement.

(f. 7).[155] Consequently, 'the very idea of it includes comparative quantity', so that 'notionally' 'body' is divisible *ad infinitum* (f. 17).

Since he believes there is 'no observation or experience to incline us to limit things' either in terms of smallness or immensity, North feels obliged to assent to the proposition that matter has parts 'infinitely small so that no space can happen but there will be bodys at hand ready to fill it'. And this assent to a universal plenum is the reason that determines his opinions against vacuity and elasticity, two concepts that also are inconsistent with his definition of body. Regarding vacuity, which in this manuscript denotes pores interspersed in body (taken as a compound), he acknowledges that 'most virtuosi' allow it in order to supply 'a seeming obstruction to motion'. But he thinks he can answer this fanciful opinion by allowing 'actual infinity', for then there is 'no impediment to motion', because 'the parts chang[e] and intermix, without clashing against the universal pressure' called gravity.

As for eternity of time, North states that it differs 'from infinity of space no otherwise then as time differs from motion', for 'it follows motion or chang[e] of position and is but the comparison of severall changes, and till a totall annihilation of body or absolute acquiescence of all the parts of it, time will be' (f. 16v). This remark follows from a digression in his comment on time as a principle (ff. 14–16), in which he states that he understands time and motion as 'one and the same observation, but to a different porpos; for in motion wee onely observe the chang[e] of position of bodys, but the notion of time is the observation or comparison of severall changes together'. North then indicates that by using this method, he was led 'to very pleasant contemplations, and perhaps not unprofitable, in relation to the most unsoluble questions of theology'. For if one lays

---

[155] RN (f. 7) uses the terms 'extension' and 'impenetrability' as synonyms for matter's space-filling property. But Descartes used the term extension on geometrical grounds, and More used the term impenetrability on the ground that 'a thing may be extended and yet not impenetrable'. What RN does not say is that More distinguished between matter as the extended impenetrable and spirit (RN's 'thing') as the extended penetrable, whereas Descartes restricted extension to matter alone.

aside the distinction of fore and after 'with body', then 'the divine *nunc stans*[156] must take place', for in the case of 'a perfect being' all knowledge which to us would be past and future is present. Accordingly, North desires that his 'thincking freinds' discard their 'prejudicated opinion, that time will remaine when body is gone'.

Note that at the very beginning of the manuscript North has raised two important issues. The first issue relates to his discussion of the nature of matter ('body'), its division and individuation. In that discussion he placed himself as a mean between two extremes, the atomists and the scholastics, who held rival metaphysical doctrines about the nature of matter.[157] According to the doctrine of the atomists, the parts of compound bodies actually exist prior to, and independently of any process of infinite division. Indeed, these parts are ontologically prior to the bodies themselves, in that the very existence of bodies is parasitic on the existence of the parts out of which they are composed. Accordingly, division cannot be considered as resulting in smaller and smaller parts without end, for then there would be nothing with independent existence out of which bodies are composed. Hence, there must be *ultimate* indivisibles—metaphysical atoms—supplying an ontological ground for the constitution of bodies out of their parts. According to the rival doctrine of the scholastics, the parts into which bodies are divisible exist only potentially, not actually, until a division produces them. Accordingly, such merely potential parts have neither independent existence nor ontological priority.

North claims that his mediating position embraces features of these two rival metaphysics. On the one hand, while he agrees with the atomists that the parts of bodies actually exist prior to, and

---

[156] I.e., (Lat.) the now that stands still. In defining eternity, Boethius, *De consolatione philosophiæ*, Bk V, Pt vi, used the phrase *nunc permanens* (abiding or permanent present). It is probable, therefore, that the phrase *nunc stans* came into Anglican theology through the influence of Thomism; see, e.g., Elders, *The Philosophical Theology of St. Thomas Aquinas*, p. 178, for Aquinas' account of Boethius' definition.

[157] For detailed treatment of these two extremes, see Holden, *The Architecture of Matter*.

independently of any process of division, he rejects their corollary of ultimate metaphysical atoms. On the other hand, while he agrees with the scholastics that matter is infinitely divisible, he rejects their potential infinity in favour of a 'notional' or 'actual infinity'. Unfortunately, in the manuscript under discussion, North provides no insight into what he means by these terms; but since I will return to this problem in the next section and, especially, in the following chapter, further discussion is deferred here.[158]

The second issue mentioned above concerns North's references to 'atomists' and 'thinckking freinds'. Who, in particular, might these be? In the case of the atomists, the reference is too general to identify any one or more persons with certainty. Since North conceives the parts of matter as absolutely hard, he may have in mind Gassendi, who worked out a Christianised form of atomism in his final exposition of Epicurean philosophy.[159] But if this part of the manuscript was written after North read the 1706 *Optice* (and I believe this is not the case), then Newton cannot be ruled out, for he had spoken of minima not only at the beginning of *Optice* but also and more extensively in its final Qu. 23, where he set out his conception of atomism.[160]

In the case of North's 'thinckking freinds', it is also difficult to determine precisely whom he has in mind. But since the reference occurs in his treatment of time and motion, it could include Isaac Barrow, who, in his *Lectiones geometricæ*, treated and defined an 'absolute time' as a quantity inaccessible to measurement except by

---

[158] See *infra* 2.2.4 and 3.3.4.

[159] I.e., *Syntagma philosophicum Epicuri*; see Jones, *The Epicurean Tradition*, pp. 176–80. In f. 208v of the manuscript discussed *infra* 2.2.5, RN refers to Epicurus' 'Epitomy in 3. letters'; and since this was included in the *Syntagma*, he must have read that work. But I have found no evidence that he read the work of Walter Charlton, Gassendi's English disciple.

[160] Initially, Newton, *Optice*, Bk I, Pt I, Def. 1, provided a hint of his atomism when he defined rays of light as its least parts; but he made it more explicit in the final Qu. 23, where he used the terms smallest particles, pp. 335–6, and more significantly, primordial particles, p. 343.

some constant, uniform motion.[161] Unlike Newton, however, Barrow observed that the word, time, denotes not 'an actual existence, but simply a capacity or possibility of permanent existence';[162] and perhaps it is this last phrase that North conceives as the 'prejudicated opinion' of his 'thinckig freinds', namely, 'that time will remaine when body is gone'. Nevertheless, because he had specifically singled out absolute time in his first note on *Principia mathematica*, here again Newton cannot be ruled out.

There is however no question whom North has in mind in the third part of his manuscript, where (ff. 21–21v) he uses the term centripetal that had been coined by Newton. For he writes: 'There may be many hypotheses or inventions for solving the planets courses'—for example, the 'Ptolomaick of old by solid spheares and epicicles' and the latest 'by reciprocall attraction, centripetall and centrifugall forces'. He adds that, 'It is strang[e] what a priveledg dogmatizing arrogates in matters not subject to immediate experiment, as in this instance of the heavens.' For without either 'direct' or 'analogicall experience', neither hypothesis can be proved true. Consequently, he favours the Copernican hypothesis, as 'unfolded more plausibly by Descartes', but with this qualification:

> I would not be understood to patronize the minute contrivances of Cartesius' first and second elements, etc. but goe with him thus far and no farther, that is[,] that the world is full of fluid matter ever in motion, and that not onely by intestine agitation, but [also by its] generally flowing in vast whirlepools about centers, which are called suns and starrs. And shewing a consequence of these girations to be a generall recess from the center of the movement, which by the laws of motion is stronger in some then in other bodys, the weaker

---

[161] See Mahoney, 'Barrow's Mathematics', pp. 204 and 245 n.36. For the treatment of time in his *Lectiones mathematicæ*, see Barrow, *The Usefulness of Mathematical Learning*, pp. 184–5, 211–18, *et passim*. Note that unlike other 'absolutists', who were more concerned with space, Barrow devoted considerable attention to time.

[162] Quoted in Burtt, *The Metaphysical Foundations*, p. 156, who also treated Barrow's conception of time at length.

give way to the stronger, and are by that means crowded towards the center, and from thence to the circumference. There resides matter of different degrees in power to recede according to distance where there is a sort of ballance of it. For if taken farther from the center, [then] it is among stronger [degrees in power], and there it must give way and come neerer, which is called gravity.

North, therefore, seeks to retain the vortex mechanism as an explanation of gravity. And he was not alone in this commitment, for as Stephen Gaukroger has pointed out, the 'resilience' of Descartes's vortex theory 'should not be underestimated', since, together with his laws of motion, that theory was 'the starting point for all serious work in physical theory in the mid-seventeenth century, Newton's included'.[163] Thus, when the latter, in *Principia mathematica*, replaced the vortex theory with the theory of universal gravitation, many philosophers were reluctant to give up the Cartesian vortex.[164] And in North's case, his notes suggest that the reasons for his own reluctance were not restricted to a concern with Newton's concept of universal gravitation but extended to his argument against Cartesian vortices in Book II.[165] For that argument was based on what North will afterwards refer to as Newton's 'doctrine of density'[166]—that the density of the medium, not the subtlety of the vortex parts, determines the resistance the medium offers.

---

[163] Gaukroger, *Descartes*, p. 383.

[164] Note that John North, who died before Newton's *Principia mathematica* was published, had written in a notebook that 'this system of vortexes [is] subject to more inconstancy and disorder then can agree with the exactness of the heavens'; see f. 216v in RN's copy of the notebook entitled 'Notes of Dr. North' (UK:Lbl) Add MS 32514: ff. 167–227v.

[165] RN would have read Book II carefully, since it includes Newton's theory of sound.

[166] See, e.g., *infra* 3.3.2: No. 9, RN→[SC 1706] (f. 384).

2.4. *ASSAYING SOME PRINCIPLES IN OPTICS AND MECHANICS*

Not long after writing the text considered in the last section, North began to revise some of his thoughts on optics and mechanics. Two early manuscripts devoted to these subjects will be considered here, in the first of which he seeks to test some concepts in Newton's book on optics, at the same time struggling with problems that would arise if he were to give up the vortex theory.[167] At the outset of the manuscript, he states:

> I doe not here intend to depreciate any of Mr. Newton's experiments, or mathematical conclusions, but owne them extraordinary, and wonderfully conducing to greater discoverys as time may produce. But so much of the phisicall part of the subject, as he hath touched, is the subject of my offence (f. 143).

Hence, he returns to, and elaborates on some points previously made in his notes concerning Newton's universal 'principles' of attraction, 'vacuum in the vast spaces', 'hardness as the only constant', as well as the theory that there is a 'stream', 'emanation' or transport of matter from the luminous source to the organ of vision. But he also adds some new points for examination—for example, that 'bodys work upon rays of light at a distance'.[168]

In the process of his examination, North rejects Descartes's theory that light is due to the centrifugal pressing of the etherial particles (subtle matter) to recede from the center of rotation of their

---

[167] RN, 'Of light and colours' (UK:Lbl) Add MS 32546: ff. 143–168v (early period, late), who provides his own numbering of the text, from which it is clear that there are a number of missing pages (pp. 11–14) between f. 147v and f. 148, perhaps because he discarded them. At f. 153 he refers to the theory of colours 'hinted' by Newton 'about 30 years since in the *Philosophical Transactions* and now publisht with the author's last compiling hand', i.e., in *Optice*.

[168] See Newton, *Optice*, Qu. 23, p. 322, where the author asks: have not the small particles of bodies certain powers, virtues or forces, by which they act at a distance?

vortex.[169] For he points out that 'Cartesius found a necessity that the matter of the world rolling about a center must recede, [and] thought that might caus in us the sense of light'; and he also used 'a new expression[,] *conatus ad motum*,[170] which is not apt, but very exceptionable'. He then states: 'Wee will ... allow it [the vortex] to be the cause of gravity but not of light' (f. 146v).

Recall that when writing *Notes of Me*, North understood the nature of light and sound differently,[171] the one transmitted instantaneously and the other, successively, that is, in time. Recall also that in the second of his three notes, he had specifically commented on Newton's proposition that light is transmitted in time.[172] In an attempt to unfold this opinion, North rejects Descartes's version of a continuum theory based on an analogy of a tennis ball hit by a racket. Instead, he attempts to apply the surface wave analogy to light as he previously had done for sound (ff. 145–147v). But the difficulties in this attempt result in an abrupt change near the end of the essay, where he repeats the distinction made previously in his autobiography—that the transmission of sound is 'by succession', but the transmission of light is 'in an instant' (f. 162). And the whole concludes (f. 168v): 'this is all at present I have to say of light and its variations by colour', although 'upon farther perusall of Mr. Newton's works and study I may add considerably'.

It is important to recognise that continuum theories of light based on the analogy of surface waves could not provide a satisfactory explanation of the rectilinear propagation of light.[173] And

---

[169] Note that RN (f. 153) accounts Descartes's theory of colours as 'exploded'.

[170] Descartes considered the centrifugal pressing as an endeavour to move (*conatus ad motum*), not as a real motion, a conception criticised by Hooke, *Micrographia*, pp. 60–1.

[171] See *supra* 2.2.1.

[172] See *supra* 2.2.2: No. 2 (f. 180).

[173] According to Shapiro, 'Kinematic Optics', this problem could not be solved until the nineteenth century.

this was one reason why Newton was a steadfast opponent of such theories, even in the papers he contributed in the 1670s to the *Philosophical Transactions*. But there also was another reason. For Newton's rays of light are discrete material entities, and this supposition committed him to an emission theory, or what North variously describes as an 'emanation' or 'stream' of such entities. Consequently, in the 1672 paper mentioned previously,[174] Newton himself used the Cartesian analogy of a tennis ball, which suggested to him that 'the rays would travel in curved lines after their emergence from the prism', whereas Descartes had not envisaged any curvature of the rays' path.[175]

Although much more could be said about this manuscript, the next one to be considered is more important, because in the course of writing it North gives reasons for concern about some of Newton's concepts and in so doing, begins to be more explicit about his own values.[176] Of its two main parts, the first (ff. 81–118v), which has the running title 'Mechanick notes' [variant 'Mechanicall notes'], consists of an introductory essay containing covert autobiographical remarks (ff. 81–88v), followed by three sections concerning the nature of matter (ff. 89–95v), its configurations or changes in motion (ff. 96–107v) and the causes and effects of these changes (ff. 108–118v).[177] The second part (ff. 119–137v), which is an untitled but incomplete continuation,[178] begins with a brief recapitulation of

---

[174] See *supra* 2.2.1.

[175] See Sabra, *Theories of Light*, p. 239 and n.17.

[176] RN, 'Mechanick notes' (UK:Lbl) Add MS 32540: ff. 81–118v, *untitled continuation* ff. 119–137v (early period, late).

[177] Since in this first part the handwriting is neat, and there are few emendations (deletions and insertions), it is possible that it represents a rewriting of the transitional manuscript discussed *supra* 2.2.3. Note that RN's understanding of configuration is combinatorial; and this led him to provide the first systematic explanation of the principle of chord inversion in music; see Kassler, *Music, Science, Philosophy*, pp. 44–5.

[178] Portions are missing after f. 137v, including a promised section on mechanical powers.

twelve principles from part one (ff. 119–120), after which follows 'A farther prosecution of a mechanicall systeme, upon the foregoing principles' (ff. 120–137v). Although diagrams appear in both parts, they are more extensive in the second part,[179] whereas the reverse is true concerning references to persons.[180]

In section one of the first part, North begins with an abstract geometrical principle that he assumes is true, because we know it 'by vniversall and quotidian experiment'. His principle is that 'body', meaning matter in general, fills space in all three dimensions 'so as no other like substance can come within the limitts of it'.[181] He thus continues to conceive space filling matter ('extension') and impenetrability as synonyms (f. 89). In stating this principle, he does not argue 'against the existence of beings, which doe not fill place'; rather he only affirms of them that 'our perception being wholly by means of body', 'wee cannot have any perception of spirits which are penetrable' and therefore cannot take them into consideration as 'a philosophicall principle' (f. 89v).

In this statement North distances himself from Henry More (without naming him), who sought to establish in the natural world the existence of spirit not only as a philosophical principle,[182] but

---

[179] Two diagrams are missing in the untitled continuation, which, as the handwriting suggests, seems to have been written in some urgency. Moreover, it is possible that this part of the manuscript represents a rewriting of the missing portion that should have followed the first part of the transitional manuscript *supra* 2.2.3.

[180] I.e., in the second part there is only one citation: 'the algebraicall work of Dr. Wallis', namely, his *Treatise of Algebra* (1685) or its augmented Latin translation in vol. 2 (1693) of his *Opera mathematica* (1693–99).

[181] See also *supra* 2.2.3 n.155.

[182] By introducing an incorporeal ('hylarchic') spirit as the operative causal agent in the natural world, More planned to solve all the phenomena of mechanics and hydrostatics, including gravity. In a long criticism of More, Hooke defended Descartes's explanation of gravity, even though he did not accept it. Indeed, in contrast to More's explanation, he described Descartes's 'Suppositions' as 'Rational and Ingenious, and so much above the Objections brought against them, and so much better than any other I have yet met with, as no wise to deserve to be

also as the instrument for implementing God's plan for his creation and, hence, to deny the world can be resolved into purely mechanical causes.[183] That he had read some of More's writings, however, appears in a postulate, reaffirmed from the transitional manuscript discussed in the previous section, that all the parts of matter which are 'single and have no pore' are 'adamantine and indiscerpable'. By adamantine North means hard; but his conception of hardness is not the same as that of Descartes, according to whom body may be pulverised.[184] Indeed, since North knows no argument to prove that 'primary matter' (body as a singularity) is 'discerpable', he can claim, first, that those primary parts are 'indiscerpable' (f. 90) and second, that they are 'the same and ever will be, as they were *ab initio*' (f. 96; see also f. 119v).

The words indiscerpible and discerpible were coined by More to mean indivisible and divisible.[185] And in several of his writings, he endorsed the actual-parts doctrine by arguing from actual parts to a determinate number of parts and then by arguing from the definiteness of parts to metaphysical atoms—'physical monads' or 'particles so minute that they cannot be further divided or discerped into parts'. However, since he conceived compound body as composed entirely of these physical monads and resolvable into the same, 'at least by divine power', he supposed that

> ...it is entirely contradictory that body cannot be actually divided to the degree that an actual division can be made in it. Since, indeed, it has been arrived at that all the parts are rightly said to be physical monads, inasmuch as in them the physical division comes to a halt, they themselves cannot be

---

esteemed *fœda deliria* [foul ravings], as the learned Doctor [More] is pleased to term them'; see Hooke, 'Lampas', *Lectiones Cutlerianœ*, pp. 182–94, p. 183.

[183] See Gabbey, 'Henry More', pp. 24–5.

[184] See Shapiro, 'Light, Pressure, and Rectilinear Propagation', pp. 243–4.

[185] See OED, where More is the first citation.

discerped into further parts, that is, [they cannot be] physically divided. [186]

North also conceives compound bodies as aggregates of parts (f. 100v) and, hence, is an advocate of the actual-parts doctrine. But he again rejects the notion that there are *ultimate* indivisibles, that is to say, that there is a limit to division, as advocated not only by More but also by Newton.[187] And his reason is the same as that given in the transitional manuscript, namely, that in supposition 'but not in fact', matter may be divided *ad infinitum* into 'an actual infinity of minuteness' (f. 107v).[188]

Having thus laid down his universal principle of matter as an hypothesis, he proceeds in a quasi-Euclidean manner with eleven other principles; these are followed by propositions,[189] which in the continuation provide the basis for the geometrical reasoning that further develops his attempt at a system of mechanics. About such attempts, he remarks in the introductory essay:

> Now you will say that at this time of the world[,] when philosophy stands refin'd by the industry of the most exquisite witts by proper and pertinent steps, and no cost nor application hath bin wanting to assist them with experiments, any man that pretends to systematize ... hath a singular confidence, and that it ought not to be offered at without some extraordinary new light (f. 87).

North often writes directly to an imaginary reader, even if the reader is none other than himself; and the 'you' in the passage just quoted

---

[186] Quoted in Holden, *The Architecture of Matter*, p. 151; see also pp. 35–6, *et passim*.

[187] For Newton, see *supra* 2.2.3 and n.160.

[188] See further *infra* 3.3.2: No. 2, RN→[SC 1706] and 3.3.3.

[189] For the three propositions that are the 'groundwork' of his system and his 'discours upon the beauty and reasonableness of them', see ff. 114–115v. He there mentions (f. 115v) the 'dedalean or mechanick powers', which is an allusion to the book by Wilkins that first introduced him to mechanics; see *supra* 1.1.3.

may be an example of this. Nevertheless, in the first part of his manuscript, there are a number of indications that the reader is not an imaginary but a real critic of systems of mechanics. For example, in a note on 'vacuity', North states: 'it appears there is no demonstration against it, and I beleev not for its actuall existence, tho Mr Newton pretends to give it' (f. 91).[190]

Although this is the only overt mention of Newton, there are a number of other references to him that are covert. For example, after stating that 'in truth the whole world is one aggregate of bodys', North comments as follows:

> I know some have supposed a fixt space in the world, wherein all bodys have a place, which he calls absolute, as if it were a stated thing [i.e., real]. Now I cannot bring my self to make any idea of space but by extension, and relation to body, whereby distance may be measured (f. 98).

From this statement we may infer two things: first, that 'some' who believed in the existence of a 'fixt' space would include Barrow, Gassendi, J. B. van Helmont[191] and More; second, that 'he' who calls it 'absolute' would be the previously-named Newton. For all these men separated space from body, whereas North holds that

---

[190] According to Bertoloni Meli, *Thinking in Objects*, p. 273, 'Newton's concerns with a demonstration of the existence of the vacuum appear so deeply rooted' in the structure of Book II 'as to affect the meaning as a whole'. But in *Principia mathematica*, Bk II, Prop. XL, Prob. X, (General Scholium) pp. 339–54, p. 346, he merely expressed a wish that experiments on air resistance could be carried out with greater precision, because on them depended the demonstration of a vacuum ('*cum demonstratio vacui ex his dependeat*'), whereas in Bk III, Prop. VI, Theor. VI, Cor. 3, p. 411, he asserted, but did not demonstrate the necessity of a vacuum ('*Itaque Vacuum necessariò datur*'). And even earlier, his suggestion for improving the air-pump experiment to test the effect of a vacuum on the sounding source was problematic and, hence, could not produce the desired demonstration; see Kassler, *The Beginnings*, pp. 94–6.

[191] I.e., the Belgian physician, Jean Baptista van Helmont; see *infra* 3.2.2: No. 1, RN→[SC 1706] (f. 83 n.67).

> ...the notion of vacuity is the same as that of space, viz., body
> under a negation of fullness or penetrability. And it must be
> allowed, that the idea of extension only gives the idea of
> vacuity, and the former we have from body; so that what D.
> Cartes says, of extension and body being all one[,] is true as to
> the idea, but not as to the supposed existence of the thing
> (ff. 91–91v).

Consequently, North 'must suppose' that space and time are 'pure
attendants of body, and that annihilated, the others vanish'
(f. 98v).[192]

On Newton's conception, however, if body were annihilated,
then absolute space and time would remain. Consequently, it would
be possible to conclude, first, that space and time are eternal and,
hence, uncreated; and second, that they are attributes of the deity.[193]
And this is the reason why, in a treatise on mechanics, North returns
to a point he had already made in the transitional manuscript
concerning the divine *nunc stans*:[194] that 'it is almost certein that
beings which perceiv not by the means of body, or extended
mensurable space, can have no sence of [the flow of] time [from past
to future], which consideration answers all the puz[z]les about
prescience, etc. and agrees with holy writt, before Abraham was I
am'[195] (ff. 104–105).

And it is for the same reason that he writes about 'that axiom,
ex nihilo nihil fit',[196] that existences are 'not spontaneous to exist and
vanish, by caprice, or that any condition of things as magnitude,

---

[192] On the same folio RN refers to an 'imagined experiment of two bodys joyned
by a diametricall-string; and turn'd round[,] should stretch the string, with force
according to the swiftness of motion'. This is the second of two experiments on
rotation intended to demonstrate *motus verus*, for which see Newton, *Principia
mathematica*, Scholium (to the Defs), pp. 9–10. A few folios later (ff. 100 and
102), RN denies that the experiment demonstrates that motion is real.

[193] For one follower of Newton who took these steps; see *infra* 3.3.5.

[194] See *supra* 2.2.3 and n.156.

[195] John viii.58.

[196] (Lat.) from nothing nothing is made.

figure, motion, rest, etc. can alter without a caus any more than the thing it self'. For

> If we would examine whence it is that this axiom hath so clear a light, [then] it will be found to proceed from the universall and perpetuall experience of all mankind. And no glimps of evidence ever appear'd to argue the contrary. That which wee beleev of the world's having bin created out of nothing and may returne vnto the same, is joyned with a beleef of an almighty power able to doe it. So when miraculous things are done contrary to the cours of nature, wee admitt the same[,] not pretending to understand but onely to admire (f. 108v).

However, since North's immediate concern is with 'the ordinary cours of things which never vary without adequate caus', he points out that miraculous cases 'are not now in question' (f. 108v).

2.5. CONCERNING NEWTON'S 'DESIGNE'

The issue of Newton's 'abstracts' is tightly linked with another that North raises in his third note—that in both his books Newton not only sought to undervalue Descartes, but also to overturn the philosophy called 'mechanical'.[197] As a consequence of this inference, North began to produce some critical reflections on Newton's 'designe', a process, which, if traced to its conclusion in 1728,[198] involves a number of inter-related themes. One theme, concerns changes in ways of philosophising from ancient times. But, as some of his other writings indicate, North's historiography is not one of linear improvement or winner takes all, for he imposes on history a pattern of cyclical change, the motor of which is a

---

[197] See *supra* 2.2.2: No. 3 (ff. 184, 186).

[198] See further *infra* 5.5.2.

combination of self-interest and power.[199] A second theme, which concerns Newton's use and abuse of Descartes, is motivated by two maxims concerning justice: to hear both sides and to give each person his due.[200] And these maxims, in turn, promote a third theme, namely, the comparative value of the philosophies of Descartes and Newton, their excellences ('virtues'), as well as defects.

All these themes are present in rudimentary form in a manuscript that dates from *c*.1707 or a little later.[201] Originally intended as a 'discussion' of philosophical authorities in history, North begins by developing a hint from his third note that Aristotle had 'overturned the learning of the ancient naturalists'.[202] But he now identifies the latter as the Greek atomists, Democritus and Epicurus, repeating from his former note that they had a better way of philosophising than Aristotle. He then adds two comments: that Aristotle's method was less good, because it subserved wrangling, not 'the ends of knowing'; and that this method was adopted by Aristotle's successors, the medieval scholastics. But if the 'disputing' method was less good, then why did it become authoritative? North's answer is that the heathen priests believed that atomism was most dangerous and Aristotle's philosophy, 'least hurtfull to their trade'. So they meddled with the natural part of his philosophy by referring his qualities to the deity, leaving the rest to be managed by God or by 'petty intelligences'. And the scholastics continued this kind of

---

[199] See, e.g., RN, *Notes of Me*, pp. 214–16, where his brief account of religion in England states that 'the reall motive of all the turnes in it' is 'secular interest'. For his innovative attempts to sketch the history of law and music, see RN, 'Of Etimology', pp. 303–37, RN, *The Musicall Grammarian 1728*, pp. 219–73 and Kassler, *HRN*, pp. 13–14, 164–78.

[200] Regarding RN's conception of maxims, see Kassler, *HRN*, pp. 44–5, *et passim*; regarding the above maxims, see pp. 113 n.152 and 119–20.

[201] RN, 'Authoritys' (UK:Lbl) Add MS 32546: ff. 207–230v (middle period, early); see also the related fragment, RN, 'Philosofy-&-fers' (UK:Lbl) Add MS 32546: f. 300 (middle period, early). Messy handwriting, inkblots on some leaves, numerous emendations (deletions and insertions) and lack of careful organisation are symptoms not only of cursory writing but also of work in progress.

[202] See *supra* 2.2.2: No. 3 (f. 185).

meddling until 'the Reformation in Europe' gave a check to it; and it was this historical event 'that let in liberty[203] and then new philosofy' which 'moves on clearer principles'.

Instead of developing the subject of new philosophy, North pauses to observe that 'in knowledg, as in faith' it is 'worldly interest' that is the motor of change in history, since that leads men to fall 'naturally into factions', which happens not only with 'the speculative sort of men' but also with 'men in publik posts', that is to say, posts in church, state and educational institutions. As a consequence, prejudices to truth begin when youth are inclined to observe, and 'perhaps in cours, come into the same interest' of one or another faction. And this thought leads him to a brief discussion about university education, in which North not only recalls some of his own experiences as a student at Cambridge, though covertly, but also proposes modifications to the curriculum in order to overcome the 'disputing method' of the old philosophy that still prevailed in the English universities of his day (see ff. 210v–211v).[204]

If this discussion about university education seems to be a digression, then it is worth re-reading that part of North's third note in which he speculates about Newton's student days at Cambridge.[205] Moreover, that same note also is the source for the next topic: that as in antiquity, so in North's day the change in philosophising was from a better to a worse way. But before that could take place, the 'disputing method' of the old philosophy had to be overturned. And this happened when certain 'brave spirits'[206] left Aristotle's

---

[203] I.e., free inquiry; see *supra* 2.2.2: No. 1 (f. 187) and n.139.

[204] Although RN admits that the old philosophy 'by maintaining the logomachy is fitter to uphold professor's places, then the new, which doth not build on words but things', yet he would not censor Aristotle's works, because 'the history of the opinions of other philosofers' is to be had 'no where els but there'. Instead, he merely proposes that the works in question 'not be crambd downe as acts of parliaments in learning'.

[205] See *supra* 2.2.2: No. 3 (f. 186).

[206] RN (f. 212v) names Bacon, William Gilbert, Gassendi, Hobbes and Descartes. Elsewhere, he includes the name of Galileo as one of his 'heroes'; see *infra* 3.3.3 n.253.

'seducements' behind, for they aimed at 'an explication of the mundane system of body and its modes', thereby returning to the way of philosophising of the ancient Greek atomists. And thus concludes his brief history of 'the authority[s] of all former ages as to naturall philosofy', after which North returns to the subject of new philosophy.

In this part of his history he focuses on Descartes, because this 'most transcendent genius ... hath by the meer strength of his mind and thought gone far beyond' all the improvements of the other 'witts' contemporary with him. Indeed, against the claim of 'an academick humoured Oxonian'[207] that Descartes's essay on geometry[208] was plagiarised from Thomas Harriot,[209] North asserts it was founded on 'all that was knowne in geometry before and advanced *de novo* with exquisite clearness and also exposing a course of discovery further to infinite, in the most abstruse doctrine of curves' (f. 213v). Moreover, that the French philosopher was 'very conversant with' the work of the ancient mathematicians and 'as nice a crittique in their severall texts and designes of them as any whatever' (f. 214) is 'enough declared' in connection with the problem of Pappus.[210]

---

[207] I.e., John Wallis, who made the claim in his *Treatise of Algebra* (1685).

[208] I.e., in Descartes, *Discours de la méthode*.

[209] Harriot contributed to a number of different branches of mathematics and made notable innovations in algebra; but his papers on this branch remained unpublished until 1631, when a limited selection of them was edited. Wallis, however, seems to have been motivated by a desire to assign the innovations in algebra to the older Harriot, not to the younger Descartes, because, the former (like Wallis himself) interpreted algebra numerically, whereas the latter interpreted algebra geometrically.

[210] For Descartes and the 'problem' of Pappus, see Crombie, Mahoney and Brown, 'René du Perron Descartes', pp. 55–8. According to Guicciardini, *Isaac Newton*, p. 80, in the essay on geometry, Descartes proposed a solution of three or four lines 'as a paradigm of the superiority of his method over that of the ancients', whereas Newton believed the method of the ancients was more elegant than that of Descartes.

Although not 'here wrighting a panigir[211] on D. Cartes', yet North values him as 'an hero', even if he was neither 'omniscient nor impeccable in philosofy'.

> But reflecting [on] what insults as well as slights are put upon him by men of academick education and hierarchicall interests[,][212] as if he were ... a shallow ringleader of a vain sect, and those that opine with him a blind unthinking but obsequious party, I could not forbear ... *to doe him right* and justify him in parity at least with any if not superior [then] to all other extant in the world (f. 215, italics mine).

As part of his justification, North lists 'some items of improvem[ent] in philosofy wholly owing to Descartes' (ff. 215–218), which in brief are as follows:

(1) his giving 'so easy and clear a method to imagin' the immensity of the world;[213]

---

[211] I.e., panegyre (OED 1, obs.) eulogy.

[212] By 'hierarchicall interests', RN here refers to the academically-trained critics in the Church of Rome hierarchy and not, as in later texts, to critics in the Church of England hierarchy. For some of the critics in the former hierarchy, see Clarke, *Occult Powers and Hypotheses*, pp. 12–13, 22–34 and 223–7. In the text discussed above, RN (f. 220) specifically mentions the Jesuit René Rapin, concerning whom see also RN, 'Inceptio' (UK:Lbl) Add MS 32546: ff. 169–173v, *continued* ff. 266–270v (early period?). From the context of both texts, it may be inferred that he is referring to Rapin's *Reflexions sur la philosophie ancienne et modern* (Paris, 1676, English translation 1678). Indeed, he once owned an edition of this work; see RN Books (1), where place and date of publication are illegible.

[213] I.e., Descartes's infinite universe, which provided the first 'kernel' for a renewed discussion of the ancient atomists' problem of a plurality of worlds (Newton provided the second 'kernel'); see Dick, *Plurality of Worlds*, pp. 106–41, 143.

(2) his declaring that 'the error is not in the sences, but that those ever informe exactly true', rather, 'it is our judgment and inferences from them that prove fals';[214]

(3) his demonstrating that 'most operations in nature are done by insensible parts, the world being, as he holds, full of bodys';[215]

(4) his solving appearances by considering 'the compass of the whole earth, what reference it may have to[,] or influence from the celestiall matter in which it is conveyed' and 'what power its motion may have to the severall parts of it, or one part of it to another';[216]

(5) his advancing 'the principle of body and its modes, or as they[217] ordinarily speak, motion,[218] and especially in one thing, that the union of parts is only from rest';[219]

(6) his 'venturing to give an account of the whole univers' and especially the celestial bodies (for, North adds, why should one not

---

[214] I.e., Descartes's sceptical epistemology—that the mind, located in the pineal gland at the base of the brain, interprets the representations transmitted in the process of sensation, whence errors arise from the interpretation, not from the representation; see, e.g., the essay on dioptrics in Descartes, *Discours de la méthode*. According to Gaukroger, *Descartes*, p. 420 n.28, sceptically-driven epistemologies were the 'dominant approach' until nearly the twentieth century. As indicated *supra* 1.1.3, RN himself had a similar approach; but his anatomy cannot be described as Cartesian; see Kassler, *Inner Music*, pp. 165–207.

[215] I.e., Descartes's universal plenum, in which spaces apparently empty of matter are filled with unobservable particles. RN (f. 219v) accepts the plenum; but, as will soon appear, he rejects Descartes's hypotheses concerning the various sizes and shapes of the particles.

[216] I.e., Descartes's vortex theory, which involves all planetary bodies, not just the earth.

[217] I.e., philosophers.

[218] I.e., Descartes's definition of motion; see *supra* 222: No. 1 (f. 174v and n.69). RN (f. 219) describes this definition as 'a superlative improvement of philosophy, viz., he hath demonstrated that motion is regulated according to the measure of quantity, so that if the state of bodys impelling each other be given [then] the consequent effect of the impuls on either is demonstrable'.

[219] I.e., Descartes's explanation of cohesion that 'rest is like glew'; see *supra* 2.2.2: No. 3 (f. 184 n.124), which RN subsequently rejects, e.g., f. 212 in RN, 'Plenitude' (UK:Lbl) Add MS 32545: ff. 210–231v (middle period, early).

endeavour to find such an account 'when the Psalmist says they have a law which they cannot pass,[220] that is, a natural reason, as any triviall thing here below');

(7) his rejection of final causes 'in all inquiries of naturall things,[221] for really it hath bin heretofore a great hindrance to the progress of knowledg, the thincking the whole univers made for the use of man and studying out the conveniences onely in reference to him';[222]

(8) his 'stupendious discovery' of the relativity of motion, which is of the greatest importance in 'all natural knowledg'.[223]

If these are Descartes's excellences, then what are the defects? North's list is rather brief and includes Descartes's *conatus ad motum*; his 'account of colours from refraction', his hypotheses concerning 'the nicetys of particules', for example, the *particulæ striatæ* and 'the anguillar forme of watery parts, and some others', which he touches on only in passing (f. 219v). But he singles out one

---

[220] I.e., Psalm 148, a song of praise for God's creation, in which v.1–6 concern the heavenly bodies and in which v. 6 states that God established them 'for ever and ever', because 'he hath made a decree which shall not pass'.

[221] I.e., Descartes's elimination of teleology both as purpose (goal or terminus) and as design (purpose preceded by an idea). But the French philosopher did not deny the presence of finality in nature. Rather, he denied the human ability to know finality and to use it in natural inquiries; see Collins, *God in Modern Philosophy*, p. 67.

[222] According to Rossi, 'Nobility of Man', the thesis of a plurality of worlds developed in polemic against anthropocentrism. For RN's acceptance of the thesis, see the incomplete essay, RN, 'The world' (UK:Lbl) Add MS 32546: ff. 33–90v (middle period, early), where he writes (ff. 71–71v) that since Saturn is not so frozen 'as they [the Newtonians] imagin', it may be a 'comfortable land to those creatures, if any are, to injoy it'; and (f. 67) 'as to inhabitants, with proper conveniences of life, whither like any here or not, wee ... say with Monsieur Hugens [i.e., Huygens], why not?', especially since (f. 76v) 'Nature hath no limits'.

[223] I.e., Descartes's principle of relativity, or in RN's terms, his conception that motion is not real, or 'nothing in it self, but as there is regard to the stations and positions of other body[s]', including that of the observer; see RN, *Notes of Me*, p. 93.

failing in particular, namely, Descartes's rule that '*in vacuo* a greater body doth not move a lesser att all, and what motion a lesser takes he thincks is from the medium'. North had mentioned this rule in his first note,[224] but he now adds that it 'is found contrary to experience' (f. 219).

Instead of pursuing this subject, he moves on to answer some of the French philosopher's critics, afterwards clarifying his own position, when he writes: 'and however I am not a Cartesian, in the sence of the academicks[,] so as blindly to idolize him, but am glad as they are (but with more civility) to use him' (f. 221v). Nevertheless, he thinks it possible to give 'a catalogue' of the 'ill usage' Descartes received from the philosophers who came after him, most of whom were his 'plagiarys'. But as that would 'swell too much for the profit it brings', North merely instances two books that are like 'trees sprang from a Cartesian root', even though the authors of them (who are not named) treat Descartes with contempt.[225]

And perhaps with this thought in mind, he now begins his consideration of 'the noble author' of *Principia mathematica*, who,

> ...*to give him his due*, is the onely person since Cartesius of a genius in the way of naturall philosofy fitt to be compared with him. And I thinck I shall doe him no wrong in making the comparison, tho like Plutark in his lives,[226] not for parralell, but [for] opposite vertues which layd together may best shew each other. And if I doe not preferr him, [then] he may excuse it, becaus he came after the other [genius] who had possest the larg plaines of discovery before him (ff. 221v–222, italics mine).

---

[224] See *supra* 2.2.2: No. 1 (ff. 176v–177).

[225] RN's reference is to John Locke, *An Essay concerning human Understanding* (1689, rev. 2/1694, rev. 4/1700) and Thomas Burnet, *Telluris theoria sacra* (Pts 1 and 2, 1681; Pts 3 and 4, 1689). Regarding the latter work, see Mandelbrote, 'Isaac Newton and Thomas Burnet'.

[226] I.e., Thomas North's translation from the French of *The Lives of the noble Grecians and Romans compared together by Plutarke* (London, 1579); see RN Books (1). RN probably inherited this first edition from his grandfather, who was the translator's great uncle.

In North's attempted comparison, excellences and defects are combined rather than separately treated, beginning with the two men's manner of expression (f. 222v), where he points out that Descartes's aim was 'to make his philosofy as comunicable to the capacitys of all mankind as was possible', whereas Newton 'keeps himself all in the dark, so that it is impossible from his wrightings to collect his general sense'. For although in his book on optics he invented

> ...the most admirable notions about light and colours and all opticall matters, as for instance that rays of light are of various colours, specifically distinguisht[,] white is an aggregate of them all blended together, but when they come to refract[,] some being in their nature refrangible to greater angles then others, as red more then blue and that more then yellow and the like, they are distinguisht by separation.

But this admirable invention 'doth not att all content the spirits, for what matters how the colours happen to appear, if wee know not what either light and colours are'.[227]

Later on, when North considers that 'all new philosofy leans on experiment[,] and that deservedly demands the greatest of authoritys', he praises (f. 230v) Newton, because his 'admirable hypothesis of light and colours' was the result of experiments from 'such ordinary things, as the froth of a barber's bason,[228] the shaddow of a moving comb and such like.'[229] Hence, like Descartes, his judgment was of greater consequence then either the contrivances or the subtlety of the experiments.[230] But this praise is not due solely

---

[227] According to RN, the aim of a 'physical solution' or natural history is to describe what, not why; see *supra* 2.2.1 and n.26.

[228] I.e., basin.

[229] Experiments 11, 12 and 14 in Newton, *Optice*, Bk I, Pt II, Prop. V.

[230] Regarding Descartes, RN mentions four 'experiments', one from *Principia philosophiæ* (a stone whirled around in child's sling) and the others from the essay on dioptrics in *Discours de la méthode*; see Shapiro, 'Light, Pressure, and

to the compatibility here between the two men, for in treating the different meanings that 'divers sorts of people' have assigned to the word 'experiment', North classes himself with those who think that 'the ordinary occurrences of life, to one that hath a sagatious sence to observe[,] affords foundation enough to guide the judgment in all generalls of philosofy' (f. 228).[231]

As for Newton's book on rational mechanics, a number of issues are canvassed. These include one raised previously in North's third note, namely, the author's 'seeming content and joy in opposing Cartesius', for 'in more places then one in his book, the conclusion comes, that this or that of Cartesius, and particularly his vortexes doe not hold water' (ff. 224v–225). Another issue concerns the re-introduction of what Descartes had previously 'disbanded', namely, vacuum and occult 'qualities' (ff. 223–223v).[232] North allows that because vacuum is 'almost impossible to demonstrate one way or other', it can be introduced 'at pleasure'. But since qualities had been exposed 'as a coven of ignorance', he finds it strange that this matter 'should gaine ground againe' in Newton's 'exquisite structure', which is built on them. For in his book 'wee have the centripetall and centrifugall qualitys, and bodys attracting each other, with force deminishing by the squares of their distance'; and what is more, the latter 'is made to solve all the mundane system' (f. 223v).

But, North asks (ff. 224–224v), how do 'the influences' of light and attraction operate through 'the immens vacuum' and how can that which is not body move bodies? He grants that Newton is 'sensible of these inquiries at first, and says he is to give mathematicall demonstration and craves no phisicall solutions may be expected'. But 'such energeticall attraction supposed', he goes on to systematise the mundane system on this principle 'with all the

---

Rectilinear Propagation', pp. 264–6, and Eastwood, 'Descartes on Refraction', pp. 487–9.

[231] For some of RN's own experiments that answer the description above, see Kassler, *The Beginnings*, pp. 111–12.

[232] See Hutchison, 'What Happened to Occult Qualities in the Scientific Revolution', especially pp. 234 and 250–3.

assurance as might be expected were it done upon clear and not precarious principles'.

> These are such flaws, that the manifest ability of the author doth not permitt me to thinck otherwise, but that he reserves to himself some systeme of phisicall knowledge that he thincks highly probable, but cannot to his satisfaction demonstrate his opinions so as to be secure against cavills. And the example of Cartesius in this may make him beware. For the world is spightfull; and if there be a lacun[a]e [then] the envious will assuredly peck there (f. 224v).

From North's point of view, Newton's 'assuming centrall powers was a peice of mathematick skill', but it was 'farr from a philosoficall aim',

> ...which in things out of our reach, by collation with other things in our view and knowledg[,] finding inductively agreements [and] analogys, concludes with sufficient probability and comands our assent. And if men are not pleased to admitt this method [then] they must lay aside a science which exercises the facultys of men in judging the beautys of the Creator's works more then all that there is besides (f. 226v).

But in writing this, he is not arguing against 'the mathematick sciences', which he admires; and he also envies those whose comprehension of them is 'vastly beyond my capacity'. Nevertheless, he holds that mathematics cannot be the standard for such branches of knowledge as theology, morality, policy, in the last of which 'some pamphleteerish writers usurp the word demonstration improperly and impertinently'.[233] And if we consider the science of physics itself, the greatest part 'must and will consist as to our skill

---

[233] RN provides insufficient information for identifying these writers. It should be noted, however, that in his *De jure belli ac pacis libri tres* (1645), Hugo Grotius was one writer who held that political philosophy ('policy') is capable of demonstration; see RN, 'Of Etimology', p. 311 and n.297; see also Kassler, *HRN*, pp. 12, 115–16.

in probabilities and that in severall degrees of more and less, so as to confine upon but not enter the lines of demonstration' (f. 227).[234]

Since Newton believed that mathematics and mathematical demonstration were instruments for delivering certainty in natural philosophy,[235] North's statement above points to an important difference between his and Newton's ways of philosophising. But from his notes on Newton's two books and from examples provided of his early attempts to unfold the meaning in them, a number of other differences have emerged. Hence, North would require more time to absorb and critically reflect on the contents of Newton's books before being able to clarify to himself the extent of their challenge. And, as will become apparent in the next chapter, in seeking clarification he did not rely solely on testing his thoughts by the method of critical reflection.

---

[234] These last remarks derive not only from RN's probabilism but also from his scepticism *supra* 1.1.4.

[235] See Leeuwen, *The Problem of Certainty*, pp. 106–20; Shapiro, 'Newton's "Experimental Philosophy"', p. 217, and Guicciardini, *Isaac Newton*, p. 385.

1706 CORRESPONDENCE WITH CLARKE
ON
'PHISIOLOGICALL MATTERS'

3.1. FINDING AN ADVERSARY

From the account in the previous chapter, we may conclude, first, that North began to test his responses to Newton's two books almost immediately after making the three notes that record his experiences of reading them; and, second, that the tests consisted of critical reflections written as a consequence of 'second thoughts'.[1] But in one of the manuscripts discussed in that chapter, North also writes: 'For want of others a man must contradict himself if he will know truth';[2] and thus sometime after reading Samuel Clarke's 1706 translation of Newton's book on optics, he sought to implement the second method of test described previously.[3] Since this method required an adversary and, in particular, one who had some involvement with Newton, it is not surprising to learn that North 'had once the favour of a short correspondence' with Clarke 'about some phisiologicall matters',[4] that is to say, matters physical or natural philosophical.

In its present state this correspondence is incomplete, so it is not possible to determine how or exactly when it was initiated. Indeed, there is no evidence that the two men were personally

---

[1] I.e., *d[e]uterai phrontides*; see *supra* 1.1.2 (f. 4).

[2] See f. 83 in RN, 'Mechanick notes' (UK:Lbl) Add MS 32540: ff. 81–118v, *untitled continuation* ff. 119–137v (early period, late).

[3] See *supra* Some Preliminaries.

[4] See *infra* 5.5.2: No. 3, RN→Bedford 1719 (f. 1v).

acquainted, even though in 1706 both were residing in different parts of Norfolk.[5] Because of this gap in our knowledge and because there was to be a second short correspondence later on, it will be useful to provide some background details about North's much younger correspondent, who was born in Norwich on 11 October 1675[6] and educated at the free grammar school between 1685 and 1690. In 1691 he was admitted to Gonville and Caius College, Cambridge, where his tutor was John Ellys, a friend of Newton.[7] After defending in public a proposition drawn from Newton's *Principia mathematica*, he graduated B.A. in 1695 and in the following year became a junior fellow of his College (1696–1700).

In 1697 Clarke and William Whiston[8] met at a coffee house in the market-place of Norwich, where one of the topics of their conversation concerned some discoveries made by Newton. According to Whiston, Clarke's knowledge of these discoveries was 'greatly' surprising, because they were 'then almost a Secret to all, but to a few particular Mathematicians'.[9] When he reported this meeting to his 'Patron', John Moore, the latter requested an introduction. And thus it was that Whiston, as Moore's chaplain,[10]

---

[5] For roads from Rougham ('Ruffham') to Norwich, see the map of Norfolk, created by Robert Morden and published in 1695, at http://nla.gov.au/nla.map-rm1508.

[6] His mother was a daughter of a merchant in Norwich; and his father, a cloth manufacturer, was an alderman in Norwich and M.P. (1700) in the last parliament of William III; see Ferguson, *ECH*, p. 1.

[7] Ellys matriculated at Caius in 1648 and graduated B.A. (1652) and M.A. (1655); he was afterwards a fellow from 1659 until 1703, when he became master of Gonville and Caius College. By 16 January 1705, when Queen Anne knighted him (as well as Newton), he was vice chancellor of the university.

[8] Whiston, a graduate of Clare College, Cambridge, had been ordained a deacon in September 1693.

[9] Whiston, *Historical Memoirs*, pp. 7–8.

[10] Whiston first met Newton in 1694, the same year that Moore appointed him chaplain. Two years later, Newton, as warden of the Royal Mint, invited him to return to Cambridge and lecture as his deputy, but he did not commence lecturing

invited Clarke and his father to the bishop's palace in Norwich.[11]

Not long after this meeting, Ellys asked Clarke to produce a better Latin version of a book in which Descartes's new philosophy was expounded and taught at Cambridge.[12] Clarke agreed to his tutor's request; and his translation was published in 1697 with a dedication to Moore and additional annotations at the end of the volume. The annotations were intended for the purpose of amending some points in Cartesian philosophy and did not incorporate much data relating to Newton's discoveries. But in the second 1702 edition, Newton's name appeared as part of the title, and the annotations (now in footnotes) increased the incorporation of material from *Principia mathematica*.[13]

On 7 August 1698 Clarke was ordained deacon; and shortly afterwards, he succeeded Whiston as chaplain to Moore.[14] He then commenced a course of theological study under the guidance of the

---

until 1701. In May 1704 he succeeded Newton as the third Lucasian professor of mathematics, but in 1710 he was expelled from the professorship and from the university for his heterodox religious views.

[11] Moore was appointed bishop in 1691 in place of the deprived, non-juring bishop, William Lloyd. Note that RN was legal advisor not only to him but also to other deprived bishops; see Lloyd's commonplace book (UK:Lbl) Add MS 40160: ff. 17–25, and Yould, 'Two Nonjurors', pp. 367–8.

[12] I.e., Jacques Rohault, *Traité de phisique* (1671), the first Latin translation of which, intended for foreigners, was published at Geneva (1674) and subsequently at London (1682). It was the latter version which Clarke was asked to improve. For Rohault and his *Traité*, see Clarke, *Occult Powers and Hypotheses*, pp. 48–50, 81–6, 149, 159–61, 171–2, 186–8, 202–3, 205–11, 213–14, *et passim*. To my knowledge, RN does not cite Rohault or his treatise either in its French version or in the Latin translations.

[13] With the 1710 edition, Clarke also included material from the book on optics, thereby combining in one text two competing interpretations of natural philosophy. For details, see Hoskin, '"Mining all Within"', who refers to the editions as 'versions' because of the extensive changes in each. According to Ferguson, *ECH*, pp. 6–7, repeated by Gascoigne, 'Samuel Clarke', p. 913, the new translation helped to replace Cartesianism with Newtonianism. For a different viewpoint, see Cajori, *Sir Isaac Newton's Mathematical Principles*, pp. 630–1.

[14] Clarke's position as chaplain to Moore ceased in 1710; see Ferguson, *ECH*, p. 43. Note that Moore had a number of other protégés who promoted

bishop,[15] whose library of books and manuscripts was extensive.[16] In the following year Moore appointed Clarke as rector not only of a small parish in the city[17] but also of Drayton in Norfolk, to which he was collated on 11 October 1699. A year later, on 17 October, his marriage to Catherine Lockwood[18] took place in St. Margaret's Church, Burnham Norton, one of the villages in the northwest corner of Norfolk. Then, on the invitation of the archbishop of Canterbury, Thomas Tenison,[19] Clarke delivered two sets of sermons at St. Paul's Church, London, the first set of sixteen in 1704, and the second set of eight sermons in 1705. These series were part of the so-called 'Boyle Lectures', instituted through the will of Robert Boyle for 'proving the Christian Religion against notorious Infidels, viz., Atheists,

---

Newtonianism; for some of these, see Gascoigne, 'Politics, Patronage and Newtonianism'.

[15] For a rather vague description of his study, see Hoadly, 'Preface', p. ii; for brief hints as to its purpose, see Whiston, *Historical Memoirs*, pp. 12–13; and for its final result and consequences, see *infra* Chapters 4 and 5.

[16] See Meadows, 'John Moore'; see also *infra* 4.4.1.

[17] Unfortunately, the specific parish in Norwich has yet to be identified. According to *The Clerical Guide*, p. 282, there were four parishes in the patronage of the bishop of Norwich—St. Margaret, Westwick; St. Peter, Southgate; St. Simon and Jude; and St. Swithin.

[18] Ferguson, *ECH*, p. 35, and Gascoigne, 'Samuel Clarke', p. 913, have assumed that she was the daughter of a 'Rev. Lockwood', rector of 'Little Maningham', Norfolk. The village name, however, is Massingham Parva (Little Massingham), where between 1664 and 1673 the rector was Robert Lockwood, who was buried there on 1 September of the latter year; see McLeod, *Massingham Parva*, pp. 114, 162. If he is the same Robert Lockwood who on 11 April 1654 was admitted to St. John's College, Cambridge, aged seventeen, then he would have been aged thirty-five at his death, making it unlikely that he was Catherine's father.

[19] Tenison's appointment as archbishop dates from 1695. In the 1680s he had worked for comprehension of dissenters with Heneage Finch, 1st earl of Nottingham, and, afterwards, with Nottingham's son, Daniel, 2d earl of Nottingham. RN's brother Francis was a friend and protégé of the first earl; see Holdsworth, *A History of English Law*, vol. 6, pp. 539; whereas RN and his siblings were cousins of the second wife of the second earl; see *infra* 4.4.5.

Theists,[20] Pagans, Jews, and Mahometans[21],[22]

On 30 December of the latter year, he read a sermon before Queen Anne at St. James's Chapel; and it was probably at this time that Moore introduced Clarke to the queen. Indeed, by her special command his sermon was published in January 1706,[23] after which he was appointed one of her chaplains-in-ordinary.[24] And, as previously indicated,[25] in April of that year Clarke's Latin translation of Newton's *Opticks* was published with a dedication to Moore.[26] Towards the end of 1706 Moore presented Clarke with the living of St. Benet's Church, Paul's Wharf, London,[27] at which time he resigned the benefice of Drayton.[28] Then, about 13 October he removed to London, where for a short time he resided in the bishop's London house in Petty-France, Westminster.[29]

Between 1705 and 1706 Clarke came to wider public notice as a result of controversies that resulted not only from his Boyle Lectures, mentioned above, but also from views expressed in a pamphlet controversy concerning the immortality of the soul and free

---

[20] I.e., deists.

[21] I.e., Mohammedans.

[22] Boyle, *The Works*, vol. 1, p. 105.

[23] Advertised as 'Just Publish'd' in the *Daily Courant* for 23 January 1706.

[24] For the termination of this appointment, see *infra* 5.5.3.

[25] See *supra* 2.2.2.

[26] According to Whiston, *Historical Memoirs*, p. 13, for this translation Newton gave Clarke £500, 'the Dr. having then five Children'.

[27] I have not discovered either the date when he was collated to St. Benet's or the date when he resigned his benefice of Drayton.

[28] It is not known whether Clarke also resigned as rector of the small parish church in Norwich. If he did not, then on his visits to Norwich (which it is known he made over the years) he must have performed his duties there, the times of his absence being filled by a deputy or assistant cleric.

[29] See *infra* 3.3.2: No. 4, SC→RN, 28 September 1706 (f. 2) and No. 6, SC→RN, 26 October 1706 (f. 1).

will.[30] But I have found no evidence in North's manuscripts that he was aware of these controversies. Nevertheless, it should be noted that although the issues raised in his 'short' correspondence with Clarke derive from North's reading of Newton's two books, yet a number of the same issues may be found in Clarke's Boyle Lectures, including, for example, the nature of matter, the absoluteness of space and time, the infinite vacuum and 'attraction' or universal gravitation. What is more, the two sets of lectures are the most important source for Clarke's thought and ways of philosophising, at least in the editions published in the same years that the lectures were given.[31] For to each edition Clarke made changes which included deletions and insertions, as well as an appendix of letters to and from his critics. Consequently, these 'editions' are better described as versions.[32]

For these reasons and also because there is some circumstantial evidence that North may have read the lectures,[33] details of the editions issued before 1713 are given below. Initially, the two series were published separately as

A Demonstration of the Being and Attributes of God: More particularly in Answer to Mr. Hobbs, Spinoza, And their Followers. Wherein the Notion of Liberty is Stated, and the Possibility and Certainty of it Proved, in Opposition to Necessity and Fate. Being the Substance of Eight Sermons Preach'd in the Cathedral-Church of

---

[30] For these controversies, see Ferguson, *The Philosophy of Dr. Samuel Clarke*, pp. 29–31, 138–51.

[31] See Ferguson, *The Philosophy of Dr. Samuel Clarke*; see also Ferguson, *ECH*, pp. 23–34.

[32] I.e., they exhibit the same tendency as his translation of Rohault (see nn.12 and 13 above). Unlike that translation, however, there has been no careful comparison between the various editions of Clarke's Boyle Lectures.

[33] See *infra* 3.3.5.

St. Paul, in the Year 1704, at the Lecture Founded by the Honourable Robert Boyle, Esq. ... London: W. Botham, for James Knapton, 1705, 2/1706, 3/1711.

A Discourse Concerning the Unchangeable Obligations of Natural Religion and the Truth and Certainty of the Christian Revelation. Being Eight Sermons Preach'd at the Cathedral-Church of St. Paul, in the Year 1705, at the Lecture founded by the Honourable Robert Boyle Esq. ... London: W. Botham, for James Knapton, 1706, 2/1708,[34] 3/1711.

Subsequently, the two series were published together,[35] with a dedication to the four trustees appointed by Boyle.[36]

As will become apparent from the correspondence itself, the issues debated included certain 'phisiologicall matters' that are tangential to my narrative. Since these relate to Books I and II of Newton's book on optics, two points about that book will be made here. First, both parts of Book I treat the author's theory of colour, the main outlines of which had been established in a series of articles published in the 1670s, at least one of which North had read.[37] As Alan E. Shapiro has pointed out, Newton's idea that

---

[34] A new 'Preface' was added to the 1708 edition, in which Clarke defended himself against an anonymous author, who had criticised the first series of lectures. One of Clarke's reasons for dismissing the critic's argument is that it depended '*entirely upon Supposition of the Truth of the* Cartesian Hypothesis, *which the best Mathematicians in the World have demonstrated to be false*' (a few pages later Newton is cited); see Clarke, *A Discourse* [1708], p. A 2 verso. Ferguson, *ECH*, p. 77, identified the anonymous author as William Carroll, who was the first critic of Clarke's Boyle Lectures and, hence, began the controversy noted above.

[35] Clarke, *A Discourse concerning the Being and Attributes of God, the Obligations of Natural Religion, and the Truth and Certainty of the Christian Revelation....* (London, [no copy of first and second edition located], 3/1711).

[36] The trustees included Tenison, then archbishop of Canterbury, and John Evelyn, the latter of whom was one of the 'ingenious acquaintances' of RN's brother Francis; see Kassler, *The Beginnings*, Appendix B.

[37] See *supra* 2.2.1.

sunlight is not simple, pure and homogeneous, but the most compound of all colors, directly opposed a millennia-old tradition of thinking about sunlight, so that the adoption of his theory demanded that an established and fundamental idea about nature be abandoned. According to Newton, it is the colors that are exhibited by the spectrum that are simple, while sunlight consists of a mixture of the innumerable spectral colors. Colors are not created by refraction of sunlight in a glass prism but are in the light all along; refraction only decomposes or separates the different colors by bending each a different amount according to their different refrangibility. Previously, colors were thought to be some modification of pure sunlight that were produced when it was refracted or reflected, such as a mixture with shadow or darkness, or a condensation. Newton replaced the concept of modification with a new way of thinking, that of decomposition or analysis.[38]

By distinguishing between simple and compound colours, Newton also introduced a fundamental, new concept to optics, that of immutable rays of a single colour.

Second, the book on optics contained 'far more' than Newton's theory of colour,[39] for Book II[40] treated colours of thin films[41] and thick plates, colours of natural bodies, and the theory of 'fits' of easy reflection and transmission.[42] And Book III included the exploratory investigation of diffraction (which, after Hooke, Newton called 'inflection'), as well as the concluding queries, some of which North

---

[38] Shapiro, 'The Gradual Acceptance', pp. 43–4.

[39] Ibid., p. 88.

[40] For details of Newton's investigations, see Shapiro, *Fits, Passions and Paroxysms*, Chapters 2–4.

[41] For RN's notice of one such effect (colours that appear when metals are heated) see *supra* 2.2.2: No. 2 (f. 180v and n.108).

[42] I.e., the theory arising from Newton's description of the periodic properties of light exhibited in both thin films and thick plates—that light rays are put into periodically recurring states ('fits') that dispose them to be easily reflected or transmitted.

had previously singled out in his notes on reading the *Optice*.[43] But in the context of the correspondence, it should be noted that Newton reserved the terms 'fringes' and 'fringed' to describe the effects of diffraction in Book III, whereas Clarke seems to have used these terms to describe borders of colours on the edges of beams refracted by parallelepipeds, triangular prisms and spheres.[44] It is somewhat difficult to determine his exact meaning (and not only in this instance), since he provided no specific citations from Newton and since North's letters on the subject are now missing, presumed lost.

3.2. THE CORRESPONDENCE: DESCRIPTION AND EDITIONS OF THE MANUSCRIPTS

As previously mentioned, the sequence of correspondence is incomplete; moreover, what now remains of North's side of the correspondence consists chiefly of fragments, only one of which is dated and identifies Clarke ('Clerck') as the recipient. These fragments, along with a few other letters in which the recipient is not named,[45] have been bound by the British Library according to their

---

[43] See *supra* 2.2.2: Nos. 2 and 3. In some later critical reflections RN singles out other features, including the various analogies to sound and musical pitch in Newton, *Optice*, e.g., Bk I, Pt II, Prop. 3, where the author divided the spectrum into seven segments that were proportional to a string sounding the seven notes of the octave tuned in just intonation. As his writings on music indicate, RN preferred meantone tuning.

[44] See *infra* 3.3.2: No. 3, SC→RN, September 1706 (f. 1, §2, f. 1v, §§3–5) and No. 4, SC→RN, 28 September 1706 (f. 1v, §7). According to Alan Shapiro (personal communication), Newton treated parallelepipeds not in *Optice* but in his unpublished optical lectures read at Cambridge between 1670 and 1672, now edited in Shapiro, *The Optical Papers of Isaac Newton*, vol. 1, pp. 563–75.

[45] See, e.g., (UK:Lbl) Add MS 32546: ff. 301–302v, which has no formal epistolary trappings, though the content, which relates to longitude, makes it clear that RN is responding to a correspondent who has sent him a paper by another author; hence, on f. 301, the heading, centred, is: 'Upon a cursory perusall of the Easy Method &c., I observe.', perhaps referring to the pamphlet by George Keith, *An Easie Method not to be found hitherto ... whereby the Longitude of any Places ... may be found* (London, 1713).

subject matter and not in the correspondence volumes. Consequently, it was necessary for me to establish at the outset rules for determining whether or not a text was part of the 'short correspondence' with Clarke. The first rule required that the subject matter concerned some content in Newton's book on rational mechanics or in the 1706 Latin translation of his book on optics. The second rule required that a text contain clear evidence that the recipient was a real person, even if that person is not named but simply addressed as 'you'. This is not a trivial problem, because this pronoun may stand for an imaginary reader or even for North's 'half freind',[46] a name that refers to his own second thoughts.[47] In the case of the 'short correspondence', the recipient ('you') was identified as a real person if the subject matter of the fragment not only conformed to the first rule but also related to subjects of on-going debate.

A slight modification to this rule occurred during work on North's fragments, since late in the course of writing, four manuscripts came to light that represent a small portion of Clarke's side of the correspondence. At the time of their receipt, North annotated them, perhaps for the purpose of providing an index that could be seen from the side of the document once it was inserted into a letter book.[48] But in annotating Clarke's letters of 28 September and 26 October, he wrote the dates of 8 September and 8 October. If those dates are not errors, then they may refer to the dates when he had written to Clarke rather than the date of Clarke's answer. Unfortunately, this possibility is called into question by North's annotation to Clarke's letter of 21 November 1706, where he reproduces the same date as that letter. Whatever the case, Clarke's

---

[46] See *supra* 1.1.2 (f. 4).

[47] See Kassler, *HRN*, p. 56.

[48] Later on, the manuscripts formed part of the collection of James Crossley that, subsequently, was auctioned by Sotheby, Wilkinson and Hodge in 1885, when they were purchased by Chetham's Library; see lot 2955 in *Catalogue of the second Portion of the ... Library of ... the late James Crossley, Esq. F.S.A. President of the Chetham Society*.

manuscripts facilitated identifying some other of North's fragments related to the correspondence that do not use the word 'you'.

In the description of copy texts below, each item is numbered; and this numbering is retained in the editions that follow. Since Clarke's manuscripts have no foliation, I have assigned one to each copy text for the purposes of cross-references in the notes to the editions, where I also gloss Clarke's foreign or obsolete terms, thereby reserving the Glossary for North's terms. Because the copy texts do not represent the complete sequence of the correspondence, that sequence has been determined chiefly, but not solely, from internal evidence.

No. 1 (Truncated copy of a letter) North→[Clarke 1706]
(UK:Lbl) Add MS 32549: in Notebook 3, ff. 81v–87v
Internal evidence indicates that this item is not the first letter of the correspondence; rather, it is an answer to the second of two items of communication that preceded it: (1) a 'paper' from North (ff. 82, 86), and (2) a criticism of it by Clarke (who is not named). Nevertheless, some of the topics discussed in that paper, as well as some insight into the criticisms of it are retrievable from the copy itself. In the latter case this is so, because North, in using em dashes ('—') as a substitute for quotation marks, has provided either direct quotations of, or paraphrases from his critic's letter.

Title. Each folio of the copy text has a running head. On ff. 81v, 84, 85v–87v the running head is 'An Answer to - - - - - - -', whereas the other folios have 'Answer to - - - - - -'. But in copying the original text into the notebook, North has dropped the usual epistolary trappings. References to 'you' are frequent; and in one place he addresses the recipient as 'my ingenious observer in his engagement against universall plenum' (f. 85v).

Physical features. With the exception of ff. 82–82v where rewriting is interleaved over crossed out sentences, there are few emendations (deletions and insertions); and the handwriting is neat. Nevertheless, in entering the text into the notebook, there are some signs of carelessness (e.g., f. 83v). Unfortunately, there is no way of knowing whether the copy in the notebook is a faithful reproduction of the original, because no previous version of the 'Answer' has been preserved. Indeed, in addition to the two rewritings mentioned above, there are other indications that there may have been some editing. For example, North began entering his 'Answer' by using the numbers 1–3 at the start of each of the first three paragraphs, whereas the fourth paragraph has number 4 added in the left-hand margin, after which no paragraphs are numbered. These numbers are retained in the edition, since numbered paragraphs occur in other communications. But I have omitted one long dash, centred, between paragraphs 10 and 11, which may have been used to indicate a new subject.

Dating. The paper of Notebook 3 is Arms/IV (1), a type of paper that North must have purchased in bulk and used over a long period of time.[49] Unfortunately, internal evidence is not helpful for determining either when the original version may have been written or when the truncated copy was entered in the notebook.

No. 2 (Fragment of a paper) North→[Clarke 1706]
(UK:Lbl) Add MS 32546: ff. 293–293v

Title. Each folio has a running head, centred, as follows: 'Actuall Infinity' (f. 293), 'Infinity' (f. 293v). Epistolary apparatus is completely lacking, but there is one reference to 'you' (f. 293v).

Physical features. Two features of this fragment suggest that it is part of a draft paper. One is the presence of a running head as mentioned above; the other is North's own numbering of each page, starting with the number '1'. Because of these features, I was uncertain whether the text met my criteria for inclusion. However, I decided in the affirmative, because the text summarises the paper on actual infinity mentioned in No. 1 (f. 85v); and consequently, in its finished form, it could have been sent with a letter to Clarke (whose name is not mentioned).

---

[49] See Chan, 'Dating the Paper', p. 49, whose dates (*c*.1708 to *c*.1726) must be taken as a guide only requiring further research for each of the entries in Notebook 3.

Both folios have some emendations (deletions and insertions), but the greatest number occur on f. 293v, the first line of which seems to be a long insertion, for it is written on both sides of the running head and afterwards proceeds about a third of the way down the left-hand margin. To the right of this insertion, some lines of the text are indented, and these, as well as some of the main text, are deleted. Besides these emendations, there are a few slips of the pen, including some omitted words. Note that I have not included North's pagination mentioned in the previous paragraph.

Evidence that the fragment served as scratch paper appears on f. 293v, for on the middle of the left-hand margin, parallel to the text, there are two sets of numbers. One set is just below the middle of the text, the first line of which starts at the margin's edge and faces the text thus: '<21 N°. 1695> 1700'. Under each date, and facing away from the text, are numbers in columns, apparently referring to years: '6 7 8 9' under the first, deleted date of 1695, and '1 2 3 4' under the second date, 1700. Just above the middle of the text, adjacent to these numbers, are other numbers that are placed downwards from the edge of the text thus: 10   C/240/20

$$\frac{12}{20}$$

20

10

Dating. Although the watermark falls in the Arms class, there is no countermark.[50] Hence, the paper cannot be dated; and internal evidence in the paper itself is unhelpful.

No. 3 (Paper) Clarke→North, September 1706
(UK:Mch) The Allen Deeds, Parcel 5 no. 5 (i–iv) [ff. 1–1v]

Title. There is no title.

Physical features. The paper is written in a neat hand, and there is only one emendation in the form of an insertion. It is structured as five numbered paragraphs relating to the theory of colour in Newton's book on optics (which is not cited).

---

[50] See Chan, 'Dating the Paper', pp. 53–5, 87.

Dating. On the first fold of f. 1v, after the end of the text, North has written a short note along the left-hand margin edge, which in transcription reads: 'Mr Clercks vindication— / here of Sr I. Newton. / Sept, 1706.'

No. 4 (Letter) Clarke→North, 28 September 1706
(UK:Mch) The Allen Deeds, Parcel 5 no. 5 (i–iv) [ff. 1–2v]

In this letter Clarke answers the objections North had made in a previous letter, which is missing, as well as the difficulties North had raised in previous 'papers' (f. 1), which also are missing, since the latter difficulties have no relation to the draft of a paper that has been preserved (see *No. 2* above).

Title. The format is epistolary, so there is no title.

Physical features. The bulk of the letter consists of thirteen numbered paragraphs. The handwriting is neat, and the emendations (deletions and insertions) are minimal, as are slips of the pen.

Dating. The letter is dated as above; but on the middle fold of f. 2v, North has written parallel to the edge of the right-hand margin (in transcription): 'Mr Clerck, / 8. [*sic*] Sept. 1706.'

No. 5 (Two fragments) North→[Clarke 1706]
(UK:Lbl) Add MS 32546: ff. 290 and 290v

The two copy texts, which are not continuous, begin with the same first sentence. Moreover, in both texts North points out that Newton's efforts will result in a new world of philosophy. Except for these similarities, the fragments have mostly different content. Nevertheless, I include them both on the assumption that they represent discards from, or drafts for a letter. Internal evidence in the first fragment (f. 290) has determined its place in the sequence; but it is not clear that the second fragment (f. 290v) belongs in the same sequence.

Title. Neither fragment has a title, running head or formal epistolary apparatus. Only the first fragment makes direct reference to 'you'.

Physical features. The handwriting of the first fragment is neat, although there are some emendations (deletions and insertions). A few spaces below the main text, several lines of text start from the edge of the left-hand margin and continue to the middle of the page. Since the first line begins 'of them in the Querys after the Opticks', it is probable that the initial part of this line was once on a previous but now missing page.

In the second fragment, the handwriting is much less neat; and there are numerous emendations, including deletions of entire lines and insertions squeezed in above the deleted line. Both features, which are signs of hasty writing, suggest that the fragment is part of a rough draft.

Dating. The paper on which the fragments are written is Arms IV (1), the type of paper that North seems to have purchased in bulk and used over a long period of time.[51] As noted above, the insertion at the bottom of the first fragment cites the 'Querys' in the 'Opticks'; but in the final paragraph of the second fragment, the context implies that North is alluding to the final three queries in Newton's *Optice*.

No. 6 (Letter) Clarke→North, 26 October 1706
(UK:Mch) The Allen Deeds, Parcel 5 no. 5 (i–iv) [ff. 1–1v]
At the beginning of this letter, Clarke indicates that he will answer briefly some issues raised by North in a previous, but now missing letter.

Title. The format is epistolary, so there is no title.

Physical features. The bulk of the letter is structured in eight numbered paragraphs. The handwriting, as well as a number of insertions suggest that Clarke was writing in haste.

Dating. The letter is dated; f. 1v has North's address and a postmark. About the middle of the same folio, on one of the folds, there are two annotations written parallel to the right-hand margin. The first, in the hand of North, reads (in transcription): 'Mr Clerk / 8. [*sic*] Oct. 1706.'; and the second, in the hand of his son Montagu,

---

[51] See Chan, 'Dating the Paper', p. 49; see also n.49 above.

reads (in transcription): 'all these Letters are transcribed by my Father / in a Book.'[52]

No. 7 (Three fragments of a letter) North→[Clarke 1706]
(UK:Lbl) Add MS 32546: ff. 280, 282 [282v is blank] and 287v[53]
These fragments, whose internal sequence has been determined by their numbered paragraphs, are not drafts of an answer to *No. 6*. Rather, the position of No. 7 has been determined by a phrase used by North in the third fragment and repeated by Clarke in *No. 8* below. As a consequence, it may be inferred that an item or items are missing between *No. 6* and *No. 7*.

    Title. There is no title.

    Physical features. All three fragments are drafts, the third with numerous emendations (deletions and insertions).

    Dating. The paper of the three fragments has been classed as 'Unidentified' and, hence, cannot be dated.[54]

No. 8 (Letter) Clarke→North, 21 November 1706
(UK:Mch) The Allen Deeds, Parcel 5 no. (iv–iv) [ff. 1–1v]
As the first sentence indicates, this letter answers *No. 7* above. Reference is made to a paper by North on attraction, now missing, but which must have been sent sometime after 26 October, the date of Clarke's previous letter (*No. 6*) and before the date of the copy text described here.

    Title. The format is epistolary, so there is no title.

    Physical features. The bulk of the letter is structured as five numbered paragraphs. The handwriting is neat, and there are few emendations. Towards the bottom portion of the letter, a tear on the right-hand margin affects some of the text.

---

[52] Judging from the rough checklist that Mary Chan made for her personal use, the letter book in question is no longer part of the collection of manuscripts held privately by RN's descendants in Rougham, Norfolk.

[53] For ff. 280v and 287, see *No. 9* below.

[54] See Chan, 'Dating the Paper', pp. 64, 87.

No. 9 (Three fragments) North→Clarke, 27 November 1706
(UK:Lbl) Add MS 32546: ff. 280v and 287
(UK:Lbl) Add MS 32547: ff. 382v–384v

Title. At the top of f. 280v, centred, is the following heading (in transcription):

'27 November 1706. /
Answer, to a letter of Mr. Clerck, dated 21 ditto /
and follows after.'

This title suggests that the first fragment is either a discard from or part of a copy of a letter to Clarke.

Physical features. The first fragment, f. 280v, has very faint pencil marks to the right of the date '27 November 1706.'; but these are not in North's hand. The copy text, which fills the entire page, ends with the word 'but' and consists of the above heading, the address to 'Sir' and then two paragraphs. The first paragraph has no emendations, whereas the second has only a few deletions and insertions. In the left-hand margin adjacent to the second paragraph, North has written '1.' to indicate the first matter for debate.

The second fragment, f. 287, is very messy, with numerous lines crossed out and insertions written above the crossed-out lines, thus suggesting that it is part of a draft of this part of North's letter.

The third fragment, ff. 382v–384v, presents similar evidence to the first fragment above (f. 280v), for its numbers (3 to 5) are also entered in the left-hand margin adjacent to certain paragraphs. Moreover, the handwriting is neat, with only a few emendations (deletions and insertions) or slips of the pen. Hence, this and the first fragment may be part of North's copy of his original letter.

Dating. The paper of all three fragments falls in the class 'Unidentified' and, hence, cannot be dated,[55] whereas the date, 27 November 1706, is found on the first fragment only.

---

[55] See Chan, 'Dating the Paper', pp. 64, 87.

No. 1 (Truncated copy of a letter) North→[Clarke 1706]
(UK:Lbl) Add MS 32549: in Notebook 3, ff. 81v–87v

[f. 81v]

An Answer to - - - - - -

1. As to [your charge against my] defining body by its nature
and making that consist in extension alone— I answer that I make no
definition either of body or space. That formality belongs not to
things unknown and sought, but to things knowne, and serves to limit
their beings whereby the discours may not possibly agree with any
thing els. For this reason the mathematicians, who dread nothing
more then ambiguity, will have the principles, tho well known yet
defined, before they will conclude any thing thereupon. But in
phisicall discourses where the nature of any thing is sought, even the
power to define it is sought; for as soon as it may be defined, the
disquisition is at an end, the definition then serving onely that it may
not be mistaken for somewhat els, or somewhat els for that. But if
there must be a definition to begin with, [then] it is [either] the
phænomena to be exposed by such ideal discours, or description as
may probably transferre what one person conceivs into the notion of
another. And in this way, I thought to hint, what to me seemed **[f. 82]**
true of body, and what I thought must also be true of space,
attributing nothing to either which I was not sure did, or rather sure
did not apparently belong to it. And the sum was that body was
extended and impenetrable, and that space also was extended, of
which the consequence is expressed in the paper,[56] and need not be
here repeated.

2. I did not pretend, nor doe I dispute the metaphisicall
determination of attributes, that such may be universall and yet agree
with different species, as warmth may belong to all animals, as to an
hors, as well as to a man. Much less doe I argue that such agreement
proves the hors and man to be all one, and to take away all difference

---

[56] I.e., the missing paper also mentioned *infra* f. 86. Clarke's subsequent criticism,
also missing, is answered in the text edited above.

of things.[57] But considering well this[58] paragraph, I find you confound species and individualls[,] for at first you speak of attributes agreeing with different species of things which may partake of attributes universall and yet have particular attributes to distinguish them, all which is very true. But at last you say that body and space have the same essential attributes (extension) and yet are different things[.][59] But that is not different species, for every individuum[,] tho alike attributed, is different from every other individuum and yet be in one and the same species **[f. 82v]** as[60] body and space is, which differ onely as other individuums within the same species doe[,] whereby you are to shew attributes that essentially agree with the one and not with the other which I allow will make them different species.

3. Solidity is set up for an attribute of body,[61] which doth not agree with vacuum. If a definition be wanted [then] I doubdt it is here. [*]*Soliditas quid*? The answer must be, that no like (species) can come in that place.[62] This is but the phænomenon. The word, solidity, is abstract, and hath no meaning but in the reall [*]essence,

---

[57] The allusion to 'the hors and man' suggests familiarity with the introduction to Aristotelian class logic, in which a 'tree' illustrates the relation of higher class or genus (e.g., species) to the lower class (e.g., individual). Written in the third century A.D. by the neo-Platonist, Porphyry, it became 'one of the great sources of logical theorizing and teaching' from the middle ages 'until well beyond the end of the seventeenth century'; see Crane, 'The Houyhnhnms', pp. 85–6, and Porphyry, *Isogoge*, pp. 35–6.

[58] From here on, the text is written above sentences that have been crossed out and which read (in transcription): '<I would / not have one species, as body & space be / made two species, becaus ye same attribute would / belong to both. as 2. eggs are of ye same species / and must not be made of 2. species becaus as you argue, becaus [*sic*] ye same attributes may belong to both, but they> / will be'. The words 'will be' are the old catchwords and should have been crossed out.

[59] I.e., RN and Clarke differ as to the constitution of the continuum; see *infra* 3.3.3.

[60] From the word 'as' to the word 'species', another interleaving occurs, where the sentences crossed out read (in transcription): '<will be and continue one & ye same species / if their natures appear to be so, however ye attributes may agree>'.

[61] For Clarke's conception of solidity, see *infra* 3.3.3.

[62] I.e., a space occupied by a body cannot be penetrated.

which consists in such consequence and nothing els. If there be ought in solidity besides refusall of place, [then] I desire to know it. —But [you say] it seems wee cannot separate solidity from body, no, not in imagination.— What wee can or cannot imagine is not materiall to the truth of things without us, which will be as they are in themselves, whither wee can imagine it or not,[63] and this setting up of imagination is no other then abetting of prejudice against truth, which is to be sought [by] abstracting all prejudice or imagination. And the contrary method[64] so much indulged is known to be the confusion of all naturall philosofy. But let solidity so understood be the intrinsick nature of body, and then with equall light I affirme it is **[f. 83]** also the intrinsick nature of space. Nothing was ever so vain, as the late chimeras about the intimate nature of body. The atheists will have it thinkable,[65] the chimists to be saline, etc.,[66] the attraction men to have [*]*vires appetendi [et] fugiendi*, etc.[67] and so with like reason, in further diversification [*]*ad infinitum* or as turnes to be served require, for when they once goe off from the undoubted phenomenon, where will they stop?

---

[63] Here, as well as *infra* ff. 83–83v, RN provides a succinct statement of his epistemological realism; see *supra* 1.1.4.

[64] I.e., Clarke's method of argument based on what one can or cannot imagine; see *infra* 3.3.4.

[65] This issue, which commenced with the Locke's suggestion that God could add to matter the power of thought, was central to the 1707–8 dispute between Clarke and Anthony Collins; see Yolton, *Thinking Matter*, pp. 24, 39–41.

[66] I.e., followers of the Swiss physician, Paracelsus, who explained the structure of matter by conceiving all substances as built from three principles—salt, sulphur, mercury.

[67] I.e., the animists who held that desire united iron and the magnet on the grounds that magnetism was a spiritual force; see e.g., Djiksterhuis, *The Mechanization*, pp. 394–5. RN may have in mind Helmont, to whom he attributes 'the plan of our new attractive philosofy' and provides a specific citation for this from a work that was once in Newton's library. See f. 88v in RN, 'Memoranda' (UK:Lbl) Add MS 32549: in Notebook 3, ff. 88–88v (undatable), an entry that follows the text edited above; see also Harrison, *The Library*, p. 158.

4. —But [you say] the idea of space is distinct from that of body, and endued with attributes distinct and contrary, so that it's impossible to imagine they can be inherent in the same subject[.]— As to ideas and all that is concluded from them I set [such considerations] apart, as determining nothing, when the question is of the truth of things without us, which[,] as I sayd before[,][68] will be the same, if no idea of them, or of any thing els were extant in the world. Whither there be any contradiction or not wee shall see; but in the mean time all [*]conceipt of possibilitys and impossibilitys fall short of the matter.

—But [you ask,] doe wee not conceiv body in its obvious appearances, moving, joyning, **[f. 83v]** separating, impelling, communicating, following, and the like?— The discussion of those mechanick laws or consequences is the work of a larger undertaking,[69] but not to the purpose here, becaus wee have onely the [*]conceipt to deal with, and if on our side, vacuum (as it is called) may be [*]conceipted to performe the same, (as I affirme wee doe) [then] there's the hare's head against the goos gibletts.[70]

—But [you say that] wee conceiv space to be that wherein all bodys are placed[;] [that it is] penetrable, refuse[s] entrance to nothing immoveable, inactive, without forme or quallity, whose parts are inseparable, [that it] is the comon measure of motions and is immovable it self. [And you add that] these so disagreeing and repugnant attributes of body and space, it's impossible [they] should belong to the same subject.— Here wants to be proved, that the truth of naturall things accords with our ideas, or imagination, whereof the contrary is found true in almost every instance of sence; for now wee are fallen from conceiving so far as in a manner to predicate the allegations as if they were true as they are conceived, which is not a fair way of arguing. For I say all **[f. 84]** those disabling ideas of

---

[68] See *supra* f. 82v.

[69] See RN, 'Mechanick notes' (UK:Lbl) Add MS 32540: ff. 81–118v, *untitled continuation* ff. 119–137v (early period, late), described *supra* 2.2.4.

[70] I..e., a rhetorical dilemma; here, a choice between space either as empty of matter or as vacuum.

space are derived by prejudice from similarity with the fals appearances of excavated bodys; and whereas body and space are here[71] taken as disparates[,] I say they are comparates and differ onely as body and body.

—But [you say] there is a demonstration, with a diagram of two sphears (required) to contain all that's body in the world.[72] And those can be tangent but in a point, whereupon the spondrills[73] must be space void of body.[74]— Is it not manifest that all this frame must dissolve, if the [*]*suppositum*, that all body is or may be conteined in two sphears, be denyed? The Cartesian says[,] the like remaines without as you put within the sphears, and to take away the matter and leav the space, they say[,] is a contradiction.[75] For all that is essentially true of body is true of space. One body cannot interfere with another, and so one space cannot interfere with another space, as the first paper[76] alledged, and so shut up all body, you must shut up all space. This is the pinch of the argument, and unless it be clearly answered[,] all that can be sayd comes to nothing. But as to this scheme of two sphears, it is a [*]*petitio* of the whole matter in controversie.

---

[71] I.e., at the beginning of the paragraph, where RN quotes or paraphrases statements from Clarke's missing letter.

[72] From this description, one might infer that Clarke had a container-contained model of the universe (space being the container).

[73] I.e., spandrels: the space included between the shoulders of two contiguous spheres.

[74] From RN's paraphrase, it may be inferred that Clarke had reproduced the geometrical demonstration of absolute space from Barrow's Lect. X (on space and impenetrability); see Barrow, *The Usefulness*, pp. 180–1 and Fig. 3 (in a plate of figures).

[75] As the next f. 84v indicates, RN here paraphrases John Keill, the first lecturer at Oxford to offer a course on Newtonian philosophy, a popular exposition of which was then published as *Introductio ad veram physicam* (Oxford, 1701); for the source of the paraphrase see n.78 below.

[76] See *supra* f. 81v and n.56.

[f. 84v]

To say truth, Cartesius used an image which proved a jest, rather then a demonstration, which was of a vessell perfectly empty, whereof the sides, nothing lying between[,] must needs touch.[77] Here wanted a fair conference in order to a clear intelligence of the matter, which (to say nothing of others) made Mr. Keil the great Cartesio-mastix and his Cartesian, like two fools, laugh at one and other.[78] For one laughs at the empty vessel, while empty space seemed to keep the sides asunder; and the other laughs becaus by that he admitts the vessell full of space called empty. What says one, full of emptyness? Yea sayd the other, for space, call it what you will[,] is something. If it were [*]*merum nihil*, [then] it were not space; and then[,] stood to it[,] the sides must touch. But waiving such logomachy, and then admitting that one space cannot become another space, what can be sayd more of body, and its solidity or impenetrability.

—Space [you say] is immoveable— As to that[,] suppose (as wee may doe in our turne) that the whole world was (in your sence) a [*]*vacuum infinitum*; and [suppose also that] the Almighty should say, let this universal space fall into infinite devision, and then to changing of distances and aspects of **[f. 85]** the parts with respect to each other, what can really be made more of the world then that? But I forbear prosecuting this fancy. Now as to the question of vacuum, it is no where so well [*]delated as in one of Dr. Barrow's lectures. He is not convinc't, but in a parity of argument makes a sort of choice to hold vacuity; for comon prejudice inclining that way in small matters turnes the equilibrium. The greatest men sometimes content themselves with words, when notions are cloudy, think[ing] to clear

---

[77] See Descartes, *Principia philosophiæ*, Pt II, Art. 18.

[78] In *Examination of Dr. Burnet's Theory of the Earth* (London, 1698), Keill attacked Descartes for initiating a trend of world-making, i.e., showing how the world might have been produced (as the title-page has it) 'by the necessary laws of Mechanisme without any extraordinary concurrence of the Divine Power'. The implication, of course, is that Descartes was an atheist.

them with synonimes. He calls vacuum a capacity,[79] that is,
[*]*merum nihil*. If a capacity (abstract still) be any thing [then] the
world made as wee beleev out of nothing,[80] must needs have bin
[*]*ab eterno*, for that nothing might be called a capacity. But in a
word if wee cannot shake off comon prejudice imbibed from
similitudinary ideas, [then] all discours and argument upon this
subject is throwne. For the sume of all that is or can be sayd for the
possibility of space in the sence of the vacuists[,] perfectly empty, is
that wee cannot imagine it should be otherwise.

[f. 85v]

But I must not baulk my ingenious observer in his engagement
against universall plenum. It's granted that the world (mostly) is
filled with fluiditys. But as to the doctrine of density, and the
generall governement of gravitation, which is a late hypothesis much
in descant, untill wee have some experiments to uncover the
elementary state of body, (which will never arrive) I shall not come
into an hypothesis of words such as density is. For I take nature
according to its undoubdted principle,[81] and then all things are dens
alike, that is, impenetrable or (in other words) that no one thing will
become any other thing, nor any two become one or one[,] two. And
as to the comon phenomena of resistance, [*]cession, gravitation, etc.
they depend on systemes of moving matter, whereby the difficultys

---

[79] In Lect. X (on space and impenetrability), Barrow, *The Usefulness*, p. 176,
stated: '*Space* is nothing but the mere *Power, Capacity, Ponibility*, or (begging
pardon for the Expressions) *Interponibility of Magnitude*'.

[80] The pronoun here could be understood in one of two ways: either as the royal
we and, hence, referring to RN's belief in creation *ex nihilo*, or as a collective
pronoun applicable to both him and Clarke. If the latter, then see Clarke,
A Demonstration [1705], Prop. X, p. 158: 'the true Notion of Creation, is not a
Forming Something *Out of* Nothing, as out of a *Material Cause*; but only a
Bringing Something [i.e., matter] into being, that before had no Being at all; or a
Causing Something to Exist Now, that did not Exist Before; or which without this
Cause, would not have Existed.'

[81] I.e., space as geometrical; and consequently, space-filling matter as extended in
three dimensions.

insisted on by Borellus,[82] Newton, and others are reconcileable, and particularly thro the simple notion of actuall infinity[,] toucht upon in a paper framed for another porpose.[83] Supposing therefore that your friend[84] will let you see it, no more need be sayd of it here.

As for the theory of [*]*motus verus & relativus*[,] you returne me an extract of the same out [**f. 86**] of the author's [book],[85] instead of an answer to what my [previous] paper containes, to which I cannot add. But it seems by your paines in describing what relative motion is, you are not aware that you are all that while proving that motion consists wholly in relation. For as that is, or may be taken, which is arbitrary, any motion is this way, that way, or any way, nay numberless ways together; and even rest it self is motion, and all these, [*]*ad infinitum* at one and the same time. Relations are infinite, whereby every thing hath infinite motions, or none at all[,] 'ene as you pleas. And [because] the distinction of [*]*verus* and [*]*relativus* out of all experiment, smells so rank of the scools who allwais take shelter in words,[86] I cannot be reconciled to it. And I wonder [how] the great author of the *Principia* should fall into so low a shift as this, which is used only to serve some turnes he wanted it for, and then seeming to support it by a pair of

---

[82] Giovanni Alfonso Borelli, Italian physician, physicist and astronomer, who conceived the effect of a continuous force as the limit of the effect of a periodic impulsive force with indefinitely increasing frequency; see Djiksterhuis, *The Mechanization*, p. 367.

[83] For a summary of this paper, see *infra* 3.3.2: No. 2, RN→[SC 1706]. The original paper may no longer be extant (it is not in the British Library collection of RN's manuscripts). For a copy of the original, see *infra* 3.3.3. and n.261.

[84] I.e., RN.

[85] I.e., an extract from Newton, *Principia mathematica*, Scholium (to the Defs), pp. 7–9, concerning the distinction between true and relative motion.

[86] I.e., the ecclesiastical schools, attached to certain abbeys and cathedrals, in which 'schoolmen' lectured on philosophy and theology. Later, such lectures were given in the medieval universitys, in which 'scholastics' from the 9th to the early 15th century were especially concerned with applying Aristotelian logic to Christian theology. But the decline of scholasticism was largely due to its own 'shelter in words'— verbal subtleties, banding about words (logic chopping), etc.

experiments[,] mere paralogismes, as to proving [*]*motus verus.*[87] It were well if he or any one could show what exists in a body, while others change distances and aspects respecting it, more then when not, or what more **[f. 86v]** then those changes the (abstract) word motion can mean. To say, continual change of place, is faulty, for place is ambiguous, and not fixable but by relation, or (what is disputable) supposition. Who can say here or there [*]*in vacuo infinito*, without some standard,[88] such at least as the reflector[89] himself (whose person ought also to be abstracted) takes himself or his station to be. But I shall be lost, if I venture further here.

Your account of time and its distinction of absolute and relative, which comes out of the same shop,[90] is yet wors, for there is not so much as an experiment alledged (as was for the other) to vouch it. Who can find a warrant to say more of time then [*]*comparatio motuum*? —But [you say] that is of apparent or relative time, subject to be incertein or incomensurate as contingent quantitys are. But[91] absolute time is never faster nor slower but allwais equable.— I wish any one would tell me how fast or how slow it is. Equallity is taken from one and the same scale, but there are different scales, and which of them is it? Till that is answered, the distinction is merely chimerique and in a word **[f. 87]** there can be no account given of time but what will fall in with space or extension, and the changes it is subject unto. And I cannot approve the veracity of pretending to give any other account of it.

---

[87] I.e., Newton's two experiments: one using a bucket of water rotated and the other, two globes connected by a cord rotated about their common center of gravity; see *supra* 2.2.2: No. 1 (ff. 178–178v) and *infra* 3.3.2: No. 5, RN→[SC 1706] (f. 290).

[88] I.e., frame of reference.

[89] I.e., the spectator who thinks.

[90] I.e., Newton, *Principia mathematica*, Scholium (to the Defs), p. 7.

[91] I.e., whereas.

I know there are some who hold that infinite space and eternity of time are necessary beings,[92] becaus the idea of them in us is inexpugnable, which would not be, if the Creator, who is no deceiver, had not planted them in us. But this failes, for the ideas are from sence, and not from reflection, as the idea of an almighty power is. And sence may deceiv in this, as in other instances, where various tryalls may not be had to correct it.[93] But this subject requires a larger tract to shew it self.

As to the mathematicall probleme, I pretend not to disprove your analytick, nor doe I see you have here disproved the synthetick, by shewing where any paralogisme lyes. But you say the lines are not defined and from thence, as I suppose[,] conclude the demonstration faulty. Now first the rectangle is certein as can be, 2 x 1 or 1 x 2, as the exposure is. Then the four similar triangles are found by a motion [*]*in hypothesi* which is a method as legitimate as any, and [f. 87v] not onely strait lined, but all curve formes are defined by it. But if the demonstration be true, [then] it is much more usefull then your analitick, for the two means may be taken in compasses and carried into practise, but your [*]resolve is in surds[94] which cannot be exposed by any numbers whereby to protract them. And I should be glad to see any other exposure answering your resolve, performed by lines strait or curve.

---

[92] I.e., have real existence or absolute being. Believers included Barrow, Gassendi, Helmont, More and Newton, whereas Locke seemed to have wavered, according to Leyden, *Seventeenth-Century Metaphysics*, pp. 235–6, 279–80. For Clarke's position and for clarification of the remainder of RN's sentence, see *infra* 3.3.4.

[93] Concerning this paragraph, see *infra* 3.3.4.

[94] I.e., irrational numbers or quantities, especially roots.

No. 2 (FRAGMENT OF A PAPER) NORTH→[CLARKE 1706][95]
(UK:Lbl) Add MS 32546: ff. 293–293v

[f. 293]

Actuall Infinity

[...] But there is another sort of infinity which I call actuall [and]
which is of greatest importance in the science of nature. Therefore I
shall take a litle more care to explaine it.

That the matter of the world, increasing from us, is actually
subdevided into small parts, every one admitts, for all motion[,]
especially that of fluidity, declares it; and wee live in and perceiv so
much of it, that nothing can be to us more notorious. And for that
reason there cannot be actuall infinity of simple magnitude in the
way of increas, but[96] by composition of divers ingredients. And then
it amounts to the idea of infinite space, and no other [idea], and here
[this increase in magnitude] doth not concerne us. So in the way of
subdevision, it is no less agreed that, mentally, every individuall
particle of body is devisible and subdevisible [*]*ad infinitum*. And
that which the ancients meant by atome or minimum is not found
[*]*in rerum natura*. This [too] is not our porpose nor how farr bodys
may [be] (if at all)[97] practically separated [by] mechanicall
application of powers and resistances that may be against them

---

[95] In his previous communication, RN writes that 'the difficultys insisted on by
Borellus, Newton, and others are reconcileable ... particularly thro the simple
notion of actuall infinity[,] toucht upon in a paper' and that supposing 'therefore
that your friend will let you see it, no more need be sayd of it here'; see *supra*
No. 1, RN→[SC 1706] (f. 85v). Based on this statement, it is reasonable to assume
that Clarke asked to see the paper and that RN sent it to him, probably with a
covering letter, although these letters, as well as the paper itself are missing,
presumed lost.

[96] I.e., except.

[97] MS has this parenthetical phrase followed by another, which I have moved to
the end of the sentence.

(which consideration belongs to another place[98]). All[99] that I here affirme is that in most places there is matter interspers't that is actually small *ad infinitum*, that is, there shall be some smaller then any magnitude assignable. And [if you] propose any magnitude of space, [then] I affirme there is matter not farr off actually smaller. As [for example] in the compass of a dye,[100] the interstices contein a matter smaller then the component parts, and those have interstices conteining farther smaller and those [yet] smaller *ad infinitum*. So that the space of the **[f. 293v]** dye is perfectly repleat with body greater and smaller and of the latter[,] actually small [*]*ad infinitum*. And you may by the same way argue against devisibility as against actuall smallness [*]*ad infinitum vitum*. For what may be mentally cannot be proved impossible to be actually; and if there be great need [then] it should be so, as I shall shew [below]. It is highly probable it is so[,] as many bellow when I shew the use of this assertion[,] which is of another place.[101]

And farther wee say this notion [of actual infinity] implying no contradiction nor any inconsistency with any knowne [*]essence or mode of naturall things, admitting also that it may or may not be[,] it cannot be by any arguments proved it is not. For there is then [*]colour of argument that way, but [in arguing] for it, there is if not direct proof [then] good reason besides the use to presume it is so[,] as I shall shew.[102] For by all that observation which shews wee can

---

[98] See RN, 'Mechanick notes' (UK:Lbl) Add MS 32540: ff. 81–118v, *untitled continuation*, ff. 119–137v (early period, late), in which separation of bodies (effect) is due to impact (cause).

[99] MS has 'And that which'.

[100] I.e., die; in reference to gaming, a single dice.

[101] For RN's use of actual infinity, see *infra* 3.3.3.

[102] I.e., in what follows.

discerne nothing smaller then some [*]*animalculi*,[103] [and] when[104] by extraordinary magnifications by glasses such appear, then[105] [they] are discerned to have animall life from their members and activitys. And yet wee must gather [*]wonderfull further minuteness in the animall spirits of those creatures.[106] And in short, dioptrick glasses by discovering no symptome of [*]minisimuses, but ever of yet further deminutions, argue that the like may[,] if art[107] could lead us farther[,] proceed to actuall infinity of smallness. And it is agreeable to the notion [of] infinity in the way of increas, for [the] one admitted seems to include the other. All termination seems to contradict infinity; and the world[,] shewing no symptome of limitts[,] it is no reason to prescribe any.[108] These I say not for demonstration but for [*]colours or probabilitys[,] such as all naturalists even the most rigorous have thought fitt sometimes to entertein them. But after all [*]*cuilibat*[109] *judere esto*.

---

[103] These one-celled organisms, later known as bacteria, were first seen by Antony van Leeuwenhoek; but RN probably learned about the 'little Animalcules' from Hooke, 'Microscopivm' [1678], *Lectiones Cutlerianæ*, pp. 297–320.

[104] MS has 'for'.

[105] MS has 'and'.

[106] For a similar inference to unobservables, see Hooke, *Micrographia*, p. 173: 'Now, if we consider [the vibrative motion of the wings of flies and] the exceeding quickness of these Animal spirits that must cause these motions, we cannot chuse but admire the exceeding vividness of the governing faculty or *Anima* of the Insect, which is able to dispose and regulate so the motive faculties, as to cause every peculiar organ, not onely to move or act so quick, but to do it also so regularly.'

[107] I.e., the dioptric art of lens grinding to assist vision.

[108] I.e., 'Nature hath no limits'; see *supra* 2.2.5 n.222.

[109] MS has 'ruilibat'.

No. 3 (Paper) Clarke→North, September 1706
(UK:Mch) The Allen Deeds, Parcel 5 no. 5 (i–iv) [ff. 1–1v[110]]

[f. 1]

1. Every single refraction necessarily produces colours by separation of the rays; but because in most single refractions the separation is made in a very acute angle, the colours are not discerned without great niceness in making the experiment, because the angle does not open enough till at so great a distance as that the colours are very languid and [with] difficulty discernible. That the colours separated by a single refraction may become visible, three things are necessary: 1st, that the incidence of the rays be very oblique, and so the refraction great; 2^{ly}[,] that the density of the mediums be very different, for the same reason; 3^{ly}, that all foreign light be excluded, which by its brightness hinders the colours from being discerned. A six-pence in a bason,[111] or an oar or staff under water, makes no discernible colours, for the third reason; namely because those bodies are not very luminous, and the feint coloured rays which come from them are obscured by the bright light of the open air. The stars appear not coloured by the refraction in the atmosphere, for the 1st and 2^{d} reason. But it is very easy to render the colours produced by a single refraction very discernible, in the following manner. Take any glass vessel with a flat bottom and with a flat cover at the top; place it full of water in a dark room, into which no light can enter but only one sunbeam through a very small hole; let this beam pass with a very oblique incidence through a small hole in the cover of the vessel, and being refracted through the water, fall upon the bottom of the vessel; there you will see the ray very visibly separated into colours by a single refraction. But if the room be full of light, they are not easily discerned. The deeper the

---

[110] On the middle of folio 1v, parallel to the edge of the left-hand margin, RN has annotated (in transcription): 'Mr Clercks vindication— / here of Sr I. Newton. / Sept, 1706.' The wrapper or covering letter for this paper, with address and postmark, is missing, presumed lost.

[111] I.e., basin.

vessel be, and the smaller the hole through which the light passes, the more distinct will be the colours.

2. The reason why two refractions with one intermediate reflection make sensible colours in a round drop of glass or water, but not in a prism, is because the incident rays are much more oblique to the two refracting superficies in one case, than they can be in the other;[112] as will easily appear by calculation. And in some cases, the two sides of a prism, by means of the intermediate reflexion, have the same effect as if they were parallell to each other. When the prism is so ordered that the rays reflected from the inward surface fall very oblique on the outward [surface], they will be fringed with colours, discernible (as before) in a perfectly dark room, but not in open day-light.

3. A transparent body with parallell sides, if it be very[113] **[f. 1v]** thick especially, and the incident rays very oblique, will make fringes of colours, discernible in a dark room.

4. The reason why the colours contract, (for they do not vanish, if the room be dark,) when the prism is held close to the candle, is because the rays are not then so oblique to the refracting superficies, as at a greater distance. Altering the obliquity, causes the colours to contract or dilate proportionably, at any distance.

5. The fringes of colours do not dilate by withdrawing the eye from the prism in a direct line, because the eye so withdrawn receives rays that emerge less obliquely; but the image cast upon a wall, shows plainly that the image dilates as the angle opens; and the eye may easily, by altering its station duly, cause the fringes of colours seen through the prism, to dilate very much.[114]

---

[112] I.e., the two transparent bodies (glass bubble and water) have different refracting densities, the greater the difference, the less obliquity is requisite to cause a total reflection; see Newton, *Optice*, Bk II, Pt III, Prop. I.

[113] Just below the word 'very', MS has: 'turn over'.

[114] MS concludes with a flourish of the pen.

No. 4 (LETTER) CLARKE→NORTH, 28 SEPTEMBER 1706
(UK:Mch) The Allen Deeds, Parcel 5 no. 5 (i–iv) [ff. 1–2v[115]]

[f. 1]

Norwich September 28, 1706

Sir

Your objections are indeed very accute and ingenious; but yet, I think, all capable of a very just answer. To an inquisitive and intelligent person, there is no need of being either long or methodical in the reply. I hope therefore you will find all your scruples answered in the following short observations, in the same order as the difficulties themselves occurr in your papers.[116]

1. Sir Isaac Newton uses no physical <u>principles</u>, no <u>data</u>, no <u>hypotheses</u> whatsoever. All his writings contain only either <u>experiments</u> of matters of fact, or <u>geometrical</u> deductions of their consequences.[117] Tis geometricall to observe the different refrangibility of rays. Tis experiment of matter of fact, that to certain degrees of refrangibility certain colours are annexed. All the various observations drawn from hence, are either strictly geometrical, or strict natural history, without any <u>hypothesis</u>, without any <u>data</u> at all. There is no one <u>supposition</u> in his whole book, on which any proposition is built.

2. Sir Isaac never recurrs (as you suppose he does, and as the antient philosophers did,) to <u>qualities</u> and <u>properties</u>. He does not mean to say that <u>attraction</u> is the <u>cause</u> of bodies coming together;

---

[115] The wrapper with the postal address and postmark has not been preserved; but on the middle fold of f. 2v, RN has written parallel to the edge of the right-hand margin (in transcription): 'M[r] Clerck, / 8. [*sic*] Sept. 1706.'

[116] These 'papers' are missing, presumed lost.

[117] In §§1–3 Clarke here, as well as *infra* 4.4.3 and n.65, has used a number of one-word generalisations (e.g., all, always, no, never), which are absolutes that run counter to human experience (hence, the problem of induction). Moreover, despite his claim ('All' Newton's writings), his description of Newton's style of exposition refers only to the book on optics. For a more accurate description, see Shapiro, 'The Gradual Acceptance', pp. 89–91.

but by <u>attraction</u> he only means to express the <u>effect</u>, that bodies of such densities and at such distances <u>do come together</u> with such a certain force. This is only <u>natural history</u>; and the geometrical consequences of this observed <u>effect</u>, answer precisely to the phænomena of nature. What is the cause of this <u>attraction</u> (that is, of this <u>effect</u>, this <u>coming together</u> of bodies,) he does not undertake to explain. This he thinks is <u>metaphysical</u>; and it would be a sufficient perfection of natural philosophy, if we could thus compleat <u>natural history</u>, though our faculties could not at all come at the <u>causes</u> of things. Nevertheless, he gives hints of this metaphysical enquiry, sufficient for the more sagacious. But to write of this <u>*ex professo*</u>,[118] would be to make an <u>hypothesis</u>, which he alwais avoids as the greatest obstacle of true knowledge.

3. Hence you see why Sir Isaac makes no <u>system of physicks</u>. He will never make any <u>hypothesis</u>; and though he has many more things <u>lying behind</u>, yet he will bring nothing forth, till it be arrived either to the certainty of <u>natural history</u>, or <u>geometrical demonstration</u>. The book of light and colours is an example in one instance, how mere <u>natural history</u> without any <u>hypothesis</u> at all, is the only way to true philosophy.[119]

4. What I said concerning <u>oblique incidence</u>,[120] is to be understood only <u>*cæteris paribus*</u>.[121] Where the refractions are such as produce colours, there oblique incidence, by increasing the refraction, makes the colours more distinct and visible. But where the refractions are such as produce no colours (or in a degree seldom sensible) as in the case of your <u>interiour surface</u>, there oblique incidence will not do. Why by the refractions in the case of your

---

[118] (Lat.) out of a declared intention; as an expert.

[119] Between the end of §3 and the beginning of §4, MS has a space in which Clarke has drawn two short parallel lines from the left-hand margin to a little past the word 'What' at the beginning of §4.

[120] See *supra* No. 3, SC→RN, September 1706 (f. 1, §§3–5).

[121] (Lat.) all other things being equal.

interiour surface no colours are produced, (because it seems to be a principal difficulty with you,) I will explain at large presently, §9.
[f. 1v]

5. Concerning the difference of the density of mediums,[122] the same is to be understood, as concerning obliquity of incidence.

6. Your distinction concerning the inconvenience of foreign light is true. I mentioned it chiefly with respect to the colours produced by single refractions. Bodies under water, are not luminous enough; and the colours of sun-beams at the bottom of water, are hind[e]red from being visible, only by foreign light: as you will see by the experiment I before proposed.[123]

7. Red and blew are not the only primary colours,[124] but the primary colours are infinite as the degrees of refrangibility. Yellow can by no conspissation[125] be reduced to red, nor blue to violet. All the colours from red to violet are constantly ranged by every refraction.[126] The cause of your whole mistake in this matter, is your not perceiving the reason why, when every refraction ranges all the colours, yet every image (as of a broad bar in the window) is fringed constantly with only red and yellow on the one side, and only violet and blew on the other. The true cause of which, is plainly this. So many differently refrangible sorts of rays as there are, so many several images of the object must you see through the prism. Suppose you look upon a little white square; all the red and least-refrangible rays will represent to you a red square; all the blew and

---

[122] See *supra* No. 3, SC→RN, September 1706 (f. 1, §1).

[123] See ibid.

[124] If RN did assert this, then it is probable he was relying on the modification theory of Hooke, who assumed that colours are not innate to white light but modifications of it and that there are only two primary colours, red/yellow and blue; see Shapiro, *Fits, Passions and Paroxysms*, p. 103, and Hooke, *Micrographia*, pp. 62–7, 69–71, 74, 77.

[125] (OED obs.) thickening; condensation.

[126] I.e., the colour of each ray is connate and immutable, which is the new concept that Newton introduced into optics; see Shapiro, 'The Gradual Acceptance', p. 15, and *supra* 3.3.1.

most-refrangible rays of the same body, will represent to you a blew square; the yellow and green, in like manner. You may conceive therefore in your imagination suppose five whole-coloured images of the same white square, one red, one yellow, one green, one blue, one violet. Now as the refrangibility differs, so these whole-coloured images, which in the white body itself are exactly co-incident, will by the refraction of the prism, be as it were <u>drawn aside</u>, one <u>from under</u> the other; and so the edges of the red and yellow image must of necessity alwais appear on one side, and the edges of the violet and blew image on the other; all the images continuing coincident in the middle part which still appears white. Innumerable experiments will demonstrate this to be the very truth, to a sagacious observer.

8. What you say concerning colours residing strangely in <u>exility</u> will not at all affect the experiment I proposed for a single refraction.[127] The exility which produces colours, is only in that one dimension of <u>thinness</u>;[128] in the other dimensions it makes no difference, except perhaps for the convenience of observation. Now your water may be as <u>thick</u> (that is, as <u>deep</u>) as you please.

9.[129] Your objection drawn from the interior surface, is founded upon a very great mistake *in tota materia*.[130] You suppose that the refractions in that case are both the <u>same way</u>, and not <u>back again</u> as in the parallel sides. But if you please to examine the following rough schemes [see *Figure I* (1–4)], you will find the direct contrary to be true, namely that the refractions are not both **[f. 2]** the <u>same way</u>, but <u>back again</u> exactly as in parallel sides.[131] In Figures 1 and 2,

---

[127] See *supra* No. 3, SC→RN, September 1706 (f. 1, §1).

[128] Possibly, a reference to colours produced by thin films.

[129] MS has in the left-hand margin: 'N.B.', i.e., *Nota bene* (Lat.) Note well.

[130] (Lat.) in the whole matter.

[131] See Newton, *Optice*, Bk I, Pt I, Defin. II, which defines the refrangibility of rays of light as their disposition to be refracted or turned out of their way in passing from one transparent body or medium into another and also defines a greater or less refrangibility as the disposition of the rays to be turned more or less out of their way in like incidences on the same medium.

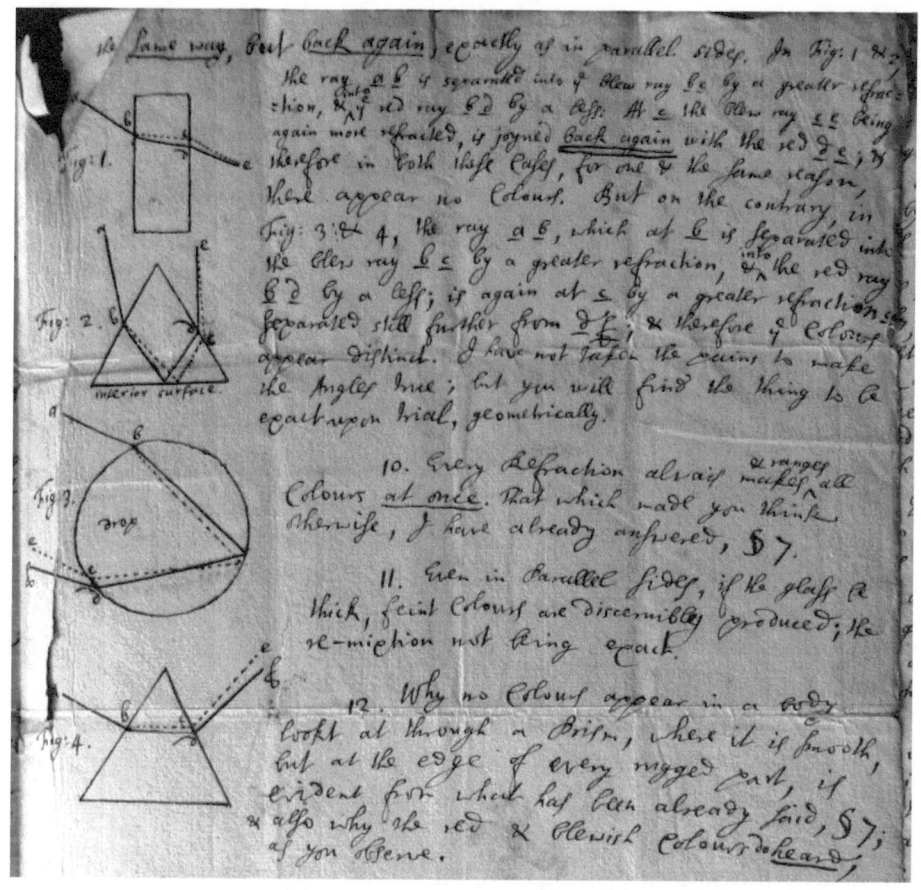

Clarke's Schemes
illustrating greater or less refrangibility of rays of light
(top of f. 2 in Clarke →North, 28 September 1706, §9)

Figure I (1–4)

the ray <u>a</u> <u>b</u> is separated into the blew ray <u>b</u> <u>c</u> by a greater refraction, and into the red ray <u>b</u> <u>d</u> by a less: At <u>c</u> the blew ray <u>c</u> <u>e</u> being again more refracted, is joyned <u>back again</u>[132] with the red <u>d</u> <u>e</u>; and therefore in both these cases, for one and the same reason, there appear no colours. But on the contrary, in Figures 3 and 4, the ray <u>a</u> <u>b</u>, which at <u>b</u> is separated into the blew ray <u>b</u> <u>c</u> by a greater refraction, and into the red ray <u>b</u> <u>d</u> by a less; is again at <u>c</u> by a greater refraction <u>c</u> <u>e</u>[?][133] separated still further from <u>d</u> <u>f</u>; and therefore the colours appear distinct. I have not taken the pains to make the angles true; but you will find the thing to be exact upon trial, geometrically.

10. Every refraction alwais makes and ranges all colours <u>at once</u>. That which made you think otherwise, I have already answered, §7.

11. Even in parallel sides, if the glass be thick, feint colours are discernibly produced; the re-mixtion[134] not being exact.

12. Why no colours appear in a body lookt at through a prism, where it is smooth, but at the edge of every rugged part, is evident from what has been already said, §7; and also why the red and blewish colours do <u>heard</u>,[135] as you observe.

13. The bar in a window is sometimes quite covered with colours, from the images of the bright glass expanding over it on each side.

If I have omitted any thing, I shall be very glad to hear further. In your next be pleased to give some directions for a certain way to send to you. We shall be going to London about the 13th of October: After next week therefore, if you cannot send sooner, be pleased to

---

[132] MS has double underlines, perhaps for emphasis.

[133] MS has barely visible a single letter written on the very edge of the right-hand margin.

[134] See mixtion (OED *sb*. obs.) the fact of being mixed or, in the above instance, remixed.

[135] I.e., herd (OED *v.*[1] 1b, rare) of things: to come together, assemble, to be assembled or associated.

inclose yours in a blank paper directed to my Lord[136] at his house in Petty-France, Westminster. If I live to return [to Norfolk] next summer, I should be very desirous of waiting upon you; but my preferment does not afford a horse. I find few people in the universitys or at London, that make half so rational objections as yours out of the country.

    I am,

Sir,

Your most obliged humble Servant

Sam: Clarke.

No. 5 (Two fragments) North→[Clarke 1706][137]
(UK:Lbl) Add MS 32546: f. 290 and f. 290v

[f. 290]

    For there is no mean, either all is perpetuall miracle, or things ordinarily[,] as reciprocall causes[,] hang one to another. And as for the hints to the sagacious you formerly touched,[138] really I find very much [that] plaine dealing is a jewell; and I never yet knew a fair reason to wrapp up phisicall truths, as is sayd of the Pythagoreans, in affected obscurity, unless it were either to advance a trade or [to] forme more then is true. Nor can I beleev, tho many by their proceedings seem to fear it, that naturall philosofy in the most perspicuous dress[139] ever hurt any sort of goodness, nor is that [style] such, whatever it be, that steals out of the way of a right understanding.

    Now having toucht on the *Principia*, and the title, it were uncivil to let it pass without a note to the matter. And out of a sea of

---

[136] I.e., John Moore, bishop of Norwich. Regarding Clarke's instruction to RN about enclosures, see *infra* No. 6, SC→RN, 26 October 1706 (f. 1, postscript).

[137] Although these fragments begin with the same words, it is not clear that they are rewritings one of the other.

[138] See *supra* No. 4, SC→RN, 28 September 1706 (ff. 1, §2 and 1v, §7).

[139] I.e., easy to be understood style.

that, I choos the first thing, absolutes and relatives, and of them motion, which is distinguish't into [*]*motus verus*, and [*]*motus relativus*; and that endeavoured to be made good by experiments, which goe no farther then the case of turning,[140] to which I say nothing at present, but desire to know your thoughts, whither a body be capable of [*]*vera quies* and *motus verus* (I use the words) at one and the same moment of time. And then I may give you a farther trouble upon that point.

[*Written near the bottom of the folio from the left-hand margin to the middle of the page*] of them in the Querys after the *Opticks*,[141] and when the great author crownes[?] or affords us hopes[,] I dare say wee shall have a new world in philosofy. But all that lys a great way behind, and it is a mountaine to be removed to come at it.

[f. 290v]

For there is no mean, either all is perpetuall miracle or things ordinary, as reciprocall causes, depend on each other. It is easy to say[,] matter is indued with propertys to any porpose, but more great and glorious [to say,] if one property be shewed truly to subserve all sensible and probably inscrutible porposes, [then it is] that of universall attraction.[142] I am sure by experiment[,][143] that to enervate

---

[140] I.e., rotation. RN here refers to the first two experiments in *Principia mathematica*, one with a bucket of water turned and the other, immediately following, with two globes on a cord revolved, from which Newton concluded that motion should be determined with respect to absolute space and not (with respect to Descartes's principle) relative to the reference frame of contiguous bodies. For RN's comment on the first experiment, see *supra* 2.2.2: No. 1 (ff. 178–178v), and for his comment on the second, see *supra* 2.2.4 n.192.

[141] After the word 'Opticks' (i.e., *Optice*), MS has deleted: '<which altogether are too much to take in hand>'.

[142] The next sentence, which refers to 'universall attraction', has only some words crossed out; but as the remaining make little sense, I have assumed that they were intended for deletion along with the sentence that follows, which has been crossed out.

[143] MS has 'experiment as may be'.

gravity may be proved to fail, which is [*]*jugulum causæ*. And it is an error of the first concoction to use a principle which any one may pleas to deny; and it is labour hazarded if not lost to superstruct elaborate demonstrations on such. I find by the *Opticks* wee may hope for an explication, or rather demonstration of the principles, either by clear sensation or [by] necessary consequences (for really less will not doe).[144] And then [we may] expect a reformed if not a new world of philosofy; when surely no notions will be current, but what are clear and distinct or necessarily deduced from such as are so. And esteaming ourselves not oblidged to go omniscient but be contented in many things to remaine ignorant[,][145] we[146] [do] not reject plain veritys, becaus they will not fitt all cases.

I find in the *Opticks*, in a [*]quere dress[,] much proposed to render the principle of attraction and some other [*]*vires* allowed plausible, but nothing to prove it.[147] But [this] is the constant method of [*]hypothecarians, that is, disabling all other solutions and applying this [one] in their room and then concluding from aptness. And however some objections are seemingly abraded,[148] as not like occult quality[,][149] possibly by pulsion (that is, it may be and not be at all, for pulsion and attraction are two notions) and the like.[150] Yet it [attraction] is grossly maintained and seems to be contradictory in

---

[144] Possibly a reference to Newton's first public statement concerning his method of analysis and synthesis, which was included in the 1706 *Optice*, Qu. 23, pp. 347–8 and not in the 1704 *Opticks*. Regarding this statement, see Shapiro, 'Experimental Philosophy', pp. 196–7.

[145] This sentence provides an instance of RN's scepticism, especially his belief that the recognition of one's own limitations is an essential mark of wisdom; see *supra* 1.1.4.

[146] MS has 'and'.

[147] I.e., In Newton, *Optice*, Qu. 21; but in the remainder of the paragraph, RN's comments also include Qu. 23.

[148] I.e., rubbed off, removed by friction.

[149] I.e., the unintelligible 'quality' of attraction.

[150] Newton, *Optice*, Qu. 23, p. 322, had stated that what he called attraction may be performed by impulse, or by some other means unknown to him.

some things[,] as one while the attraction is according to density and then small bodys, etc.

No. 6 (Letter) Clarke→North, 26 October 1706
(UK:Mch) The Allen Deeds, Parcel 5 no. 5 (i–iv) [ff. 1–1v[151]]

[f. 1]

London October 26, 1706

Sir,

Your last[152] is so diffusive, that I dare not presume to be certain of giving a satisfactory answer to it: But, being sensible of your candour, I shall endeavour to touch briefly upon the several particulars.

1. According to your definition of a physical hypothesis, that it is <u>an arguing from some things palpable, to others out of our reach</u>, we agree that hypotheses <u>may</u> and <u>must</u> be used. Thus, that <u>all the</u> matter in the universe is impenetrable, is an <u>hypothesis</u>. But such hypotheses, as Cartesius makes, that <u>all</u> space is matter, without proving that <u>any</u> is so; and that the pores of <u>all</u> bodies are full of subtle matter, without proving that the pores of <u>any</u> are so; and that <u>all</u> light consists of <u>globules</u>, without demonstrating that <u>any</u> does so; such hypotheses, do only hinder knowledge.

2. It is a mistake, what you affirm, that <u>the *anomala*[153] of the moon, and some other cases, are ascribed to some attractions not yet discerned</u>. All these minute cases could not be solved at the time when the <u>*Principia*</u> were wrote, only for want of sufficiently exact

---

[151] Folio 1v has the postal address (and postmark) as follows: 'For Mr North at his house / at Rougham near / Swaffham / in / Norfolk / by way of Thetford / (J Norwich)', i.e., John Moore, bishop of Norwich. About the middle of this folio, on one of the folds, there are two annotations written parallel to the right-hand margin. The first is in the hand of RN: 'Mr Clerk / 8. [*sic*] Oct. 1706'; the second is in the hand of his son Montagu: 'all these Letters are / transcribed by my Father / in a Book.'

[152] RN's previous communication is missing, presumed lost.

[153] (Lat.) irregularities.

observations of the phænomena; but they are since solved equally by the one general law of gravitation; without supposing any new vires;[154] which would be an hypothesis indeed, in the sense we disclaim.

3. It is likewise a great mistake to say that comets move in parabolas, and by some unknown *vir attractiva*[155] are reduced to an elliptick course. On the contrary, tis evident they move in ellipses by the common force of gravitation, as other planets do; only their parabola is a less errour on one hand, than any species of ellipsis we can assigne, is on the other.[156]

4. Colours, if not very strong, are not so easily discerned by simple vision, as by reflexion: nay[,] the stronger colours, as when you look upon the sun through a prism, are not discerned half so well, as when they are cast upon a white wall. Tis no wonder at all, that you cannot see the colors made by single refraction, unless you please to try some such experiment as I before mentioned.[157] Yet I have sometimes in the open day-light seen a bright shilling at the bottom of a pail of water fringed with colours.[158]

5. As to your choosing rather to think that colours are only rags[159] or edges, than whole images; to what was before said,[160] I can only add this: that in a sun-shine-day, by looking through a prism with the greatest possible obliquity upon a small dazy,[161] illuminated with the strongest light of the sun-beams, I can easily draw out not

---

[154] (Lat.) forces.

[155] (Lat.) attractive force.

[156] MS has the continuation on the same leaf but written parallel to the first part of the letter.

[157] See *supra* No. 3, SC→RN, September 1706 (f. 1), where Clarke has described how to render 'the colours produced by a single refraction very discernible'.

[158] I.e., the six-pence in a basin; see *supra* ibid.

[159] (OED *sb.*[1] 5, rare) a jagged projection (perhaps RN's term).

[160] See *supra* No. 4, SC→RN, 28 September 1706 (f. 1v, §7).

[161] I.e., daisy.

only the edges, but even almost the whole entire coloured images from under one another, like five coloured flowers in a row.

6. Colour is no further a diminution of light, than as tis a passaging[162] of it. All the prismatick colours drawn into a focus, make as bright a ray, as the original one, excepting for the rays reflected by the glasses.

7. Thin transparent plates of <u>exactly equal spissure</u>,[163] (as was observed by Mr <u>Hook</u> with microscopes,[164] as well as by Sir Isaac Newton)[165] yield the brightest colours in the world.

8. The colours you expect after decussation of the rays[166] emerging out of a prism after the reflexion, are much less discernible than in single refractions, and for the very same reason. The angles should be more acute than I have drawn them.[167]

9. I find no other difference in the colo[u]rs by the object approaching the prism, than what plainly arises from the different obliquities of incidence, or from the great <u>divergence</u> of the rays proceeding from a neer object.

I am, Sir,

Your most humble servant
S. Clark.

If you please to inclose a letter to me at any time, in a blank paper directed to my Lord himself[168] at his house in Petty-France, Westminster, you need pay no postage.

---

[162] (OED *v.*[2] 1. *intr.*) to move across, pass, cross (first citation is 1824).

[163] Not in OED; but see spissness (OED *adv.* obs.) thickness, density.

[164] See Hooke, *Micrographia*, Obs. IX and X, pp. 47–79, in which he proposed an hypothesis to explain why thin films, as well as other transparent bodies produce coloured rings (now called 'Newton's rings').

[165] For thin films, see Newton, *Optice*, Bk II, Pts 1–3.

[166] I.e., crossing of the rays to form a figure like the letter 'X'.

[167] See *supra* No. 4, SC→RN, 28 September 1706 (f. 2) and *Figure I (1–4)*.

[168] I.e., John Moore, bishop of Norwich.

No. 7 (THREE FRAGMENTS) NORTH→[CLARKE 1706]
(UK:Lbl) Add MS 32546: ff. 280, 282, 287v

[f. 280]
[...]In astronomy, that force [of attraction,] tho it commands all the planets in our system[,] is barely to be observed in one effect pointing to the north. But by handling the [lode]stone and iron [*]obnoxious to the same force,[169] wee understand somewhat more tho[170] not enough yet to explaine it, but wee hope for the farther discovery from experiments upon earth, then from astronomical observations.

2. Whither the [*]*vires*, supposed, are not fundamentall in the structure of the planetary system, and the latter recurring to the [*]*vires sit liber index*; it seems to me a circle. Supposing such *vires*, the planets must have such cours, they have such cours, therefore the *vires* supposed are true. So in time of yore, supposing orbs and epicicles solved the phænomena and that went for proof of reall orbs, etc. So [also] went the disputes about the time of new philosophy first professed. And now is it not the dispute which way will and which will not jump to an inch? All is understood and nothing left in the dark.

3. The earth cannot accelerate by passing between Venus and Mars; for they are not as shoars to a current but as bodys conformably floating with others[,] which makes no coarctation[171] at all amongst them. And so vast a body as that [*]*c[a]elum*, with the earth and its vortex if any such be[,] may require ages to accelerate[172] or retard them in motion so as to be sensible to us,[173] [...] **[f. 282]** If wee thinck to have any rule for the heavens, [then] wee shall certeinly be mistaken; for the motions of the several distances of the

---

[169] I.e., magnetic attraction.

[170] MS has this word crossed out, but it should be retained for sense.

[171] I.e., compression.

[172] MS seems to have 'xrate'.

[173] MS has the next three words crossed out: '<If there with>'.

vortexes are not upon the same center, but vary infinitely. The planets are not all, and perhaps none, in the largest circle, if you consider the vortex as a uniforme sphear, but they are towards it. The moon does not move exactly in the equinoctiall of the earth, but varys much from it, which is more visible to us then the variations of the other planets, because it is vastly neerer. Neither are the circles they move in neer perfect but rather ellipticall, as appears by the apo-and-periges.[174] Nor doe I thinck the centers remaine the same, but chang[e] every hundred years; a short time to such magnitudes as appears by the recorded accounts of Ticho Brahe[175] and the best astronomers. Besides I cannot be perswaded that without a manufacturer, any thing in nature except the very laws, which are very plain and short, is regular. Tho our ages are so short wee cannot perceiv any great difference of any thing but what is neer us. Otherwise wee look upon the world to be sphear and circles, as it is painted to the eye, and that serves the turne well enough, until our magnitude is such that we may inspect the univers, as wee doe a whirle pool. [**f. 287v**] [...] altho wee allowed (as wee doe not) some caus for it. Much less in such a point of time (comparably) as that aspect continues.

4. Saturne is allowed to be, in our sence, cold enough[176] and that reflection of heat from the flat ring might warm it and that [may be] justly esteemed a single caus of the [*]*ansae*.[177] And Mercury

---

[174] I.e., apogee or most distant point in an earth orbit; perigree or the point in the orbit of the moon that is nearest the earth.

[175] I.e., the sixteenth-century Danish astronomer, Tycho Brahe, whose access to improved or new instruments and his 'unparalleled observational talent' helped to raise 'observational astronomy to a level which was unprecedented before his time and was not reached again before the invention of the telescope'; see Djiksterhuis, *The Mechanization*, p. 300.

[176] MS has deletions and insertions, making the order of the rest of the sentence difficult to determine.

[177] I.e., handles, an astronomical term applied to the appendages of Saturn's ring, first observed by Galileo. The epoch-making treatise on the subject of ring theory was the *Systema Saturnium* (1659) of Huygens, who supposed that there was a single ring, that it was solid and that its appendages were also solid. In 1675 the

may be also hott, and both[,] as all the rest of the planets[,] be compound of materiall apt for the position. Saturne may[,] thinck wee[,] burne as wee thinck[178] Mercury [does]; the one may be unctuous and the other gold, while wee in a medium are water, earth and stone.[179] But heat and cold are not as distance from the sun, but as reflection and warrantable stuff occasion.[180] So that I conceiv all the arithmetick bestowed in the calculates of heat near the sun is lost.

5. Why should not the irregularitys of the moon and the rest of the planets be reall irregularitys[,] fortuitous, and inexplicable?[181] Tell me any one naturall thing in particular that is not so. Perhaps wee hold, as Aristotle, [*]*c[a]elum immutabile* and thinck[182] there is a decorum in a rule more then ordinary place.[183] There may be generall rules certein enough. But in application to particulars in practice everything hath unaccountable and inexplicable accidents and measures. And why great things should be mathematically exact and small ones, which wee better know, not so, I cannot answer. Nor why casualtys about the planets must needs be explicable, or that an

---

French astonomer, Jean Domenique Cassini, determined that Saturn's ring was composed of multiple smaller rings with gaps between them. But in 1709 Clarke still held to the single ring theory, and as late as 1713 it seems RN did also; see Whiston, *Historical Memoirs*, pp. 14–15, and *infra* 5.5.2: No. 1, RN→SC 1713 (ff. 19v, 24). For details of the developing understanding of Saturn between 1610 and the first decades of the eighteenth century, see Alexander, *The Planet Saturn*.

[178] MS has this word crossed out.

[179] I.e., the inhabitants of Saturn have an astronomical perspective different from the inhabitants of Earth, an allusion to Huygens' *Cosmotheoros*; see *supra* 1.1.3 and n.107. This work presented a more developed treatment of the astronomy of the inhabitants of Saturn for which the *Systema Saturnium* (1659) was only a prelude.

[180] Concerning this topic, see *infra* 3.3.4.

[181] For the reply to this statement, see *infra* No. 8, SC→RN, 21 November 1706 (f. 1, §1).

[182] MS has in left-hand margin opposite this word which begins the line: [*]'quære'.

[183] I.e., position?

hypothesis, becaus it fitts them, must be true.[184] Therefore as to your precisely regular,[185] I beg pardon, scarce thinking[186] [...]

No. 8 (LETTER) CLARKE→NORTH, 21 NOVEMBER 1706
(UK:Mch) The Allen Deeds, Parcel 5 no. 5 (i–iv) [ff. 1–1v[187]]

[f. 1]

London November 21, 1706

Sir

In answer to your last,[188] I have only the following remarks to make.

1. All that you say about bodies, [either] <u>attracted by the center</u> or <u>equilibrated in a fluid</u>, is a number of gross mistakes. The proposition that the planets are <u>attracted by the center</u>, does not depend upon the *vires centripetæ*[189] to be first granted; but the <u>phænomena</u> evidently prove it. The earth moving <u>slowest</u>, where its vortex (if there was any such thing[190]) is most compressed between the heavens of Mars and Venus, is a demonstration that it is not carried by a fluid, which alwais moves <u>fastest</u> in the narrowest part of its passage. Again, if the planets were equilibrated in a fluid, then Saturn being furthest from the center, would be <u>densest</u>, and Mercury <u>rarest</u>; which is the absurdest thing in the world. For so Mercury

---

[184] RN's stance on irregularities is not restricted to astronomy; see *infra* 2.2.1 n.1, 4.4.5 and Kassler, *HRN*, pp. 139–40, *et passim*.

[185] See *infra* No. 8, SC→RN, 21 November 1706 (f. 1, §1).

[186] MS has this word as the catchword.

[187] Folio 1v has the address (with a postmark): 'For Mr North at his house / at Rougham / near Swaffham / in / by way of Thetford Norfolk / Free / J Norwich', i.e., John Moore, bishop of Norwich. On the middle fold of the same folio, RN has written parallel to the edge of the right-hand margin (in transcription): 'Mr Clerck 21 N^r 1706'.

[188] See *supra* No. 7, of which only fragments now remain.

[189] (Lat.) centripetal forces.

[190] Cf. *supra* No. 7, RN→[SC 1706] (f. 280, §3).

would be <u>burnt up</u> in a moment, and Saturn <u>frozen</u> to a stone. Again, if equilibrated in a fluid, then the irregularities (as they[191] call them) of the moon particularly, would be <u>really</u> irregularities, fortuitous and inexplicable;[192] when on the contrary they are precisely regular and demonstrable from gravitation.

2. The increase of the motion *in directum*[193] will not (unless it be increased to a very great degree,) ever carry off a planet, but only lengthen its ellipsis, nor will jogging of it in, or increasing the attractive force, (how much soever) ever drive it into a <u>spiral</u>, but only shorten the ellipsis. This is geometrically demonstrable. And all that you say upon this head, is errour in mathematicks.

3. Equall areas described in equal times, does not at all depend upon the figure; but the proposition is universally true in the circle, in all ellipses, parabolas, hyperbolas and all the figures in the world, as Mr. Newton has demonstrated[194] geometrically.[195]

4. The arguments for a vacuum are not supposalls, as you call them, but demonstrations;[196] and to try whether the pores of bodies are full of subtle matter, or not, you may indeavour, if you please, to

---

[191] I.e., the astronomers.

[192] See *supra* No. 7, RN→[SC 1706] (f. 287v, §5): 'Why should not the irregularitys of the moon and the rest of the planets be reall irregularitys[,] fortuitous, and inexplicable? Tell me any one naturall thing in particular that is not so?'

[193] (Lat.) in a direct line.

[194] MS continues on the same leaf but is written parallel to the first part of the letter.

[195] See Newton, *Principia mathematica*, Bk. I, Sect. iv (finding eliptic, parabolic and hyperbolic orbits from one focus given), pp. 61–9.

[196] Like Newton, Clarke believed in the existence of the vaccum; but in §§4 and 5 above he cited no chapter and verse from the *Principia* to indicate specifically where Newton had actually *demonstrated* the vacuum. And the same is true of his claim in the first series of his Boyle Lectures, where he stated: 'if Gravitation be an Universal Quality or Affection of *All Matter*; then there is a Vacuum; (as is abundantly demonstrated by Mr. *Newton*:)'; see Clarke, *A Demonstration*, Prop. III, p. 49. Whether Newton ever provided the demonstration he so desired is questionable; see Bertoloni Meli, *Thinking in Objects*, pp. 272–3, and *supra* 2.2.4 n.190.

evade this one plain argument. Two pendulums of equal length, the one having its knob[197] of wood, the other of gold, make your vibrations in equal times in an exhausted receiver. Now, say I, this would happen quite otherwise, if there were any considerable quantity of subtle matter (for some little there is, as light and magnetick particles, etc.) pervading their pores. For the pores of wood being vastly more than the pores of gold, the pervading subtle matter must of necessity make much more resistance to the numerous superficies's of the solid parts of wood, than to the fewer superficies's of the pores of gold; and consequently make the vibrations of the two pendulums unequal. The fineness of the matter, will not prevent this; because how fine soever it be, yet if there be a plenum, tis denser than quicksilver. Nor will any supposed quickness of its motion signify any thing; for as many of its part[s] moving one way as another, the resistance of some of it will as [much][198] be increased by its motion, as that of other parts be diminishe[d] and so upon the whole it must make a resistance as great as if i[t] [w]as[199] a quiescent fluid denser than quicksilver.

    5. As to your paper about attraction;[200] if there be a vacuum, (as is I think sufficiently proved) then tis plain the effects we ascribe to attraction, cannot possibly be mechanical. And if our philosophy cannot go beyond natural history, (which yet is infinitely various and pleasant) who can help it, that we have no more faculties than we have?

I am Sir

Your most humble servant

S. Clarke

---

[197] I.e., bob.

[198] MS is torn on the right-hand margin, and this affects the end of this and the next two lines.

[199] MS seems to have a mark above and to the right of 'as', possibly the bottom of a letter from the above obscured line ending.

[200] This paper is missing, presumed lost.

No. 9 (THREE FRAGMENTS) NORTH→CLARKE, 27 NOVEMBER 1706
(UK:Lbl) Add MS 32546: ff. 280v and 287
(UK:Lbl) Add MS 32547: ff. 382v–384v

[f. 280v]

27[201] November, 1706
Answer to a letter of Mr. Clerck, dated 21 ditto
and follows after.

Sir,

I received the favour of yours, of the 21st instant, and admire at your lenity in not chiding, as so many mistakes deserve not for the thing (for what imports it one way or other?) but for your owne sake, having your time invaded, and also matters, with you, setled upon reasons invincible and authority inexpugnable.[202] And to say truth, such basket-attaques[203] ought, as children's impertinences[,] to be snapt up short, or like incroaching weeds[,] be cropt up all at once. For enforcing which point of discretion, I cannot give you better encouragement, then by the sequel here, that may perhaps amount to a necessity of mustring a good stock of acrimony against next occasion, for prevention of more such [*]inconveniences.

1. I deny that most astronomicall observations prove any [*]*vires*, for in that imensity [of celestial space] causes may be such as no observation can discover. Wherefore I distinguish between observation and experiment. The former perceives a [single] thing of divers, but one way, that is, distant view; but experiment perceivs the same thing divers ways. As the force of a magnet is proved variously, but [...] **[f. 287]** [...] thincking the best astronomers will pretend it. But [since] all other measures take up with a litle more or less[,] they[204] never expect the most correct table they can make shall

---

[201] MS originally had '29'; but RN altered the '9' by writing an em dash just below the closure of the circle that makes the upper part of '9'.

[202] I.e., the authority of Newton.

[203] I.e., the quantity that fills a basket. The phrase was used as a measure of an uncertain amount, here, the quantity of Clarke's 'attaques'.

[204] MS has 'and'.

last [a] long time. But in turne as hath betoken to others before them[,] [their tables] grew as these now are[,] fals. And whereas[205] you alledg all the cours and aspects of the moon are resolved by gravity onely, wee say the same but ours ownes an experiment,[206] yours[,] none.[207] Wee think wee understand somewhat of it [i.e., the moon's motion], if not all. You disowne to understand any thing [about its irregularities, whereas], wee leave much to ignorance and accident. Your machine is mathematically exact, even as the orbes of Ptolomy; and there is the state of this [our] difference.

2. I may not enter into the [*]*penetralia* of geometry to be judg of the demonstration about the reciprocall efficacy of the supposed [*]*vires*; or [to] resolve why water under great pressure and issuing in a small thredd is so soon drawne downe by gravity, and the [*]*vis impressa* of the planets to move equably [*]*in directum* not unlike the other center in a parabola, and the *vis impressa* of the planetts to move *in directum* not unlike ships, but as the medium hinders,[208] and to be held, as a stone in a sling, by a counterworking of those *vires*, and not goe on, tho the centripetall force determines both continually. But in the room of them grossly fancy that (admitting all), if[209] a ballance of forces be[,] as I think it is[,] supposed, and the planet by accident (as hitcht a litle remoter from the center) gets an advantage, [then] it must needs be gone in the way of [*]*directum*,

---

[205] MS may have this word crossed out.

[206] I.e., RN's hydrostatic account of gravity is based on the Cartesian vortex mechanism, a motion of the minute (unobservable) matter that is necessitated by a universal plenum. It is probable, therefore, that by 'experiment' he means the rotation of a stone in a sling which provided the conceptual model for Descartes's vortex theory; see *supra* 2.2.2: No. 1 (f. 177 n.82), 2.2.3, *et passim*.

[207] Later on, RN mentions that 'attractionists' had tried an experiment with falling bodies of different specific gravities in an exhausted receiver of the air pump, but he denies that it amounted to a demonstration of 'an universall power of discent in body'; see ff. 65v–66 in RN, 'Of Gravitation, general and particular' (UK:Lbl) Add MS 32548: in Notebook 3, ff. 65–74 (middle period?).

[208] MS has 'medium hinders' in the left-hand margin.

[209] MS has, opposite this first word in the line, what appears to be a little drawing, possibly scrubbed out.

and on the other side for alternate cause possibly[?] towards the center. And you seem to joyne with me in this, by a parenthesis [unless it be increased to a very great degree].[210] I ask[,] by what degree? Every degree (mathematically) determines an [*]*equilibre*. And when comes a new force to recover it? and [yet] it seems the planets' courses are precisely mathematical [...] **[f. 382v]** [...] to pretend the contrary is to contradict our genius. But all persons' relish is not the same, tho all have appetites. Some are satisfyed with comon places, as attraction, and the like; and as those places fill [so they] thinck their knowledg increaseth. Others, like the ancients, if they can bring unusuall effects to be like usuall, tho alike understood, are satisfyed. Others that affect more penetration are less happy, but either in reason or imagination have their share. And no man's head is broken, of which I will not excuse most of the ordinary and pevers enterteinements of life.

Sir,

I thinck these causes are tryed,[211] at least as farr as I may sift 'em, and if you pleas [then] wee will, for exercise and variety, call [up] another [matter]; and that may be the opinion at the beginning of the *Principia*, that distinguisheth [*]*tempus absolutum & relativum*, and [*]*motus verus & relativus*, which I cannot yet digest. I shall say litle of the former, onely that *tempus absolutum* moves equably, that equably is a relative term, and I would gladly know, [relative] to what? If you say, it self, [then] I ask what that is. The author says wee[212] confound things with their measures.[213] Wee know the measures, but are at a loss for the thing. And if that happen to be no other then the very measure, [then] there is no danger of confusion;

---

[210] MS has brackets, which enclose Clarke's words; see *supra* No. 8, SC→RN, 21 November 1706 (f. 1, §2).

[211] I.e., put to the test.

[212] I.e., those who adhere to Descartes's principle of relativity of motion.

[213] See Newton, *Principia mathematica* (Scholium to the Defs), p. 7, according to whom the duration, or perseverance of the existence of things, remains the same, whether motions are swift or slow or none at all; consequently, this duration ought to be distinguished from what are only sensible measures thereof.

but if wee referr the measure to something, but wee know not what
and cannot express otherwise then by shifting synonymous words[,]
then I thinck wee are in a way of confusion.

[f. 383]

But to tell you the truth, my dullness is such, as to have no
opinion of geometry, or demonstration as it is called, of matters that
depend on event. For, whatever may be, it seldome happens that the
subject matter, as wee may call the principles, are certein enough to
warrant that method.[214] Nor will the demonstration be compleat
without tryall. Mathematick principles therefore are usually
supposititious, as being naturall possibilitys, but scarce ever
[*]obnoxious to sence or practise. The Grecian Venus is in every
stone, but no carver since Phidios[215] could find it. And even in
mechanicall cases, which depend on the events of impulses, and
those [impulses] governed by quantity, which in simple and stated
cases are a subject of demonstration[,] yet that knowledg entered by
tryall and is then approved. The masters of computation have
exercised their witts in diversifying propositions. But in cases
infinitely complex, which[216] cannot be clearly and distinctly stated,
but are derived upon occult[217] [*]essences called [*]*vires*, which are
not and cannot be defined, but only described by blind resemblances
to the confused phænomena of complex and unknowne motions, I
say[,] none will trust conclusions from such processes without tryall,
nor receive the value of 5 pounds, on the credit of them. And the
rather becaus the principles must not onely be certein, but exclude all
other causes and incidents whatsoever, which in generall cannot be,
and in particulars is not in the subject now in view. And this makes
[f. 383v] me content to stand on one side, if not superior to the

---

[214] I.e., mathematical demonstration cannot be the standard for 'matters that
depend on event'; see *supra* 1.1.4 and 2.2.5.

[215] I.e., Phidias or Pheidias, considered as the greatest sculptor of classical Greece,
superintended all the works of art which were erected at Athens during Pericles'
administration, including the Parthenon.

[216] MS has 'and that'.

[217] I.e., unintelligible.

thornes and bryars, and from thence view the vaine intricacy of them, rather then enter, where I thinck there is more of affectation and enigma then of natural knowledg. This[,] you will say[,] is the bolt of one uncapable, [*]*et habes confitentem reum.*

3. The areas may, as divers other propertys[,] be common to all conick sections, but I see not the connexion of them with the progression of a planet.[218]

4. As to the two pendulums,[219] I have bin told,[220] it is not in the power of art to forme two pendulums that shall continue to swing together, but shall soon devide, which I suppose holds in the receivor,[221] as well as without[222] it. And then where is the argument? And I have no means to make the experiment but should be glad to hear of any way to adjust pendulums so exactly, as to swing together, which would be an exquisite demonstration of isocronisme, or equation of time.[223]

But in the meantime desiring not to elude nor be [*]illuded, I will deal plainely as to the matter, and say [1.] that permeable bodys

---

[218] See Newton, *Principia mathematica*, Bk I, Sect. vi (how motions are to be found in given orbits), pp. 104–14. That RN did not understand the connection of conics with the orbits of a planet may be due to his reliance on Descartes's 'doctrine of curves', with which he was quite familiar; see, e.g., f. 66v in RN, 'Preface to a philosofick essay' (UK:Lbl) Add MS 32526: ff. 64v–68 (early period). That doctrine excluded mechanical curves (cone sections) from geometry, an exclusion which Newton defeated by subsuming geometry under mechanics; see Guicciardini, *Isaac Newton*, pp. 294–308 *et passim*.

[219] See *supra* No. 8, SC→RN, 21 November 1706 (f. 1, §4): 'Two pendulums of equal length, the one having its knob of wood, the other of gold, make your vibrations in equal times in an exhausted receiver.'

[220] Since Hooke's connection with some members of the North family can be dated from 1676, it is possible that he may have been RN's informant; see Kassler, *The Beginnings*, pp. 27–8, 72–3, *et passim*. See further n.223 below.

[221] I.e., that part of the air pump in which air is evacuated.

[222] I.e., outside the receiver.

[223] Hooke showed once and for all that, *physically*, no pendulum is truly isochronous, whereas Newton, in working out the problem of the dynamics of the pendulum, showed that, *ideally*, the cycloidal pendulum is an isochronous system; see Ariotti, 'Aspects', pp. 409–10.

carry the included [parts of] matter with them; and such hath efficacy with the rest, onely upon stops, not being shutt in, may pass on. An hollow tower[,] shak't with the wind by the [*]*vis impressa* of the included air, shall be thrown downe if vent windoes are not made to eas it. And clouds, that consist of devided[224] **[f. 384]** minute dropps, pass afore the wind with the intermediate air all in one lump. So the wood with the included air or subtile matter is resisted in a body, as gold is. The difference appears upon impulses, for the latter coheres, and the other devides, so that the fancy of the subtile matter pervading the wood as it swings may be layd aside. This is some answer consistent with the doctrine of density.[225]

But 2. I must deny all that [doctrine] and the puz[z]les that are wiredrawne[226] from it.[227] I beleev you cannot charg us with maintaining that matter is less solid, for being more minutely devided.[228] But this wee say, that a fluid doth not resist a compound [body], in the manner as compounds resist one and other, quantity against quantity. For, the latter being equall, one in an instant stops the other. But a fluid permitts a compound to pass many and many diameters, before the motion wast[e]s, and this more or less, not from solidity[,] for that wee say in all space is the same, but from the fluidity, which makes the difference. So air is less impediment then water, that less then quicksilver, and that [less] then melted pitch. This is proved by the experiment of burnt alabaster or [*]iczen, which is a stone pulverised and set over a fire in a ketle. However

---

[224] MS has 'minute' as the last word, with 'drops' as the catchword; whereas f. 384 carries over both words.

[225] I.e., Newton's doctrine; see *supra* 2.2.3.

[226] I.e., drawn out to great length with subtle ingenuity.

[227] I.e., the puzzles either in Newton, *Principia mathematica*, Bk II, the whole of which was devoted to motion in a resisting medium, or in Newton's, *Optice*, Qu. 20, in which consideration was given to fluid media such as air, water, quicksilver, etc., and their power of resistance that arises from the density and tenacity of those media.

[228] I.e., the two men have competing conceptions of the nature of matter; see *infra* 3.3.3.

fine the powder is, it is no fluid, untill the heat is advanc't to actuate all the granules and then it takes the forme of fluidity. Before[,][229] a stick with a button[230] could scarce be moved **[f. 384v]** by the hand thro it, but when fluid, it will swing in it as a pendulum. For passing [the stick through] a fluid is not so much creating, as determining motion in the parts of it, and they are left in the same order as they were with litle disturbance, but as accident and friction occasion. For a fluid is a perfect ballance in it self, and in perpetuall agitation. And to adumbrate the nature of this action, let a weighing beam have 200 pounds in each scale, and so stand [*]*in equilibr[i]o.*[231] Thro[w] a ginny[232] into either, and the whole 400 weight is moved and that considerably swift. This motion is not imparted by the ginny, but determined [by it]. Such is the difference between minute bodys in the state of powder and in the state of fluidity.[233] And it is I thinck a fallacy that a compound, in the latter,[234] moves impulsively,[235] its equall quantity, in every diameter of the progression, as I take the doctrine of density to maintaine.

    5. I am glad of your declaration, that you goe no farther then naturall history,[236] and it had bin well all the titles of late books and chapters were as explicite, so as to give no occasion to the

---

[229] I.e., in its previous state.

[230] I.e., a knob on a handle (e.g., the knob of metal at the breech end of a piece of ordnance).

[231] Note that Newton's achievement in hydrostatics lay in providing a theoretical formulation of that science based not on weight, as in RN's adumbration, but on pressure; see Shapiro, 'Light, Pressure, and Rectilinear Propagation', p. 266.

[232] I.e., guinea.

[233] See further *infra* 3.3.3.

[234] I.e., in the fluid.

[235] I.e., like the swings of a pendulum.

[236] See *supra* No. 8, SC→RN, 21 November 1706 (f. 1, §5): 'And if our philosophy cannot go beyond natural history ... who can help it, that we have no more faculties than we have?'.

[*]misprising an occult[237] filosofy, and there wee rest. And none can idolize naturall history more then I doe, but yet thinck the pleasure is not merely as it is [a] relation of sticks and stones, but as [it] brings some light, as wee cannot but fancy, to our understanding of efficient causes, and to[238] [...]

3.3. CONSTITUTING THE CONTINUUM

As pointed out previously,[239] there is no evidence to determine the exact date when the correspondence with Clarke began; and the same is true concerning its conclusion. In addition, from internal evidence in the foregoing editions, it is clear that a number of letters or papers are missing. Nevertheless, from what now remains, it is apparent that in its initial stages, the tone of the correspondence was friendly; but from the end of October there is a subtle change of tone as Clarke becomes irritated by North's adherence to the vortex as a mechanism of gravity. That Cartesian concept requires an infinite plenum, whereas the Newtonian concept of attraction requires an infinite vacuum. And, according to Clarke: 'if there be a vacuum, (as is I think sufficiently proved) then tis plain the effects we[240] ascribe to attraction, cannot possibly be mechanical'.[241] Moreover, as Clarke indicated in an earlier letter to North,

> What is the cause of this <u>attraction</u> (that is, of this effect, this <u>coming together</u> of bodies,) he [Newton] does not undertake to explain. This he thinks is <u>metaphysical</u>; and it would be a sufficient perfection of natural philosophy, if we could thus compleat <u>natural history</u>, though our faculties could not at all

---

[237] I.e., unintelligible.

[238] Catchword is 'to'.

[239] See *supra* 3.3.1.

[240] I.e., Newton and Clarke.

[241] See *supra* 3.3.2: No. 8, SC→RN, 21 November 1706 (f. 1, §5).

come at the <u>causes</u> of things. Nevertheless, he gives hints of this metaphysical enquiry, sufficient for the more sagacious.[242]

If vacuum, in the sense of infinite space, is conceived as an attribute of the deity, as some philosophers previous to Clarke had already claimed,[243] then the cause of attraction is the deity. And here we find the main reason why North has described, and will continue to describe attraction as 'occult' (unintelligible). For although he acknowledges God as an almighty power or first cause, his aim as a philosopher is to provide a natural, not a supernatural explanation of physical events.[244]

Now, the problem of the existence of a vacuum is closely connected with the problem of matter, its division and individuation. And as previously indicated,[245] North adopts a mean between two extreme conceptions of material structure, the one, a geometrical conception that continuous magnitude is infinitely divisible,[246] the other, an atomist conception that continuous magnitude has actual parts that are only finitely divisible. The first conception seems to mandate infinite divisibility, for the mere fact that a body occupies space seems to guarantee that 'it has spatially distinct parts whose separation is logically possible'. But the second conception, in allowing that bodies are only finitely divisible, seems to preclude infinite divisibility.[247]

As a consequence of these opposing metaphysical positions, there were various responses to the problem of material structure,

---

[242] See *supra* 3.3.2: No. 4, SC→RN, 28 September 1706 (f. 1, §2). For hints concerning 'this metaphysical enquiry', see Newton, *Optice*, Qu. 23.

[243] See *supra* 2.2.4.

[244] RN is also the first writer on music to ascribe its origin to the voices and language of humans and not, as was then the norm, to the gift of God; see Kassler, *HRN*, pp. 173–8.

[245] See *supra* 2.2.3.

[246] See, e.g., Hooke, *Micrographia*, p. 2: 'certainly the *quantity* of extension of any body may be *Divisible in infinitum*, though perhaps not the *matter*.'

[247] Holden, *The Architecture of Matter*, p. 2.

which Thomas Holden has grouped into four philosophical 'factions': (1) those who maintain, with Aristotle, that body is infinitely divisible and reject the actual-parts doctrine; (2) those in the tradition of Epicurus who reject infinite divisibility of matter and uphold the actual-parts doctrine; (3) those who endorse the actual-parts doctrine and infinite divisibility, maintaining that the two theses can somehow be squared with one another; and (4) those who endorse both the actual-parts doctrine and infinite divisibility, but then admit that the two doctrines are incompatible and, hence, an antinomy of reason.[248]

From what we learned in the previous chapter, North's position is the same as that of the third faction.[249] But this faction had a number of sub-groups, in one of which are the philosophers who seem unaware of any supposed tension between the actual-parts doctrine and infinite divisibility. This sub-group includes Barrow, who accepted an actual-parts analysis but treated all infinities as potential.[250] But since North treats infinities as actual, he does not fit into this sub-group. Rather, he belongs to the sub-group who accepted that body (taken as a compound) is indeed constructed from actual infinities of ultimate parts, each body being a structured aggregate built up from a greater-than-infinite number of actual first parts.[251] According to Holden,[252] the main inspiration for this

---

[248] Op. cit., pp. 51–73.

[249] See *supra* 2.2.3 and 2.2.4.

[250] Holden, *The Architecture of Matter*, pp. 56–7; see also Barrow, *The Usefulness*, pp. 142, 148, 150–1, 162. Some historians of mathematics have suggested that Barrow's traditionalism was not so clear cut; see, e.g., Hill, 'Neither Ancient nor Modern'; Malet, 'Barrow, Wallis and the Remaking of Seventeenth Century Indivisibles'; and Mahoney, 'Barrow's Mathematics', p. 185.

[251] According to Holden, *The Architecture of Matter*, p. 62, Henry More, in his later writings, 'seems' to have adopted this solution. But this statement is questionable, because in the case of matter (but not spirit) More posited ultimate indivisibles and, hence, was a finitist; see *supra* 2.2.4.

[252] Ibid., pp. 59–62.

particular conception was the *Discorsi e dimostrazioni matematiche, intorno à due nuove scienze* (1638).

Written by Galileo Galilei, the founder of the mathematical theory of the infinite,[253] the book is structured as four dialogues, the first of which contains an extensive treatment of the concept of infinity[254] in the context of what Homer Le Grand has described as 'a highly original matter theory'.[255] In this treatment, Simplicio represents the Aristotelian tradition of potential infinity, whereas Salviati represents the author, who seeks to overcome the paradoxes involved in admitting actual parts, on the one hand, and potential divisibility on the other. Using the *persona* of Salviati, Galileo also made two claims that are important for our purposes: first, all continuous quantities have actual parts; and second, their infinite divisibility entails that they are constructed from *actual* infinities of metaphysical *indivisibles*.[256]

Compare these two claims with the one made at the outset of North's fragment on actual infinity, where he writes:

> That [*claim 1*] the matter of the world, increasing from us, is actually subdevided into small parts, every one admitts, for all motion[,] especially that of fluidity, declares it; and wee live in and perceiv so much of it, that nothing can be to us more notorious. And [*claim 2*] for that reason there cannot be actuall infinity of simple magnitude in the way of increase, but by

---

[253] See Galileo, *Dialogues concerning Two New Sciences*, pp. 30–52. For Galileo as one of the 'heroes' who overcame the disputing method of Aristotle and his followers, see *supra* 2.2.2: No. 3 (f. 187) and RN, 'Præfanda' (UK:Lbl) Add MS 32526: f. 2 (early period).

[254] I.e., indefinitely continued divisibility, as well as continuity and its paradoxes.

[255] See Le Grand, 'Galileo's Matter Theory', p. 198.

[256] Ibid., pp. 33–4. Note that these two claims imply continuous atomism (i.e., there is no ultimate indivisible), perhaps somewhat similar to the ancient Stoic quantum model of the continuum reconstructed by White, *The Continuous and the Discrete*, pp. 284–326, whose work Holden does not cite.

composition of divers ingredients. And then it amounts to infinite space.[257]

He then states that this mathematical argument of increase in magnitude does not concern him. Moreover, the physical argument for divisibility—that 'bodys may [be] (if at all) practically separated [by] mechanicall application of powers and resistances'—is not to his purpose either. Rather, all he affirms is that 'in most places there is matter interspers't that is actually small *ad infinitum*, that is, there shall be some smaller then any magnitude assignable'.[258] He supports this affirmation by recourse to two other arguments (ff. 293–293v). The first is mathematical: that the very concept of extension includes the notion of sub-parts that are themselves extended and, hence, anything extended is formally divisible to infinity. This argument follows from his consideration of 'the components' of a die ('dye') used in gaming. The second is physical, for it follows from observed subtle matter, the *animalcula*, discovered by investigations of others with the microscope.[259]

Whereas the latter argument requires no comment, the former, from a single die, is insufficiently elaborated. To discover more

---

[257] See *supra* 3.3.2: No. 2, RN→[SC 1706] (f. 293).

[258] As will become apparent below, RN's phrase, 'smaller then any magnitude assignable', means that metaphysically *indivisible* parts are extensionless (zero size) and not that such parts are infinitesimal, i.e., neither finitely extended nor altogether unextended.

[259] In *Recherce de la vérité* Nicolas Malebranche argued that although the infinite divisibility of matter can be demonstrated mathematically, what is revealed in microscopes confirms our notion of God's purpose—that He can create, and in fact has created, smaller and smaller worlds to infinity. To illustrate this, he used the example of seeds that are encapsulated to infinity, as in the dormant seed of a tulip bulb, which, when examined with a magnifying lens or even with the naked eye, can be seen to contain an entire tulip; see Malebranche, *The Search after Truth*, pp. 26–7, 118, 353, 741–3. That RN had read one of the early French editions of this work appears from his mention of its author in a number of texts; see, e.g., f. 68 in RN, 'Preface to a philosofick essay' (UK:Lbl) Add MS 32526: ff. 64v–68 (early period), where he describes the Frenchman's philosophy as 'so awd with holy church, as it wants the freedome such a designe should have'.

detail, therefore, we need first to examine the 'paper' to which North refers in both the first and second fragments of his correspondence with Clarke.[260] For in this paper he records his speculations on the interaction of 'body' and 'spirit', the latter conceived as material and as the residence of the immaterial soul, the location of which is 'somewhere about the brain'.[261] And he also stated that his speculations 'depend wholly upon the supposall of actuall infinity in the minuteness'.[262] To make it probable that 'material influences of organized matter may by direction be drawne into such a point where some matter is also infinitely reduced',[263] he briefly addressed the ways in which the ratio of solidity and superficies are varied,[264] using the example of a die to show that 'increasing the substance lessens the ration of the superficies and decreasing the substance adds to the superficies'. If this increasing or decreasing is carried on 'in progression to infinite', then 'on the one side an infinitely extended body (or space as is the supposed world) has no superficies or limit at all' and 'on the other side a body deminished *ad infinitum* is all superficies and no body at all'.[265] Unfortunately, the mathematical basis of this method is assumed, not explained, but a manuscript

---

[260] See *supra* 3.3.2: Nos. 1 and 2, RN→[SC 1706] (f. 85v and f. 293v).

[261] RN, 'Essay on the reciprocall forces of body & spirit influencing each other' (UK:Lbl) Add MS 32549: in Notebook 3, ff. 77–81 (middle period, early). Although RN's speculation may have been a response to his experience of reading More, *The Immortality of the Soul* (1659), his approach is different not only regarding the seat of common sense (for More, the fourth ventricle of the brain), but also regarding the conception of spirit (for More, immaterial). See More, *Philosophical Writings*, pp. 270–1, 307–8, for an outline of his doctrine of the relation between body and spirit.

[262] RN, ibid., f. 81.

[263] Ibid., f. 80.

[264] Ibid., f. 77v, where RN refers to 'isoperimetry', i.e. that branch of geometry which deals with figures having equal perimeters and the problems connected with them. For one such problem, see Galileo, *Dialogues concerning Two New Sciences*, pp. 57–60. Note that the OED gives 1706 as the first citation in English for the term 'isoperimetrical' and 1811 for the term 'isoperimetry'.

[265] Ibid., f. 78.

from North's late period provides an explanation relating to the 'mind having power to move the body', part of which is reproduced in *Figure II.*[266]

There are two points to note about this explanation. First, for the word 'dye' used not only in his paper but also in the fragment relating to actual infinity, North substitutes the words 'Cubes or Solids'. Consequently, his numerical scheme seems to represent what, according to Galileo, 'geometry teaches us', namely, that 'in the case of similar solids, the ratio of two volumes is greater than the ratio of their surfaces'. For this statement is followed by an illustration of 'a cube two inches on a side so that each face has an area of four square inches and the total area, i.e., the sum of the six faces, amounts to twenty-four square inches'.[267]

Second, the discursive part of the explanation seems to be a consideration of the infinite from the Galilean point of view of multiplicity and, as a consequence, the possibility of an actual infinity with respect to multiplicity or, in modern terms, infinite aggregation.[268] And it is this mathematical method that North seems to apply to the problem of 'body-spirit' interaction, for he has translated a solid with a similar surface into mathematical symbols and then proceeded to the limiting situation obtained by indefinitely exhausting its dimensions,[269] until at last it is possible, as he claims

---

[266] See the section 'Of spirituality' that begins on f. 252 in RN, 'Physica' (UK:Lbl) Add MS 32544: ff. 1–274v (late period), where RN describes his explanation as 'a meer conceipt'. Note that this manuscript, which lacks diagrams, is in the hand of Montagu North with RN's corrections. For the original from which the copy was made, see RN, 'Phisica being a congeries of discourses upon most subjects in natural philósofye' (UK:Nr) Rougham MS (late period).

[267] See Galileo, *Dialogues concerning Two New Sciences*, pp. 89–90.

[268] See Boyer, *The History of the Calculus*, pp. 115–6, 270, 291.

[269] In his paper (see n.261 above), RN assumed, f. 78v, that 'others' will object that solidity and superficies 'must subsist and expire together', so he answers that 'if it be considered that in this progression, the ration between them becoming expanded to infinite (these speculations admitt no stop or stay) so that the solidity with respect to the superficies becomes next to nothing ... [then it] will amount nearly to the same' as that of indefinitely shrinking the ratio to infinite.

...it is in mathematiques agreed, that superficies increase as squares, and solids as cubes of the root; whereby circles or sphears are of the squares, and globes [are of the] cubes of the diameter [which in] figures [are] thus—

Roots	Squares	Cubes or Solids	Faces or Superficies	A
1	1	1	6	6 to 1
2	4	8	24	3 to 1
3	9	27	54	2 to 1

And the reverse of this ... proceeds in the like proportion, the solidity decreasing, the superficies (in proportion declared) increaseth.

North's numerical Scheme
showing how the Ratio of Superficies and Solids
increase or decrease ad infinitum

Figure II

in his paper, to conceive 'spirit' (which is material) as 'a mathematical point' with no magnitude.

This suggests that North's infinitely small constituents of geometric magnitudes are indivisibles—points, lines, surfaces—not infinitesimals.[270] If so, then he is relying on the method of indivisibles, which he could have learned from reading either Galileo's book or the works of his pupil, Bonaventura Cavalieri.[271] However, there is some evidence that rules out the latter possibility. For in his 1706 correspondence with Clarke, North raises two points relating to Newton's doctrine of density, the second of which begins: 'I beleev you cannot charg us with maintaining that matter is less solid, for being more minutely divided. But this wee say, that a fluid doth not resist a compound [body], in the manner as compounds resist one and other, quantity against quantity.'[272] And then follows his attempt to show this difference between compounds and fluids from the example of 'minute bodys in the state of powder and in the state of fluidity'.

It is to be noted, therefore, that Galileo had recourse to these two states of matter in an examination that represents a conjunction between the mathematical and the physical.[273] For he attempted to show 'a remarkable property' that explains 'the vast alteration and change of character which a fine quantity would undergo in passing to infinity'. In brief, his argument is that if continuous magnitudes are made up of indivisibles and if the number of parts is infinite (as he had previously claimed), then the aggregation of these is not one resembling a very fine powder but rather a sort of merging of parts into unity, as in the case of fluids.[274]

---

[270] See n.258 above.

[271] See Malet, 'Barrow, Wallis, and the Remaking of Seventeenth Century Indivisibles'.

[272] See *supra* 3.3.2: No. 9, RN→SC, 27 November 1706 (ff. 384–384v).

[273] See Le Grand, 'Galileo's Matter Theory', pp. 202–4.

[274] Galileo, *Dialogues concerning Two New Sciences*, pp. 38–41.

And thus we return to the main problem that seems to underlie the issues raised in the first fragment, where North responds to Clarke's missing critique concerning divisibility and individuation. From the foregoing comments concerning infinite aggregation, it is clear that North uses the term 'solidity' in its geometrical sense to mean the amount of space occupied by a solid body—its volume, cubic or solid content. But since Clarke 'set up solidity' as an attribute of matter[275], his use of the term means a body's relative density (specific gravity). And since he denied that matter's space-filling properties are geometrical, we may infer that he fits into Holden's second faction, that of the Epicureans (e.g., Gassendi), who uphold the actual-parts doctrine but reject infinite divisibility of matter in both its forms—the potential of the Aristotelians (e.g., Barrow) as well as the actual of the Galileans (e.g., North). Rather, for the Epicureans, the divisibility of matter is finite, since they conceive least parts or atoms as ultimate metaphysical indivisibles.[276]

3.4. UNCOVERING IRRECONCILABLE DIFFERENCES

But there is more to be learned from the fragments that are the sole evidence of North's side of the correspondence. To begin with, recall that most of the fragments are drafts or discards, the exceptions being the first communication and part of the last,[277] which are copies from an original. Since the latter preserves the name of the recipient and the date the original letter was sent, it represents a copy perhaps intended for a letter book. But the former was not so intended, for North entered a copy of his original communication in a bound notebook, along with other notes that he wished to preserve as a

---

[275] See *supra* 3.3.2: No. 1, RN→[SC 1706] (f. 82v).

[276] For a hint of Clarke's finitism, see the negative arguments in *A Demonstration* [1705], Prop. I and Prop. III, pp. 20–3, 72–4, that all the metaphysical difficulties arise from applying 'the measures and relations of things finite to what is infinite and from supposing finites to be *aliquot* parts of infinite'.

[277] See *supra* 3.3.2: Nos. 1 and 9.

record of his thoughts. It is clear, therefore, that he considered the substance of his original letter of some importance. Yet in copying from the original, he removed all epistolary apparatus, instead providing a title: 'An Answer to ------'. It might be asked, therefore, why did North omit Clarke's name?

In terms of the 1706 correspondence I cannot answer this question with certainty, although I shall return to it later on.[278] But from what now remains of both sides of the correspondence, it is possible to infer that the two men had irreconcilable differences not only concerning certain issues in the philosophy of nature, but also in terms of their overarching philosophical positions and values. We have already encountered one instance of this in connection with the problem of material structure. But the truncated copy of his letter referred to above is helpful in teasing out another instance, for there, in response to Clarke, North states: 'What wee can or cannot imagine is not materiall to the truth of things without us, which will be as they are in themselves, whither wee can imagine it or not', for truth is to be sought by 'abstracting all prejudice or imagination' and not by 'the contrary method' that leads to 'the confusion of all naturall philosofy'.[279]

There are two parts to North's reply, just quoted, the first of which is a succinct statement of his epistemological realism described in Chapter 1.[280] But what now remains of the 1706 correspondence is silent about Clarke's philosophical position. To discover this, it is necessary to turn to the first series of his Boyle Lectures,[281] the aim of which was to demonstrate the existence of God using 'the Rules of strict and demonstrative Argumentation'.[282] In the chain of reasoning that follows, there are twelve linked

---

[278] See *infra* Epilogue.

[279] See *supra* 3.3.2: No. 1, RN→[SC 1706] (f. 82v).

[280] See *supra* 1.1.4.

[281] See *supra* 3.3.1.

[282] See Clarke, *A Demonstration* [1705], 'The Introduction concerning the Causes of Atheism', pp. 15–17, p. 15.

propositions which are supposed to demonstrate the existence of God as well as his 'omnipresence, omnipotence, omniscience, infinite wisdom and beneficence ... as Euclid demonstrates the equality of the angles at the base of an isosceles triangle'.[283]

As James Ferguson has pointed out, Clarke's demonstration is a composite one, since not only does it utilise two patterns of reasoning, both scholastic, but also several different arguments for the existence of God.[284] In regard to the patterns of scholastic reasoning, one is *a priori*, which moves from principles, reasons or causes to their consequences or effects, whereas the other is *a posteriori*, which moves from consequences or effects to their principles or causes. As for the arguments themselves, Clarke relies primarily on the cosmological argument, an *a posteriori* argument that employs some *a priori* principles, as, for example, the logical principle of non-contradiction, which states that a proposition and its negation cannot both be true.[285]

There are four points to note about Clarke's version of the cosmological argument. First, it begins with an existential premiss—that a being exists—and afterwards proceeds to deduce certain attributes classed as natural and moral. Second, it uses indirect knowledge, or what Clarke calls 'metaphysical reasoning', to prove that a being is the sufficient reason of the cosmos in a temporal sense as well as in rank, so that in this context sufficient reason is understood as the principle of causality.[286] Third, it is based on the assumption that although errors may occur in the deductions,

---

[283] Stephen, *History of English Thought*, vol. 1, p. 121.

[284] See Ferguson, *The Philosophy of Dr. Samuel Clarke*, pp. 22–7.

[285] In his Boyle Lectures and other writings, Clarke's use of the term contradiction, as well as such cognate terms as absurdity and impossibility is symptomatic of his reliance on this principle; see, e.g., *supra* 3.3.2: No. 8, SC→RN, 21 November 1706 (f. 1, §1).

[286] The principle of causality implies that the sufficient and only reason for an event may be in the will of God; but this contradicts a second version of the principle, in Clarke, *A Demonstration* [1705], Prop. XII, pp. 233–59, that God always has reasons for his acts (will). There is some irony in this oversight, since one of Clarke's principal tactics is to expose contradictions in the work of others.

metaphysical reasoning itself, like mathematical reasoning, can yield certainty, because the propositions themselves are true.[287] Fourth, the cosmological argument is supplemented by interspersed teleological arguments or arguments from design, which seek the reason or cause of the universe as a cosmos from empirical proof of its regularity[288]—its 'order, beauty, harmony'—and from an emphasis on 'design' or adaptation of means to ends.

In one of these interspersed arguments, Clarke drew his empirical proof from the 'late' discoveries in natural and experimental philosophy and astronomy, including the

> ...exact *Accommodating* the *Densities* of the Planets, to their Distances from the Sun, and consequently to the Proportion of Heat which each of them is to bear respectively, so that neither Those which are nearest to the Sun, are destroyed by the Heat; nor Those which are farthest off, by the Cold; but each enjoys a Temperature suited to its proper Uses, as the Earth to ours[.][289]

Since this particular 'discovery' was raised in the 1706 correspondence,[290] it is important to recognise that the passage quoted above is a paraphrase of the only passage that mentions God in the first edition of Newton's *Principia mathematica*.[291]

---

[287] For others who held this assumption concerning metaphysical and mathematical reasoning; see Leeuwen, *The Problem of Certainty*, especially pp. 101, 124.

[288] I.e., uniformity, concerning which RN and Clarke have opposing positions; see *infra* 3.3.5.

[289] Clarke, *A Demonstration* [1705], in Prop. XI, pp. 230–1.

[290] See *supra* 3.3.2: No. 7, RN→[SC 1706] (f. 287v).

[291] According to Newton, *Principia mathematica*, Bk III, Prop. 8, in Cor. 5, p. 415, the deity positioned planets at different distances from the sun so that they would receive heat from it according to the proportion of their densities. In 1692 the same passage was paraphrased by Richard Bentley, the first of the Boyle lecturers; see Cohen, *Introduction*, pp. 155–6.

In addition to the cosmological and teleological arguments, two others were included in the early propositions. One, the so-called ontological argument, is an *a priori* argument from the nature of absolute and antecedent necessity,[292] whereas the other, an argument from the nature of space and time, is an *a posteriori* argument based on attributes or consequences or effects of God's existence, an existence that is prior to, and presupposed by them. It is the latter argument that is particularly revealing for the meaning of the second part of North's reply, quoted at the outset of this section. For Clarke had proceeded on the assumption that if we try to suppose there is no necessarily existing being, then we always find we cannot remove from our imagination the ideas of immensity or infinite space and eternity or infinite duration.[293] In short, he depended on what one can or cannot imagine as the basis for his assertions; and it is this kind of procedure which North criticises[294] and describes as 'the contrary method',[295] that is to say, a method contrary to his own conception of space as an idea abstracted from his definition of body as the geometrically-extended impenetrable.[296]

But further, Clarke's ideas of infinite space and eternity of time are ideas of attributes or modes and, hence, must inhere in some

---

[292] Clarke relied on the ontological principle that God exists, because as a necessary being, the supposition of non-existence would be self-contradictory. Pretending to distance himself from Descartes's ontological proof, he still employed the French philosopher's procedure, although 'by no means improving on it'; see Ferguson, *The Philosophy of Dr. Samuel Clarke*, pp. 24–6.

[293] Clarke, *A Demonstration* [1705], in Prop. III, particularly pp. 41–2, where he argued that since he cannot in 'my Imagination' take away the idea of God as a necessarily existing being, then 'the Certainty of the *Existence* of that Thing' is the same 'and stands on the same Foundation, as the Certainty of ... the Relation of Equality between twice two and four', for the latter has 'no other Certainty but this, that I cannot, without a Contradiction, change or take away the Idea of that Relation'.

[294] See *supra* 2.2.2: No. 1 (ff.174–174v and n.68) and 3.3.2: No. 1, RN→[SC 1706] (ff.82–83v, 85).

[295] See *supra* 3.3.2: No. 1, RN→[SC 1706] (f. 82v).

[296] See *supra* 2.2.3 and 2.2.4.

substance. As Ferguson remarked, in admitting the necessity of their existence, Clarke at the same time admitted the necessary existence of the substance possessing them, so that such a necessarily-existing substance, possessing the attributes of infinity and eternity can only be God.[297] But what did Clarke mean by the term 'substance'? At the time of the 1706 correspondence, there were two rival definitions of the term, one, as that in which properties ultimately inhere but which itself is not a property or quality, and the other, as that which is independent or self-sufficient, not requiring anything else in order to exist.[298]

   That North follows the first definition is clear from the manner in which he treats material substance ('body') as a universal principle. In his usage, which is empirical, substance merely means that 'there are "thats" which have qualities and relations' but which are 'not divisible into substratum and attributes'. By contrast, Clarke's usage, which is metaphysical, involves 'the passage to an unobserved substratum', which bears 'the attributes but which is not itself a part of or the whole of the class of attributes'.[299] Indeed, he stated that immaterial substance is incomprehensible[300] and that the same is true of material substance, because we have no appearance of it in our imagination.[301]

---

[297] Ferguson, *The Philosophy of Dr. Samuel Clarke*, p. 26.

[298] See Holden, *The Architecture of Matter*, pp. 112–3.

[299] Morris, *Six Theories of Mind*, p. 7. Newton's usage seems similar; see Harman, *Metaphysics and Natural Philosophy*, Chapter 2; see also *infra* 3.3.5.

[300] Clarke, *A Demonstration* [1705], Prop. IV, p. 74: what the substance of the necessarily-existing being is 'we have no Idea, neither is it possible for us in any measure to comprehend it'. Clarke's conception here is in line with a voluntarist and essentialist approach to the infinite God, an approach also found in Descartes, according to Collins, *God in Modern Philosophy*, p. 61.

[301] See Clarke, *A Demonstration* [1705], in Prop. X, p. 164. In a later edition, he made this claim more explicit; see ibid. [1832], in Prop. X, p. 79: 'of the *Substance of Matter* itself, (its Simple *Substance*, considered as abstract from, and as the Foundation of That Essential *Property of Solidity*,) we have no Idea'.

In these statements Clarke reveals his commitment to the independence thesis that is central to realism, namely, an ontological commitment to entities that exist independently of our knowledge or experience of them. Recall, however, that the independence thesis can be held in different ways. In North's case, his ontological commitment is epistemological.[302] But in Clarke's case, it is metaphysical, because his ontology is indebted to the Aristotelian concept of primary substance (*ousia*)—that which exists in and of itself[303] and is the foundation of other, derivative existence and causes itself. Accordingly, in his existential premiss Clarke argued that because there is something, there must be something ultimate, that is to say, something uncaused or self-caused or necessary. And this argument follows from the belief, first, that a 'ground' of being is required for what there is and, second, that this ground or ultimate existent is real, that is, literally true.

Bearing in mind, then, that North and Clarke have different versions of the independence thesis and, hence, different conceptions of space and time, it is possible to understand a passage in the first fragment, where North states:

> I know there are some who hold that infinite space and eternity of time are necessary beings, becaus the idea of them in us is inexpugnable, which would not be, if the Creator, who is no deceiver, had not planted them in us. But this failes, for the ideas are from sence, and not from reflection, as the idea of an almighty power is. And sence may deceiv in this, as in other instances, where various tryalls may not be had to correct it.[304]

---

[302] See *supra* 1.1.4.

[303] According to Grayling, *Truth, Meaning and Realism*, p. 23, the label, metaphysical realism, is most aptly applied to the Aristotelian conception of substance as those things which exist, and can only be understood as existing, 'in and of themselves'.

[304] See *supra* 3.3.2: No. 1, RN→[SC 1706] (f. 87).

Since Clarke had asserted that infinite space and eternity of time are necessary beings and hence, uncaused,[305] he would be included among those who had made a similar claim.[306] But what about the remaining clause in North's statement, that the same 'some' had based their claim on the 'idea' of space and time as 'inexpugnable' because implanted by God? For this clause, with its reference to God as 'no deceiver' and hence, as guarantor of an innate idea of his nature, suggests a reference to part of Descartes's conception of God.[307] Is it possible, therefore, that North understood Clarke as promoting a similar conception? Perhaps, but only if he had read the first series of Clarke's lectures, for there he would have encountered not only a version of Descartes's ontological argument,[308] but also the assertions concerning the impossibility of imagining God as not existing,[309] as well as those concerning 'All and Each of those Powers and Faculties' in 'Things' which are *derived wholly from* [God] *himself*.[310] Whether or not he had read any of the foregoing,

---

[305] Clarke, *A Demonstration* [1705], in Props. III, p. 32, and IV, p. 79.

[306] For the names of some who made such claims, see *supra* 2.2.4. and 3.3.2: No. 1, RN→[SC 1706] (f. 87 n.92). To these names RN later adds the name of 'Parker', i.e., Samuel Parker, author of *Tentamina physico-theologica de Deo* (London, 1665). For RN's references to this book, see f. 90 in RN, Untitled (UK:Lbl) Add MS 32549: Notebook 4, ff. 89–102v (middle period), and RN, 'Of Etimology', p. 253.

[307] See Collins, *God in Modern Philosophy*, pp. 58–62; see also Popkin, *The History of Scepticism*, pp. 143–57, concerning Descartes's attempt to objectify subjective certitude by attaching it to God as the source of an idea and as the guarantor of its truth.

[308] See n.292 above.

[309] See, Clarke, *A Demonstration* [1705], e.g., in Prop. III, p. 32: 'the idea' of a necessarily existing being is 'the First and Simplest Idea we can possibly frame; or rather which (unless we forbear thinking at all) we cannot possibly extirpate or remove out of our Minds, of *a most Simple Being, absolutely Eternal and Infinite, Original and* Independent'.

[310] Ibid. [1705], in Prop. XI, p. 224. Regarding his belief that God also '*implanted*' the intellectual faculty (understanding, reason and judgment), see Clarke, *A Discourse* [1706], Prop. I, pp. 66–7.

North still rejects innate ideas, at the same time indicating to Clarke that 'this subject requires a larger tract to shew it self'.[311]

Not long after the 1706 correspondence came to a conclusion, North made a number of attempts towards such a tract by elaborating on the reason he had previously hinted for his rejection of innate ideas, namely, that 'ideas are from sence, and not from reflection, as the idea of an almighty power is'.[312] Part of this elaboration appears in a fragment that includes a defence of Descartes's ontological argument for the existence of God,[313] in which the French philosopher utilised a way of philosophising that, in North's words, begins with a method of doubt, which is reduced to a 'rule of thincking'; and then from the truth of his own existence and from human defects, he determined the notion of 'a better being', a notion that he called an innate idea.

> But give it what name you pleas, the thing is the same, viz., mankind is sensible of want, *ergo*, he is sensible of a better being, or one which doth not want, as he himself would be if he found such perfection in himself, as to want nothing. Now that this notion is incident to humane nature and grows up into strength and action, as a man grows in body and the use of thincking, is most certein. Why then may it not be called innate? (f. 271).[314]

Since this rhetorical question requires no answer, North turns to the problem of the formation of the idea of God in infancy; but because

---

[311] See *supra* 3.3.2: No. 1, RN→[SC 1706] (f. 87).

[312] Ibid.

[313] See RN, '2 Cartes ... A' [*sic*] (UK:Lbl) Add MS 32546: ff. 271–172v (middle period, early). Descartes's argument is found in a number of his writings; see, e.g., the fourth part of the *Discours de la méthode*, pp. 27–33, and *Principia philosophiæ*, Pt I, Arts 14–21; see also Gaukroger, *Descartes*, pp. 195–9, and Collins, *God in Modern Philosophy*, pp. 61–3.

[314] Note that on f. 271v RN has a short answer to Locke's notion that 'men have no ideas att all but from sence and then secondarily [from] reflection'; see Locke, *An Essay concerning Human Understanding*, Bk II, Ch. 1, pp. 105–18.

the text is a fragment or discard, it contains only a few lines towards a solution.

Fortunately, however, a more complete solution may be found in another manuscript, written as part of an answer to some critics of Descartes's ontological argument—that necessary existence is contained in the idea of God. The critics had claimed that the French philosopher's argument proceeds from an innate idea that 'could not forme it self' and, hence, is 'not concluding'.[315] But North wonders if perhaps such critics did not consider that this was 'the first attempt from pure reason' to prove a deity, so that its 'expressions may not be so well adjusted' and consequently, 'men may take them with some difference in intellection'. Nevertheless, if one examines the idea of God, then it will be found 'in truth to be onely an idea of our owne wants and imperfection, for it's plain [that] wants are familiar even with infants'. And 'what is the idea of want or defect, but that of the alternate'?

According to North, this was Descartes's meaning: that as infants are 'sensible of themselves', so they are 'sensible of their owne defects, that is, life and desire are coetaneous'. And reflecting on that desire, 'which Cartesius holds did not forme it self, wee from an innate principle [desire] must of necessity conclude in God' not only as creator of us but also of 'our desires or natural appetites'.[316] Although 'no formed image enters by sense', yet even new born infants are 'sensible of a God'. But there can be no introduction of the attributes until 'reason and memory are ripe to digest and retein the acquired steps'. And, then, what is the difference in meaning

---

[315] Note that Clarke, *A Demonstration* [1705], in Prop. III, pp. 37–42, had criticised Descartes's ontological argument (without naming him), apparently on the grounds that its 'Obscurity and Defect' arises from 'the *Nominal Idea* or *Definition* of a Self-existent Being', that does not 'with a sufficiently evident Connexion refer and apply that *Nominal Idea, Definition,* or *Notion* which we frame *in our own Mind,* to the *Real Idea* of a Being *actually existing without us*'.

[316] See ff. 220–221 in RN, 'Authoritys' (UK:Lbl) Add MS 32546: ff. 207–230v (middle period, early). See also Kassler, *HRN,* pp. 86–9, 156, *et passim.*

between 'innate ideas' and 'formed images, such as wee have, and generally invest in language, which is a memoriall of them to us'?[317]

North, therefore, does not reject Descartes's version of the ontological argument. What he does reject is the interpretation of it as based on an innate *idea*. And the reason he gives elsewhere is that the belief that the idea of God is innate 'doth not agree altogether with the notions of creation out of nothing, and annihilation (which is comprized in the idea of an almighty power)'.[318]

3.5. CONCERNING CLARKE'S BOYLE LECTURES

In the introduction to the first series of his Boyle Lectures, Clarke addressed those whom he considered deists and whose philosophising he considered materialistic and, hence, atheistic. Although he singled out Hobbes and Spinoza for particular criticism, he also included the Cartesians, because they were 'driven to that most intolerable absurdity of asserting matter to be a necessary being', that is to say, a real (absolute) entity.[319] As a counter to their supposed atheism, Clarke's two sets of lectures present his philosophy of God, which included his beliefs in an unknowable substratum, in absolute space and time as necessary attributes of that substratum, in the axiomatic principle that every event must have a cause, in a strong (libertarian) theory of free will and in moral

---

[317] For RN's conception of human development (i.e., how the foetus eventually acquires a human nature), see Kassler, *HRN*, pp. 86–9.

[318] See f. 75v in RN, 'Of axiomes, and the foundations of them' (UK:Lbl) Add MS 32548: in Notebook 3, ff. 75v–77 (middle period).

[319] Clarke, *A Demonstration* [1705], Prop. III, p. 33. The names of those who hold such an 'intolerable absurdity' are not given; but it is possible that Clarke's source was the seventh edition of Stillingfleet, *Origines sacra*, to which was added new material, including a long critique of Descartes's philosophy and an answer to the objections of some French Cartesians (who are named), as well as some deists and atheists (Hobbes and Spinoza are named).

principles that are necessarily true.[320] Since the lecturer also believed that reason was a more reliable way to knowledge than the senses, his philosophy has been described as 'in the rationalist style' and his arguments, as 'out of the line of development of British empiricism'.[321]

Had North read the first series of Clarke's lectures, he would have discovered that some issues raised in the 1706 correspondence appear there as well, including 'the metaphysical determination of attributes',[322] an issue that was treated publicly at length by Clarke but not at all by Newton. However, North's brief mention of this issue is insufficient to determine with any certainty whether or not he had read Clarke's lectures. Consequently, in this section some further evidence will be examined that might increase the possibility that he did. Before doing so, however, it is necessary to revisit briefly the reasons for North's concern about abstract terms. One reason was noted in the previous chapter, where I quoted him to the effect that such terms are not reducible to test or experiment, and as a consequence, they 'ever were and will be obnoxious to various fancys and opinions'.[323] But another reason for his concern is the danger of reification; for if abstract terms are used 'substantively', then we are deceived 'into an opinion, there is a reallity where there is none'.[324]

This deception occurs when abstract terms are materialised, that is to say, mentally converted into things or real entities, each with its absolute, though empirically indiscernible, meaning. For North, however, 'abstracts' are neither constituents of things nor real;

---

[320] For RN's conceptions of free will and morality, see Kassler, *HRN*, pp. 97–9 and 44, 86, 100, 111, 156, 164, 179–82.

[321] Armstrong, *Metaphysics and British Empiricism*, p. xv.

[322] See *supra* 3.3.2: No. 1, RN→[SC 1706] (f. 82).

[323] See *supra* 2.2.3.

[324] See f. 51 in RN, 'Abuses of words' (UK:Lbl) Add MS 32548: in Notebook 2, ff. 50v–53v (undatable), a short critique of abstract terms used by philosophers, entered as a commonplace note, possibly for further development.

rather, they are relational and ideal. Hence, he requests that such terms be 'laid aside, and the language fall upon realities', such, for example, as 'this history is true, this action wise, this ordinance politick, this resolution virtuous, or this argument reasonable'. For when 'meer words' are taken for things, 'and without any signification really defined', they 'pass in discourses as axiomata, and serve in the quality of principles to sustain certain pretended demonstrations'.[325]

Since North singled out Newton's 'absolutes' as indicative of this latter practice,[326] and since he again raised the issue of 'absolutes' and other 'abstracts' in the 1706 correspondence with Clarke, it is worth pointing out that in the Boyle Lectures Clarke exhibited a tendency to hypostatise abstractions—for, example, in the first series he conceived antecedent necessity as a real entity that precedes the existence of God;[327] and he also concluded that there is 'such a thing as fitness and unfitness, eternally, necessarily, and unchangeably in the nature and reason of things'.[328] Moreover, in his argument from space and time, he used these terms to signify not only the attributes or necessary consequences of the divine nature but

---

[325] See f. 120 in RN, 'Reason' (UK:Lbl) Add MS 32526: ff. 120–123v (late period), which is in the hand of his amanuensis, Ambrose Pimlowe, and dated 1732. The original version may be found about three quarters of the way into RN, 'Physica' (UK:Nr) Rougham MS (late period), where it is followed by another essay entitled 'Duty'. Note that this information corrects an error in Kassler, *HRN*, pp. 414–5.

[326] See Strong, *Procedures and Metaphysics*, p. 226, who pointed out that by absolutising a philosophical concept into a principle, that principle 'is assumed to have a sanction more ultimate and priveleged'; and 'Newton's doctrine of absolute space and time seems to be an apt illustration of this'.

[327] Clarke, *A Demonstration* [1705], in Prop. III, pp. 27–8; see also Ferguson, *The Philosophy of Samuel Dr. Clarke*, pp. 24, 94.

[328] Clarke, *A Demonstration* [1705], in Prop. XII, p. 235. In Clarke's usage, 'fitness' (also termed 'rules, relations, proportions, agreements') denotes God's reasons, which are antecedent to his acts and, hence, determine those acts. As such, Clarke must then prove that reasons, so conceived, are consistent with God's free will. For the role of the fitnesses in the second series of Clarke's Boyle Lectures, see Ferguson, *ECH*, p. 28–9.

also the omnipresence ('immensity') and continued existence
('eternity') of the unknowable substratum.[329]

Not long after the 1706 correspondence concluded, North
apparently realised that more was at stake than he had previously
recognised,[330] for in a draft essay he returns to the abstract terms,
space and time. in order to defend and clarify some aspects of his
own position. The substance of his reflection is that if space and time
are set up as real entities, then it might be possible to infer that the
deity coincides with them and, consequently, becomes
materialised.[331] Or in his words, if absolute space, 'meaning here or
there absolutely and eternally fixt', has necessary existence, then it
comes 'in being, like the devinity immortal'. And if absolute time
also has necessary existence, then it too is 'like the deity, immortall
and eternall'. But as

> ...devine authority which in words and sence prove the
> contrary—before Abraham was I am,[332] that oracle[,] from a
> sublime example[,] inferrs beings not dependant on body and
> (not like us creatures having all their sensations and most of
> their thoughts from corporeall impressions) [for immaterial
> beings] know no [flux of] time....[333]

The concerns expressed by North could have arisen when reading the
relevant parts in Newton's *Optice* concerning space as God's organ

---

[329] Clarke, *A Demonstration* [1705], in Prop. IV, p. 79: 'Infinite Space, is nothing
else but an abstract Idea of Immensity or Infinity; even as Infinite Duration, is of
Eternity ... Indeed they seem Both to be but Attributes of an Essence
Incomprehensible to Us'. Cf. ibid. [1832], in Prop. IV, p. 40.

[330] See, e.g., *supra* 2.2.4.

[331] For a clearer statement of this concern (where the danger of materialism is
called 'Hobbisme'), see f. 133v in RN, '...a dissertation of the new and moderne
(new) philosofye....' (UK:Lbl) Add MS 32514: ff. 62–125v (late period).

[332] John viii.58.

[333] See ff. 97v and 98v in RN, Untitled (UK:Lbl) Add MS 32545: ff. 92–151v
(middle period, early); see also *supra* 2.2.3 regarding the divine *nunc stans*.

of perception.[334] And they also could have arisen if he had read the first series of Clarke's Boyle Lectures. But these are merely conjectures.

To return, therefore, to the manuscript in which the above-quoted passage appears, North planned to treat two main subjects: 'the substance of the corporeal world' and the principles of motion and impact. In the course of writing, he made numerous alterations, including moving various pages (possibly taken from elsewhere) to some new locations. In this altered form the draft begins with some critical reflections on the state of natural philosophy, which reappear in other places of the text. And in one of these he considers the concept of universal plenum, a subject raised in the 1706 correspondence.[335] Moreover, Clarke had previously criticised this concept in the first series of his lectures, where he relied on the principle of non-contradiction to oppose two central Cartesian premisses concerning motion in an infinitely divisible material plenum.[336]

In the manuscript under consideration, however, North simply reaffirms that as to universal plenum he sides with Descartes against Newton's 'vacuity' (the immense space void of resistance), though he is otherwise critical of a number of Descartes's particularities, which he rejects. He also includes remarks concerning certain unnamed persons who have nothing good to say about Descartes, even though 'to him, wee must owne the invention of our moderne philosofy, whatever cry hath bin raised, and yet continues to the prejudice of his works, as if he were an heretick and author of a pernicious sect rather then a philosopher' (ff. 94v–95).[337] Following

---

[334] See Newton, *Optice*, Qu. 21, p. 315, and Qu. 23, p. 346.

[335] See *supra* 3.3.2: No. 1, RN→[SC 1706] (f. 85v).

[336] Clarke, *A Demonstration* [1705], in Prop. III, pp. 46–7.

[337] RN may refer here to criticisms of Descartes either as a promoter of atheism or as a promoter of the religion of the Church of Rome. For the latter charge, see the 1674 critique of Thomas Barlow which was published posthumously in *The Genuine Remains of that Learned Prelate Dr. Thomas Barlow*, pp. 157–8. Note that this book was once in RN's library; see RN Books (1).

from these thoughts and, in particular, from those regarding universal plenum, North reiterates his conception of matter as the geometrically extended impenetrable. He also points out the relevance of this conception to the argument from design: that 'this extended mass, so disposed with the adjunct of sensation given to creatures, produceth that image of beauty as well as order in the world wee dayly observe, and ever invites to religious dutys' (f. 108). Yet, he wonders (f. 108v)

> ...what makes men fancy this opinion to be against religion, which according to right reason, as I judg it, doth more eminently promote it, or what need is there and why should folks strain as they doe to obviate the expectation some curious persons have of gaining a satisfaction concerning naturall things out of the corpuscular hypothesis?

And he continues with the reflection that

> ...the world is much changed; now policy, I would not say pious fraudes, succeed. What els can be the meaning that the ecclesiasticall order make it a buissness to batter an hypothesis merely philosoficall, as if it were an heresie. I am sure the making matter of truth and falsehood in phisicks to come under theologicall government is superstition or wors (f. 109).

In the first passage quoted above, might we read the words 'some curious persons' as signifying North? And in the passage next quoted, might we read the words 'the ecclesiasticall order' as signifying Clarke? Perhaps we might, for in a further digression on the state of natural philosophy, North indicates that, originally, he had supposed that the 'generall and affected swerving from a plaine and just method of philosophy had bin the result of humane infirmity' (f. 110), but

> ...of late I find, and am therein guided by the excellent abilitys of the persons, that all is a meer confederacy to depose not onely the plain principles of philosofy that Cartesius useth, but

indeed all naturall philosofy,[338] and to reduce all speculation to the compass of geometry. And there [in geometry] men are encouraged to expatiate as they pleas[e]. And this I beleev may be done by many sincerely, as they thinck for the improvement of religion, and out of an opinion that philosofy leads men to atheisme, then which in my opinion nothing is more fals. And they must not thinck as I doe, if they can conceiv the mundane system without adoring [its Creator] (f. 111).

After considering 'the notorious practise of the Roman hierarchy which will let no new philosofy (as it is called) be read nor by their good will privately studied', he notes that that same hierarchy had gone 'a great way, where they have power[,] towards wholly suppressing it'. He then adds: 'some amongst us, inclined to the other extreme [Protestantism,] are practising the same thing' (f. 111v).

As to the Protestant 'some' who seek to suppress 'new philosofy', it should be noted that Clarke's Boyle Lectures included criticisms aimed specifically at Descartes, including one which asserted that he had given 'a most impossible and ridiculous account' as to how the world might be formed by the necessary laws of motion alone', even though many 'ancient and modern writers' had provided learned arguments for the existence of God.[339] Indeed, as a means of subverting new philosophy, Clarke had drawn on elements of Newton's newer philosophy. For example, after his criticism of the Cartesian concept of the plenum, noted above, he had stated that the 'Idea of Gravitation is separable from that of Matter, and Matter may be conceived without it,' and that 'if Gravitation be an Universal Quality or Affection of *All Matter*; then there is a Vacuum; (as is abundantly demonstrated by Mr. *Newton*:)'.[340] And later on he

---

[338] Perhaps a reference to the aims of Newton and Clarke.

[339] Clarke, *A Demonstration* [1705], in Prop. VIII, p. 119. For another criticism of Descartes, see *supra* 3.3.4 n.315.

[340] Ibid. [1705], in Prop. III, p. 49. The assertion in parentheses is questionable; see *supra* 2.2.4 n.190.

also had recourse to universal gravitation as an argument 'both from Experience and Reason that there are such things as Immaterial Substances'.[341]

Clarke's longest use of Newton's discoveries occurs in the context of one of the interspersed teleological arguments or arguments from design.[342] In his version of this argument, there are two principal assumptions that can be stated as axioms from which investigations proceed. For, on the one hand, there is the assumption that insofar as experience extends, there is uniformity in nature, that is to say, 'order, beauty, regularity'; and from this assumption an empirical principle is derived. And, on the other hand, by assuming a final cause or superintending providence understood not as ruling over events but as the unknowable substratum underlying them, his arguments are based on a metaphysical principle. Consequently, Clarke's version of the teleological argument may be described as physico-theological.[343]

North's assumptions are different from those of Clarke. For finding in his inquiries only a high degree of probability that certain natural events will occur with a constant uniformity, he comes to regard the principle of causality as the sufficient reason for such events and, hence, not only as the principle on which all his inquiries should rest but also as the final result towards which his study of nature tends. For, as he tells us in a draft preface, he chose natural philosophy for 'diversion, as well as study', because it alone is a science of truth; and truth is 'what nature leads directly to, and

---

[341] Op. cit. [1705], in Prop. X, p. 164.

[342] Ibid. [1705], in Prop. XI, pp. 225–33, which includes, pp. 230–1, the paraphrase from Newton, *Principia mathematica* cited *supra* 3.3.4 and n.291.

[343] For the development of this type of argument, see McAdoo, *The Spirit of Anglicanism*, pp. 240–315. For a critical examination of how these arguments have been misinterpreted in recent socio-historical speculation, see Young, *Religion and Enlightenment*, pp. 83–9, 99–100, *et passim*.

advanceth in us by continuall degrees more or less from the first opening our eys in the world, to the final closing of them again'.[344]

Because seeking truth is a lifelong process, North, in several manuscripts, expresses a desire to replace the physico-theological argument with one of a different kind. For example, in one manuscript fragment consisting of two brief notes or memoranda,[345] he states that arguments demonstrative of a deity 'are of such various natures, that each state and capacity of mankind hath them'. Two examples of such varieties are given, the first is that of the 'gentleman, by the beauty and order of the world', the second is that of the 'philosofer, by the connec[t]i[on] of soul and body, which is the most cogent [argument] of all'.[346] Since the argument of the gentleman is the type of argument used by Clarke and since the argument of the philosopher is different, what was the purpose of North making this note?

An answer to this question appears from another manuscript, the text of which records some of North's responses to philosophical positions in five separately titled commonplace notes that fill an entire notebook.[347] The fourth note, entitled 'Change of philosoficall methods' (ff. 94–95), refers to the argument from design and then ends with the statement: 'I should unwillingly part with the argument for a deity and providence, from the order and beauty of the world if I had not a better, of which elswhere' (f. 94). It is of interest, therefore, that the fifth, final and longest note, 'Objections to these

---

[344] See ff. 194–194v in RN, 'Prefaces' (UK:Lbl) Add MS 32546: ff. 191–194v (middle period).

[345] The other note states: 'In circular forme, every point of the perifery is also a point of a tangent state[?] and also a point of all possible curves that can be tangent to it. Then which shall a body take of all these? I say the strait, viz., tangent[,] for what should set it into one or other of the rest?'

[346] See RN, Untitled (UK:Lbl) 32546: f. 291 [291v is blank] (middle period, early?). Chan, 'Dating the Paper', p. 53, identified and dated the paper as Arms/DL, 1706–10, so it is possible that its two notes may be related to the first of the two fragments *supra* 3.3.2: No. 5, RN→[SC 1706] (f. 290).

[347] See RN, Untitled (UK:Lbl) Add MS 32549: Notebook 4, ff. 89–102v (middle period).

theoremata' (ff. 95v–102v), must have been added sometime later, since in that note he writes:

> Those [physio-theologians] who ascribe all sensible formes to the nature of the objects and not to the working of the imagination charge us with supplanting what all the world hath agreed upon to be the incontestable proof of the deity, viz., the order and beauty of the universe. ... [Hence, it] concernes us to supply[348] the proof pretended of a deity by what, as I think, and have promised, is much better (f. 95v).

As might be expected from an epistemological realist for whom external reality is 'simple extension, infinitely discerpted and moved' (ff. 97–97v), North's argument turns on the proof that ideas are in the mind and not in the external object. Or, as he states elsewhere: 'really and truely, there i[s] no naturall beauty or decorum in [external] things, but it ariseth from our senc[e] and judgment'.[349]

Unfortunately, his proof is a disordered miscellany of notions that are more clearly treated in other manuscripts. Fortunately, however, it is possible to discern the basis of the argument in his theory, discussed in Chapter 1,[350] that nature has limited our sensori-motor capacity for information processing to what we can physically or mentally number or count. North first developed this theory in relation to his work on sound and music, which, in short, is that the auditory system, at both subconscious and conscious levels, computes periodicities: it 'measures'—that is, counts—vibrations arithmetically in time. When periodicities are too fast for conscious counting, the auditory system organises the input sequence into larger units compounded of frequencies, for example, pitches

---

[348] I.e., to put or appoint as a substitute; see OED *v.*[1] 6b, obs.

[349] RN, 'Some Notes upon an Essay of Musick', p. 188.

[350] See *supra* 1.1.3.

('tones') singly taken or compounded together in concords, discords and larger structures.[351]

A concise version of this theory appears in North's statement that if 'the facultys of the body are outdone by the celerity of sensations, as in musicall tones and accords and by mixtures of colours', then 'there emergeth new ideas which are not *in rerum natura* extra to the mind' (f. 101). And because, for him, it is 'more reasonable to assign the variety of ideas upon the modes of the mind, occasioned by the reall modes of body, rather then upon externall or ... corporeall qualities', he rejects the opinion of 'the Newtonians' that 'light in its body compriseth all the qualitys in reall essence,[352] as the images of light are red, blew, etc.' (f. 98),[353] as well as their version of the teleological argument from the uniformity of the external world.

But there may have been another reason for North's dissatisfaction with the physico-theological version of the teleological argument, namely, Clarke's claim, in the first series of his Boyle Lectures, that it was 'Necessary to the Order and Beauty of the Whole', as well as 'for displaying the Infinite Wisdom of the Creator, that there should be different and various degrees of Creatures, whereof consequently some must be *less Perfect* than others'; and hence 'necessarily arises a Possibility of Evil, notwithstanding that the Creator is infinitely Good'. In this context, what Clarke called 'evil' is a '*Want of certain Faculties and*

---

[351] If acoustic information is communicated as a composition of periodicities, how is musical knowledge possible? For RN's answer, see Kassler, *The Beginnings*, pp. 118–20, with reference to the text cited in n.349 above.

[352] I.e., as real properties.

[353] According to Newton's theory, sunlight consists of a mixture of innumerable spectral colours; see *supra* 3.3.1. These colours are transmitted to the organ of vision, the retina, as an emanation or stream of atomic rays that have a disposition for colour-making. This theory implies, first, that the atomic rays enter the eye, and, second, that in perception the immediately given (colour) is external, not internal to experience. According to RN's theory, information is transmitted as frequencies; and perception of the immediately given is internal because inferred from limited sensory experience.

Excellences which other Creatures have'. Accordingly, he pointed out that, unlike natural and moral evil, this kind of evil is not an evil, properly speaking; rather it is *'an Evil of Imperfection'*.[354]

In the few pages in which the above quotations occur, Clarke attempted to explain the origin of evil and to vindicate the goodness of God from the imperfection of the created world. In so doing, he alluded very briefly to a principle whereby 'Creatures' are graded according to a scale of perfection,[355] that is to say, they have different degrees of worth or value. North, however, would not have accepted this principle. Indeed, in a brief assay concerning various arguments for the existence of God, he writes that although there are degrees of power, the greatest of which is 'almighty power', there are 'no degrees in existence, or between something and nothing'. And his reason is: 'Everything that exists is perfect, or substantive, equally with the greatest'.[356]

---

[354] Clarke, *A Demonstration* [1705], in Prop. X, pp. 218–9; repeated ibid. [1832], in Prop. X, pp. 107–8, and, hence, also in the intervening editions.

[355] For the classic study, see Lovejoy, *The Great Chain*, pp. 163, 352 n.42. By 1715–16, however, Clarke apparently had adopted a. different conception of perfection; see Priestley, 'The Clarke-Leibniz Controversy', pp. 43, 48, 51–2.

[356] See f. 140v in RN, 'Philosofye not demonstrative of religion' (UK:Lbl) Add MS 32548: in Notebook 5, ff. 139v–150 (late period, early).

CHANGES IN THEOLOGICAL METHOD

4.1. HEADING TOWARDS CONTROVERSY

In January 1709, on the recommendation of Moore, now bishop of Ely,[1] the queen appointed Clarke as the fifth rector of St. James's, Westminster, a position he maintained until his death. Before taking up this appointment, he returned to Cambridge to obtain a doctorate of divinity, which was awarded after successfully defending two propositions: first, that all religion supposes the freedom of human actions; and second, that the Christian religion contains nothing contrary to reason.[2] Then, as Dr. Clarke, he gave up the benefice of St. Benet's Church and removed to the parish church of St. James's, one of the most influential churches in London. Hence, a few words should be said about the church and its parish, the former of which had been consecrated on 13 July 1684 and the latter, created in the following year by the statute of Jac. II. c. 22, where its outlines are given.

The church itself occupied a site, the front of which faced Piccadilly and the back, Jermyn Street. At the time Clarke took up his appointment, the church had a vestry room, where its affairs were organised;[3] and from the time of the first rector, it became the custom

---

[1] According to Ferguson, *ECH*, p. 43, Clarke's chaplaincy to Moore ceased in 1710.

[2] Ibid., pp. 40–1; see also Whiston, *Historical Memoirs*, p. 18, and Hoadly, 'Preface', p. vi.

[3] See Ferguson, *ECH*, pp. 160, 196–209. For business details, including sources of income for the church and its rector, see pp. 160, 196–209, where, p. 198, it is noted that carvings by Grindling Gibbons were donated by 'Sir R. Geere', i.e., RN's subsequent father-in-law, Sir Robert Gayer. Note that during the period the

to hang portraits of departing rectors on the vestry walls.[4] The site itself included a churchyard, as well as a rectory, watch house and engine house. The rectory, situated on the western corner of the boundary with Piccadilly, was completed in 1686 as the residence for the rector. The watch house, located on the east side of the churchyard gates fronting Piccadilly, served for the detention of prisoners. In 1704 a school room was built over it for the education of a small number of poor boys, a pew in the north gallery of the church being allocated for their use. Since in 1690 Joseph Williamson had provided the parish with two fire engines,[5] the engine house was built sometime afterwards to house them.

Members of the congregation included tradesmen who lived or worked in the parish,[6] as well as a number of wealthy aristocrats. Indeed, in 1703 John Evelyn wrote that a sermon he heard on costly apparel would have been more appropriate for 'St. James's or some other of the Theatrical Churches in London, where the Ladys and

---

church was designed and built, its architect, Christopher Wren, was also supervising the schedule of payment to the workmen on the Temple gateway, designed by RN after the old structure was destroyed by fire in 1679; see Kassler, *HRN*, p. 22–3. Note also that in 1691 the organ presented to the church by Queen Mary was installed by the organ maker who in 1701 built RN's organ at Rougham Hall; see Lambert, *An Illustrated Guide*, p. 5, and Kassler, *The Beginnings*, pp. 115–17. For a brief geography of the parish, see Chancellor, *Memorials*, pp. 34–5; and for a survey of the church itself, see British History Online: St. James's Church, Piccadilly.

[4] Tenison, the first rector, was appointed in 1685. In 1692 he offered the post to William Wake; who declined on the grounds of ill health, whereby the existing assistant preacher, Robert Moss, filled the position (1692–4) until Wake was well enough to take it up (1695–1706). In 1706 he was succeeded by Charles Trimnell, who in 1708, on Tenison's recommendation, was translated to the diocese of Norwich in the room of Bishop Moore.

[5] Williamson, who had been called to the bar from the Middle Temple (1664), was second president of the Royal Society and an amateur viol player; see Evelyn, *The Diary*, vol. 3, p. 601, vol. 4, pp. 39, 277, and RN, *The Musical Grammarian*, p. 262.

[6] E.g., owners of taverns, coffee houses and shops, particularly in Pall Mall, the heart of residential St. James's parish; see Phillips, *Mid-Georgian London*.

Women were so richly and wantonly dressed and full of Jewells'.[7] Prior to Clarke's arrival, the future queen, Anne, had attended a service when living at Berkeley House;[8] and it is plausible that from time to time other royalty did likewise, thus adding to the theatrical spectacle. However, two residents of the parish deserve particular mention. One was North's nephew, Francis, 2d baron Guilford,[9] who between 1698 and 1724 lodged nearby in Arlington Street at the end of Piccadilly.[10] Presumably, when in town he attended service at St. James's.

The other resident was Newton,[11] who from 1710 for the next fifteen years occupied 35 St. Martin Street off Leicester Square[12] and, hence, was within walking distance from the rectory of St.

---

[7] Evelyn, *The Diary*, vol. 5, p. 542 (18 July 1703). According to Ferguson, *ECH*, p. 200, the right to have a pew was purchased, and the cost was nearly equivalent to that of attending the theatre.

[8] Chancellor, *Memorials*, p. 76.

[9] I.e., Francis North, eldest son and namesake of RN's brother, who died in 1685, whence RN became executor of the will and guardian of his brother's three children. Young Francis was educated at the free grammar school at Bury St. Edmunds; on 20 February 1689 he matriculated at Trinity College, Cambridge, and on 11 December 1690 graduated M.A. According to Evelyn, *The Diary*, vol. 3, p. 347, he left shortly afterwards 'in order to accompany the king [William] in his journey to Holland'. In 1694 he took his seat in the House of Lords and served as lord-lieutenant of Essex (1703–5), privy councilor (1711), lord of trade (1712–13) and first lord of trade (1713–14). In 1695 he married (1) Elizabeth, third daughter of Fulke Greville, baron Brooke of Beauchamps Court; and in 1703 (2) Alicia, second daughter and co-heir of Sir John Brownlaw. Francis North's name appears on a number of subscription lists for books and music, interests that he shared with his uncle. See Korsten, *Roger North*, Appendix II, for a list of some of the correspondence between RN and his nephew.

[10] Chancellor, *Memorials*, pp. 81–2.

[11] Ibid., p. 73, according to whom Newton twice occupied a house in Jermyn Street, initially in 1689 and again between 1697 and 1709, whereas Westfall, *Never at Rest*, pp. 556, 580, 594–5, 710, mentioned only the latter period.

[12] Late in 1709 Newton had removed from Jermyn Street to Chelsea for a period of about nine months and from thence to the house in St. Martin Street; see Westfall, *Never at Rest*, pp. 670–1, 710 (Fig. 13.2.), *et passim*.

James's.[13] Indeed, Newton served as a vestryman of the church,[14] as well as governor or trustee of a chapel-of-ease, the Golden Tabernacle, in King Street, Golden Square, where Clarke was the incumbent vicar.[15] From the proximity of their residences and their mutual connection with both the church and the chapel, it can be inferred that the two men met with some frequency in order to discuss business. But it cannot be inferred, for example, that during their meetings they also discussed other subjects of mutual interest, since before 1712, when Clarke was to make his intentions public, there is scant evidence on which to base such an inference.

What evidence we do have for the period 1709–12 indicates that Newton and Clarke had some common ground in relation to church and chapel, so it is important to note that despite his numerous public duties as rector of St. James's and as incumbent vicar of the Golden Tabernacle, Clarke was continuing his investigations into the doctrine of the Trinity. I say continuing for two reasons. First, he told a cleric that his conclusions concerning this doctrine were arrived at after a period of long reflection,[16] which makes it probable that he began his investigations in Norwich when studying theology with Moore. But at that early stage in his clerical career, it is unlikely that he would have possessed a library adequate to such investigations and, hence, the importance of unrestricted access to the prelate's collection of patristic and Anglican writings

---

[13] Ferguson, *ECH*, p. 215.

[14] Ibid., p. 200, who named only a few of the vestryman, including John Roberts, 1st earl of Radnor, who had been a member, with RN's brother Francis, of the reorganised privy council (1679). The earl's son, Francis Roberts, a politician, virtuoso and musical amateur, was an acquaintance of RN; see Kassler, *The Beginnings*, pp. 101–2, *NP 2*, pp. 28–35, and RN, *The Musicall Grammarian 1728*, pp. 165, 262, 265.

[15] Ibid., pp. 197, 201–2, 216; see also Westfall, *Never at Rest*, pp. 594, 778, 815, 828. From *c.*1701 the music tradesman, Ambrose Warren, kept the chapel's accounts, and in *c.*1707 he was appointed as tuner and keeper of the organ at St. James's; see Kassler, *The Science of Music*, pp. 1045–8.

[16] See Ferguson, *ECH*, p. 77.

(manuscripts and books).[17] Indeed, even if Clarke did form and later add to his own library (and there is no evidence for this), he still had access to Moore's collection after his removal from Norfolk to London in October 1706, when he resided with that prelate in Petty-France, Westminster. And even after 1707, when Queen Anne translated Moore from the diocese of Norwich to that of Ely, Clarke still would have had access to the collection, since the bishop's house was not in Ely itself but in Ely Place, Holborn.[18]

Second, some results of Clarke's Norwich investigations were included in his Boyle Lectures, although these results are mainly hints of the doctrine he later made public. Nevertheless, the hints suggest that Clarke was in the process of developing a philosophy of the Godhead that was inconsistent with the consubstantial doctrine that there are three co-equal and co-eternal Persons in one unity, a doctrine that was normative in the Church of England.[19] In the 1970s James Ferguson drew attention to two such inconsistencies,[20] one of which is found in Proposition VII of the first series of the Boyle Lectures.

In this proposition Clarke asserted that God as the '*Self-Existent Being, must of Necessity be but One*', that is to say, '*Absolute in itself*',[21] because whatsoever exists necessarily is 'the

---

[17] At Moore's death the library consisted of nearly 29,000 books and 1,790 manuscripts; see Meadows, 'John Moore'. Note that, when in London, Clarke also could have had access to the library in Lambeth Palace, since one of his other patrons was the archbishop of Canterbury, Tenison.

[18] Moore died on 31 July 1714. As a tribute to his patron, Clarke edited some of Moore's sermons. Initially published in one volume in 1715, there was a second issue, also under his editorship, that was published in two volumes in 1724.

[19] I.e., it constitutes Article 1 of the so-called 'Thirty-nine Articles', subscription to which was part of the requirement for membership in the Church of England. Dating from the end of the sixteenth century, the Articles had been written as an attempt to establish the uniqueness of the Church of England by defining Anglican doctrine as it relates to the practices of the Church of Rome; see *Articles ... touching the true Religion* (unpaginated).

[20] Ferguson, *ECH*, pp. 47–8.

[21] Clarke, *A Demonstration* [1705], Prop. VII, p. 93.

One Simple Essence of the Self-Existent Being: and whatsoever differs from that, is not Necessarily-Existing: Because in absolute Necessity there can be no Difference or Diversity of Existence'.[22] He then went on to argue, in a brief consideration of the Trinity, that as to 'the *Diversity of Persons* in that One and the same Nature', the co-existence of two other distinct beings of necessary and independent existence was impossible.[23]

The other inconsistency noticed by Ferguson occurs in Proposition XIII of the second series of his lectures, where Clarke again considered the oneness of God—his 'absolute' supremacy and independence—not only as 'the first and principal Article of the Christian Faith'[24] but also as a belief necessary for salvation. While agreeing with the established doctrine that 'the fullness and perfection of all divine Attributes' dwells in the Son, he went on to exclude from that fullness absolute supremacy, independent existence and self-origination.[25]

At the time when Clarke's Boyle Lectures were first published, there seems to have been little recognition that such statements were heterodox; but this situation was about to change.[26] For in July 1712 the result of his long investigations, *The Scripture Doctrine of the*

---

[22] Ibid., p. 94. According to Stead, *Doctrine and Philosophy in Early Christianity*, I, p. 180, this doctrine of God as absolutely simple derived not from Scripture but from mathematically-minded philosophers such as Pythagoras and his followers, who supposed that just as the number 1 was the source of all cardinal numbers, so too the 'monad' was the source of all rational order in the universe—or conversely, that the creative power behind that order had the characteristics of the 'monad'.

[23] Clarke, *A Demonstration* [1705], Prop. VII, p. 95.

[24] I.e., in contradistinction to Article 1 of the 'Thirty-nine Articles'; see n.19 above.

[25] Clarke, *A Discourse* [1706], Prop. XIII (on the necessity of believing doctrines only if they are necessary to salvation, the first of which is that there is a God who is 'One'), p. 291, repeated p. 292.

[26] In subsequent 'editions', however, Clarke made his philosophy of the Godhead increasingly clear; see, e.g., Wiles, *Archetypal Heresy*, p. 119 n.295, regarding the 1711 and 1719 versions.

Trinity, was announced as 'in the press';[27] and after its publication, Clarke became involved in a controversy with various critics that continued up to the time of his death. At least one careful reader grasped that an underlying problem posed by Clarke's book concerned the nature and conditions for religious truth and, hence, the theological method by which that truth was to be sought. Since the reader in question is Roger North, this chapter departs from the format of the previous ones in order to function as an introduction to Chapter 5, which concerns his correspondence and related material pertaining to Clarke's book.

4.2. DETERMINING FUNDAMENTALS

During the seventeenth century theologians in the Church of England accepted the predominant but rarely challenged theory that scriptural texts had been written under the inspiration of the holy Spirit and, hence, they imparted God's revelation as infallible, immutable truths.[28] As a consequence, special authority was given to specific texts in Scripture that had become the main source of authentication for determining 'fundamentals', that is to say, questions concerning ecclesiastical polity and liturgy.[29] However, textual interpretation had two constraints. One, an external constraint, was the appeal to tradition, defined as the 'church' in the years immediately after its foundation, when the fathers were contemporary with the Apostles.

---

[27] *The Spectator* (31 July 1712), no. 445, p. 2. Note, however, that in the previous month, an anonymous writer provided an abstract of the book in which he interspersed some flattering remarks about Clarke's 'great Modesty', 'exquisite Judgment' and 'universal Learning'; see [Michael de la Roche], *Memoirs of Literature* (Monday, 2 June 1712), vol. 2, no. 22, pp. 169–74.

[28] See Ferguson, *ECH*, pp. 17–18, 30, 52 and 69. The main challenge to the theory of inspiration came from the deist movement at the end of the century; and this may partly explain Clarke's riposte to certain deists in the first series of his Boyle Lectures, for he himself held the theory of inspiration; see *infra* 4.4.3.

[29] I.e., form of organisation or government and form of public worship, including the collection of formularies (e.g., creeds) for its conduct.

Accordingly, tradition referred to patristic texts as the 'deposit of faith' from which doctrine was to be derived.

Derivation itself required a second, internal constraint, the appeal to reason. And the Anglican conception of this appeal developed in reaction to the scholasticism and intellectualism not only of Roman Catholicism but also of Calvinism. In both, natural reason was conceived as a type of divine reason; and reasoning was modelled on Aristotle's doctrine of the syllogism. Consequently, the appeal to reason meant an appeal to *a priori* categorical propositions from which doctrine was deduced.[30] By contrast, Anglican theological method conceived reason not as transcendent but as immanent, a 'liberal' conception, according to Henry McAdoo, because it holds that natural reason is competent not only to deal with questions of ecclesiastical polity and liturgy but also to be in itself an ultimate factor in theological method.[31]

If natural reason is conceived as subjective and self-sufficient, then individual interpretations of Scripture would in fact become so many different authorities, thereby leading to never-ending disputes concerning fundamentals. Hence, the need for a norm or standard of reference as a balance against reason; and this was met by tradition— the deposit of faith which had been laid down during the first centuries. But just as the appeal to reason could and did lead to matters of dispute, so too could the appeal to tradition, for some were to define the deposit of faith as that which had been laid down by the Apostles and Greek fathers during the first three centuries, whereas others defined the deposit to include the early Latin fathers.

---

[30] For a brief critique of Aristotle's logic, especially its over-estimation of deduction as a source of knowledge, see Russell, *A History*, pp. 195–202.

[31] McAdoo, *The Spirit of Anglicanism*, pp. 309, 311, 315. This now classical study provides essential background for understanding the various external and internal pressures on Anglican theological method as it developed during the seventeenth century, as well as the reasons for its decline during the eighteenth century. For pressures due to the conflict between conserving and reforming aspirations, see, e.g., Oelsner, 'Tradition as a Theological Problem', Cornwall, *Visible and Apostolic*, pp. 11–18, *et passim*; and Reedy, *The Bible and Reason*, pp. 119–41.

Regardless of definition, the appeal to tradition was generally applied within the context of continuity—the church as a living embodiment of the deposit of faith.[32] And this application had two important consequences, for if truth is conceived not as immutable but as an historical development, then reason becomes trans-subjective—the 'life' of the church. It is remarkable, therefore, that this particular development in Anglican theological method paralleled a concurrent development in the English common law. For, as previously indicated,[33] Edward Coke first propounded the conception of procedural fairness as the 'life' of the law and the law itself as the artificial perfection of reason. And these conceptions were afterwards developed by John Selden, Matthew Hale and other common lawyers, including North himself, for whom the subjective grounds of individual experience and reasoning were to be established on, and confirmed by the objective grounds of 'common sense' defined as legal reason that is historical and, hence, trans-subjective.[34] Accordingly, both law and theology shared a similar conception of method as an ongoing process of adapting past solutions to present circumstances in the interpretation of fundamentals; and thus, continuity in change was the norm.[35]

During most of the seventeenth century, applications of this method maintained a balance between Scripture and the two constraints on interpretation: tradition and reason. For the method was flexible; for example, it could give different weight to one or the other of its three constituents; and it could incorporate different developments which emerged in that century, including developments in natural philosophy.[36] But there were limits to its

---

[32] See McAdoo, *The Spirit of Anglicanism*, pp. 316–414.

[33] See *supra* 1.1.4.

[34] See Kassler, *HRN*, pp. 110–28.

[35] Ibid., pp. 113–28. To my knowledge, this parallel has not been recognised; but it would make a fascinating topic for investigation.

[36] For the incorporation of developments in natural philosophy, see McAdoo, *The Spirit of Anglicanism*, pp. 240–315.

flexibility; and these were reached during the first decades of the eighteenth century, when there was a gradual impoverishment of the conceptions of reason and tradition that had been developing earlier.

In the case of reason, the general trend was towards a narrowing of the meaning of rationality,[37] a trend that began with Hugo Grotius, who reformulated classical natural-law theory by separating reason from its theological basis, redefining rights as part of a people's rational nature and correlating the problem of reason with that of mathematics. Subsequently, his successors, which in England included Thomas Hobbes and John Locke, created a new version of natural-law theory, in which reasoning involves moving from a system of internal mental representations ('phantasms', 'ideas') to determinate referents which are regarded as lying outside the system of mental representations. Accordingly, a premium is placed on the mind's clear and distinct ideas, as well as on words as signs of ideas, because definitions are to serve as principles for rigorous logical argument.[38]

Initially, this trend developed as a response to what theologians and some philosophers grouped under the heading of 'enthusiasm', that is to say, the 'Spirit-formed and Spirit-led' sects: from the 1640s the Familists, Grindletonians, Seekers, Diggers, Muggletonians, Ranters, Quakers and the more radical Puritans; from the 1650s the Calvinists; and after 1660 the 'fancies' of Dissent.[39] Over time, therefore, the response of Anglican divines to the different kinds of enthusiasms led to greater weight being given to reason as the main constraint on biblical interpretation. By the end of the century,

---

[37] I.e., narrowing from a conception of reason which recognises its own limitations unless fulfilled and extended by divine grace; see, e.g., Frye, 'Reason and Grace', and Dockrill, 'Spiritual Knowledge', pp. 147–8.

[38] See Kassler, *HRN*, pp. 114–17 and sources cited there.

[39] In the latter part of the century responses were written by such divines as Isaac Barrow, George Hickes, Simon Patrick and Edward Stillingfleet. But, according to Dockrill, 'Spiritual Knowledge', it was Locke's chapter on enthusiasm, added to the fourth (1700) edition of *An Essay concerning Human Understanding*, which sums up the epistemological and theological critique of the sectaries.

Locke's *Reasonableness of Christianity*, published anonymously in 1695, contributed significantly to this general trend, since the author conceived reasoning as deductive from propositions derived not *a priori* from speculation, but *a posteriori* from textual analysis.

In a brief assessment of this book, Gerard Reedy concluded that for its author

> ...the 'reasonableness' of Christianity meant ... what is plainly, simply, and uncontrovertibly taught by the four Gospels and Acts, the principal historical sections of the New Testament. 'Reasonableness' thus means the *ratio* or essence of the object in view, unencumbered by the presuppositions sectarian commentators bring to it.[40]

But, with apology to Reedy, no teaching is uncontrovertibly plain and simple, and no '*ratio*' (meaning or logic) is unencumbered by presuppositions.[41] We might ask, therefore, what meaning Locke would place on the scriptural texts?

An answer to this question is found in his book just mentioned, where he defined '*Natural Religion*' as a 'Body of the *Law of Nature*', that is to say, a system of morality,[42] and where, in some scattered remarks, he suggested applying a threefold approach to construct such a system. First, a collection is to be made from biblical texts of 'all those Rules of *Morality*, which are to be found in the Christian Religion'.[43] Second, since those texts were written on different occasions, it would be necessary to observe in them what is

---

[40] Reedy, *The Bible and Reason*, p. 141.

[41] For some of Locke's presuppositions, see Cornwall, *Visible and Apostolic*, pp. 21–4.

[42] For his empirical version of natural-law morality, see Locke, *The Reasonableness of Christianity*, pp. 226–41; for the development of this kind of morality in the seventeenth century, see Schneewind, *The Invention of Autonomy*; and for its relation to natural-law jurisprudence, see Kassler, *HRN*, pp. 115–17, *et passim*. Note that Locke's reduction of natural religion to morality evades the problem whether humankind can know anything about the deity by the exercise of the natural faculties of sense and understanding (i.e., intellect).

[43] Locke, *Reasonableness of Christianity*, p. 230.

principally aimed at; what the argument is and how it is managed, for these observations would assist in discovering

> *...the true* [i.e., literal] *Meaning and Mind* [i.e., intention] *of the Writer*; for that is the Truth which is to be received and believed, and not scattered Sentences in a Scripture-Language accommodated to our Notions and Prejudices. We must look into the Drift of the Discourse, observe the Coherence and Connexion of the Parts, and see how it is consistent with itself, and other Parts of Scripture, if we will conceive it right. We must not cull out, as best suits our System, here and there a Period or a Verse, as if they were all distinct and independent Aphorisms; and make these the Fundamental Articles of the Christian Faith, and necessary to Salvation, *unless God has made them so.*[44]

Finally, it would be necessary to build the system 'upon Principles of Reason, self-evident in themselves' and deduce 'all the Parts of it from thence, by clear and evident Demonstration'.[45] And, as he elsewhere made clear, by 'Principles of Reason, self-evident in themselves', he meant 'self-evident Propositions', from which 'by necessary Consequences, as incontestable as those in Mathematicks, the measures of right and wrong, might be made out'.[46]

Locke himself never constructed a system along the lines suggested in his book; but, as will appear in the sections that follow, at least one writer applied the same threefold approach to expound a philosophy of the Godhead. It should be noted, therefore, that since Locke's approach took no account of tradition, his book is also

---

[44] Op. cit., p. 249 (italics mine).

[45] Ibid., p. 232.

[46] See Locke, *An Essay concerning Human Understanding*, p. 549. Note that Locke recast the meaning of morality as an assent to propositions. According to Cornwall, *Visible and Apostolic*, p. 23, he also recast the meaning of faith in the same way. For details of what Locke signified by a demonstrative science of morality up to and including his *Essay*, see Mattern, 'Moral Science'. RN, of course, denies that moral truth can be demonstrated 'with that rigorous certeinty as mathematitians pretend'; see RN, *Cursory Notes of Musicke*, p. 33.

symptomatic of the increasing isolation of the appeal to antiquity 'from the element of reason and from the ambit of liberality'.[47] As a consequence, its sphere of application became narrowed, for example, by transferring the appeal into the context of ancients *vs.* moderns;[48] and in some instances, tradition was rejected altogether, leaving reason as the only constraint on interpretation. As a consequence, a rupture in the balance of theological method was coming about and party theology was emerging.[49] McAdoo assigns the commencement of this rupture to Benjamin Hoadly's rejection of the appeal to antiquity.[50] But an earlier rejection, at least, rhetorically, was made by Clarke in the 'Introduction' to his 1712 book, mentioned previously,[51] where he stated:

> Now in *the Books of Scripture* is conveyed down to us the Sum of what our Saviour taught, and of what the Apostles preached and wrote: And were there as good evidence, by any certain means of Tradition whatsoever, of any other things taught by Christ or his Apostles, as there is for those delivered down to us in these Writings; it could not be denied but that such Tradition would be of the same Authority, and in every respect as much a part of the Rule of Truth, as the Scripture itself. But since there is no Tradition (and indeed in the nature of things there can be no such Tradition) at this distance of Time; therefore *the Books of Scripture* are to Us Now not only *the Rule*, but *the Whole and the Only Rule of Truth* in matters of Religion.[52]

---

[47] McAdoo, *The Spirit of Anglicanism*, p. 397.

[48] Ibid., pp. 346, 398, 405–6, 409.

[49] I.e., the theologies of high, broad and low church; ibid., pp. vi, 356–8, 394–8, 405.

[50] Ibid., p. 399. According to Cornwall, *Visible and Apostolic*, p. 24, Hoadly's ecclesiology, as also that of other followers of Locke, 'come much closer to Nonconformity than to traditional Anglicanism'.

[51] See *supra* 4.4.1.

[52] Clarke, *SDT*, Introduction, pp. iv–v (see also p. xx): 'The Church of *Rome* indeed requires men to receive her particular Doctrines (or Explications of Doctrines) and Traditions, as part of the Rule itself of their Faith: And therefore

In short, Clarke not only claimed that the source of authority was Scripture alone—in his book, just the New Testament—but he also rejected the external constraint of tradition—or at least, he appeared to have done so. But a few pages later, he introduced his own conception of tradition as the deposit of faith of the 'Church from the Beginning', when the fathers had selected from Scripture 'those plain fundamental Doctrines, which were there delivered as of necessity [for salvation] to be known and understood by all Christians whatsoever'.[53]

Repeating the myth, first proposed by Eusebius of Caesarea,[54] of a morally pure early church that was corrupted over time, Clarke asserted that during the period of the first three centuries, Christianity was 'perfect' but that soon afterwards 'needless Contentions' began to arise, and 'Faith became more intricate', untill 'at last ... it was swallowed up in the great Apostacy',[55] from which it began to recover at the Reformation, 'when the [scriptural] Doctrine of Christ

---

with Them no good Christian can possibly comply.' By contrast, 'the Protestant Churches, utterly disclaiming all such Authority' require 'men to comply with their [liturgical] Forms, merely upon Account of their being agreeable to Scripture'. Since these assertions represent post-reformation developments, the term 'Protestant Churches' implies the reformed (Calvinistic and Lutheran) churches and excludes the Church of England, which, 'even in pre-reformation times', differed not only from the Church of Rome but also from the continental (reformed) churches; see Oelsner, 'Tradition as a Theological Problem', pp. 151–7.

[53] See Clarke, *SDT*, Introduction, p. vii.

[54] I.e., the fourth-century church father, who enunciated the myth in his *Vita Constantini*, a work that Clarke cited, ibid., Pt III, Ch. 2, p. 473, although he did not acknowledge Eusebius as the source of the myth.

[55] Ibid., p. vii. Note that here, as well as in the Introduction to *SDT* [1719], Clarke gave no date for the 'great Apostacy'; but in the Introduction to *SDT* [1732], p. iv, he attributed the cause to '*Metaphysical Uncertainties*' in the creeds of the fourth-century. According to Wiles, *Archetypal Heresy*, p. 91, Newton assigned several dates for the 'great Apostacy', but most particularly 381, the date when the Council of Constantinople was summoned and various heresies, including Arianism, were anathematised.

and his Apostles was again declared to be the Only Rule of Truth'.[56] But the recovery did not last long, for after the Reformation the *sola scriptura* rule was not steadfastly adhered to;[57] and as a consequence, disputes as well as schisms prevailed among Protestants. Accordingly, Clarke's express purpose was to resolve religious disputes in order to reinstate the (Platonic) ideal of 'Catholic Unity', the source of which was to be found in 'the Apostolical Form of sound words', that is to say, words free from logical defect or error.[58] But 'for the sake of Peace and Order', he also required that

> Every sincere Christian, assenting ... to the Use of any Forms of Words [in the liturgy]; must take care to assent to them in such a Sense, as may make them consistent with the Scripture; (otherwise he assents to what is False:) and in such a Sense, as may make them consistent with Themselves (otherwise he assents to Nothing.)[59]

To grasp the implications of Clarke's demand that the meaning of words must be consistent with Scripture and with themselves, it is necessary to examine more closely his 1712 book.

4.3. ANCHORING TEXTUALISM TO A DOCTRINE OF ORIGINAL INTENT

In the course of tracing the intertwining and developing strands of the three elements used in biblical interpretation—Scripture, tradition and reason—Henry McAdoo pointed out that none of these, taken

---

[56] Clarke, *SDT*, Introduction, p. ix. For Scripture as 'the Only Rule of Truth', see also pp. v, x, xx, xxi.

[57] According to one commentator, 'the Reformation produced far more doctrinal chaos than the church had experienced since the second century'; see Phan (ed.), *The Cambridge Companion*, p. 141.

[58] Clarke, *SDT*, Introduction, pp. viii, ix, xxviii. The irony, of course, is that instead of ending disputes, Clarke's book generated them.

[59] Ibid., p. xxv. Note that the entire paragraph in which this passage occurs was removed in Clarke, *SDT* [1719].

separately, are 'the prerogative of any period or of any ecclesiastical interest'; rather, the three elements taken together are the method 'which constitutes the Anglican approach to theology'.[60] Clarke, however, signalled his disagreement with that method when he seemed to reject one element of it in his 1712 *Scripture Doctrine of the Trinity*. Nevertheless, like many of his day, he accepted the theory of inspiration[61]—that 'the *Scripture* itself' was '*given by inspiration of God*';[62] therefore, the 'Words of God are the most proper Signification of his Will, and adequate expressions of his own Intention'.[63]

On the assumption that scriptural texts represent God's intent, it will follow that the first task of an interpreter is to discover that intent. But to accomplish this task, what method is to be used? Clarke's answer, hinted in various parts of his book, suggests that he had applied to his subject matter the method outlined previously in Locke's *Reasonableness of Christianity*.[64] For first, he collected 'ALL *the Texts*'[65] that relate to his subject and 'set them in One View, with such References and Critical Observations' that might be of use towards understanding 'their true Meaning'.[66] This collection

---

[60] McAdoo, *The Spirit of Anglicanism*, p. 319.

[61] For Newton's acceptance of this theory and its consequences for his biblical hermeneutics, see Appendix A in Manuel, *The Religion of Isaac Newton*, which includes, pp. 116–25, a transcription of Newton's 'Rules for interpreting the words and language of Scripture' and 'for methodising the Apocalyps'.

[62] Clarke, *SDT*, Pt III, p. 449.

[63] Ibid., p. viii.

[64] See *supra* 4.4.2. See also Clarke, *A Discourse* [1708], unpaginated Preface, where he stated that from Locke '*I neither cited any one Passage, nor (that I know of) borrowed any Argument from him*'. This seems to be the only reference to Locke in the early versions of his Boyle Lectures.

[65] Clarke's recourse to one-word generalisations is found also in his other writings; see, e.g., Clarke, *A Discourse* [1706], pp. 298 ('all the Philosophers of all Ages'), 300 ('all Antient Philosophers and Poets'). For the problem with such generalisations, see *supra* 3.3.2: No. 4, SC→RN, 28 September 1706 (f. 1, §1, n.117).

[66] Clarke, *SDT*, Introduction, p. xvii.

and its adjuncts form Part I of his book, the four chapters of which digest the data into several sections each with different headings. The data itself consist of 1,251 numbered passages or phrases extracted from verses in the New Testament, which Clarke ordered so as to reveal 'the whole Tenour of Scripture'. But a digest of biblical texts cannot reveal the entire scope of Scripture. What it may achieve is consistency with an hypothesis that guides a selection of texts. In Part I, however, nothing is said about an hypothesis or about the purpose of the adjuncts.

Disregarding his statement that the selection includes 'ALL' relevant texts (for this could be tested by recourse to a computerised concordance), the most we can say about the four chapters is that Clarke organised his texts as they relate to the Father, the Son and the holy Spirit, mentioned separately and then mentioned together. But more can be said about his adjuncts, which provide some insight into the method he used for determining the meaning of the passages and phrases, as well as the sources that guide his interpretation of certain controversial texts. To begin with the 'Observations', these take the form of glosses and emendations, the purpose of which is to indicate either the most literal ('natural', 'proper', 'true') sense of a passage or phrase, or the possible literal sense when expressions are either ambiguous[67] or figurative ('unnatural').[68] For these cases, Clarke relied chiefly on grammatical construction or comparing a passage or phrase with its recurrence in other 'parallel' places within Scripture.

---

[67] For obscurity, see op. cit., e.g., Pt I, Ch. 1, Text No. 264; for ambiguity, see e.g., Pt I, Ch. 2, Text Nos. 539, 614, 634, 663, 690, 706, 710, 713, 722, 725, 726, 731.

[68] For 'unnatural' expressions, see ibid., e.g., Pt I, Ch. 2, Text Nos. 535, 546, 550, 574, 591, 607, 934, in which those guilty of utilising such expressions are the Sabellians and Socinians; see also Clarke, *SDT* [1832], p. 149, where the terms Sabellianism and Socinianism are said to be synonymous. Possibly, therefore, he conceived the two movements as versions of modalism, a doctrine that affirms the unity of the Father, Son and Spirit by holding that there are three 'modes' or ways in which the one God acts in history. Note that the opposite extreme is tritheism, a doctrine that privileges God's plurality by conceiving the Persons of the Trinity in terms of distinct entities rather than in terms of their relations. For orthodoxy, therefore, the problem was to find the mean between those two extremes.

In addition, he drew on his knowledge of Greek and Latin in cases where, he claimed, translators had corrupted or mistranslated a text, which then is to be read in the way suggested in the annotation.[69] For example, one of his suggestions appears as a bracketed phrase '[*or* through]' inserted after the word 'by'. Initially, this distinction occurs concerning biblical passages which declare that the world was made by the Son.[70] But since the bracketed phrase then appears throughout the rest of the book in relation to these passages, it is important to note that the words 'made by' imply a first or efficient cause, whereas the words 'made through' suggest a second or instrumental ('Ministring') cause. Consequently, in this very gloss Clarke intimates the hypothesis that guided his selection of texts: that there is no co-equality of the Persons of the Trinity, because the Father as efficient cause is superior, whereas the Son as instrumental cause is subordinate.[71]

As for the second adjunct, the 'References', these serve to indicate that Clarke's various glosses and emendations are consistent with the 'reading' of the ancient ('primitive') Christian church, as he defined it.[72] When passages and phrases are controversial, he added citations from both ancient church fathers and modern theologians, the citations from the ancients tending to favour Eusebius of Caesarea, Justin Martyr, Novatian and Origen of Alexandria, all of whom adumbrated versions of a subordinationist theology.[73] Like his other citations in Part I, these function as proof texts to support his

---

[69] See Clarke, *SDT*, e.g., Pt I, Ch. 2, Text No. 940.

[70] Ibid., Pt I, Ch. 2, Text No. 546.

[71] For the first explicit statement of the Son's subordination, see ibid., Pt I, Ch. 2, sect. 5, p. 144.

[72] See *supra* 4.4.2.

[73] The citations to these authors increase considerably in Clarke, *SDT* [1719] and [1832].

case, although it should be noted that in removing a citation from its textual, as well as historical context, meaning can be distorted.[74]

Clarke, then, had an agenda, the nature of which is revealed in Part II of his book (Locke's stage two). For here, his inductions from the data of Part I are unified by a universal proposition, the hypothesis previously mooted but now made increasingly explicit: that in the '*Monarchy* of the Universe',[75] the Father alone is 'absolute' and supreme and the Son and holy Spirit are 'relative' and subordinate. And it is this doctrine that enables Clarke to rationalise his generalisations into a system of fifty-five 'methodical' propositions concerning

> ...the distinct *Powers and Offices* of each of the *Three Persons* [of the Trinity], in the *Creation, Government, Redemption, Sanctification*, and *Salvation* of man; and the proper *Honour* due consequently from Us to each of Them distinctly; [which] is the great Foundation, and the main Oeconomy of the Christian Religion....[76]

Although to each of these propositions he annexed testimonies of some ancient as well as some modern writers, the evidential value of these testimonies is only probable, because the evidence collected is incomplete and, therefore, inconclusive. And this may account for Clarke's statement that the testimonies were alleged not 'as *Proofs* of any of the Propositions, (for *Proofs* are to be taken from the *Scripture alone*,) but as Illustrations only'.[77]

But he clearly indicated that these testimonies were chosen from the majority of writers before, and at the time of the Council of Nicaea who were 'really of that Opinion ... which I have

---

[74] Although RN will hint that such distortion occurs in Clarke's citations from George Bull, other critics were more explicit; see *infra* 5.5.2: No. 2, RN→Hickes 1713 (f. 35v and n.234) and Ferguson, *ECH*, pp. 61, 68–70 and 72–3.

[75] Clarke, *SDT*, Pt II, Prop. IV, p. 243, *et passim*.

[76] Ibid., Introduction, pp. xxvi–xxvii.

[77] Ibid., p. xvii.

endeavoured to set forth in those Propositions'. For 'as to the Writers
after that Time, ... many Passages [are] not consistent with (nay,
perhaps contrary to) those which are here cited'.[78] In short, Clarke's
Eusebian myth of a morally pure primitive Church extends only to
325, when the Council of Nicaea not only condemned the unequal
Trinity doctrine of the Alexandrian presbyter Arius, but also
introduced the term *homoousios*—of the same substance or
consubstantial—in reaffirmation of the Godhead as three Persons in
one unity.[79] Although Clarke dismissed this doctrine as
'metaphysical',[80] yet he never acknowledged that in substituting
subordinationism for consubstantialism, he was replacing one
metaphysics with another.

The concluding Part III consists of two chapters, in the first of
which Clarke presents a 'great number of Passages out of the *Liturgy
of the Church of England,* wherein the Doctrine set forth in the
former Parts is expresly *affirmed*'.[81] The digested passages are
organised in five sections, each with different headings, the final two
sections of which list passages that testify to the subordination of the
Son and holy Spirit. These sections, then, function in a similar way to
the data presented in Part I. However, unlike that part, Clarke did not
make clear when he had interpolated his own gloss into a quoted
passage or phrase.[82] For example, two passages from the Athanasian
creed read: 'The *Son is of the Father* alone—*begotten*';[83] 'The *Holy
Ghost is of the Father* and of the Son—*proceeding*'.[84] As a

---

[78] Op. cit., p. xviii.

[79] For details, see Ayres, *Nicaea and its Legacy.*

[80] Clarke, *SDT*, Pt II, Prop. IV, p. 243, *et passim.*

[81] Ibid., Introduction, p. xix.

[82] In ibid., Pt I, Clarke tended to distinguish his additions by inserting them in
square brackets within a quoted scriptural passage or a quoted exegetical writing.
But this is not always the case, for sometimes his insertions appear in round
brackets or even without any brackets at all.

[83] Ibid., Pt III, Ch. 1, sect. iv, Text No. 3.

[84] Ibid., Pt III, Ch. 1, sect. v, Text No. 6.

consequence, an unwary reader might not realise that the words which follow the em-dash are interpolations.

The second, more substantial chapter consists of a digest of 'the principal Passages, which may seem at first sight to *differ* from the Doctrine set forth in the former Parts', the doctrine that the Father is 'absolute' and the Son and holy Spirit, 'relative'.[85] Despite this seeming, Clarke promised to show how the passages in the digest

> ...may be understood in a Sense consistent both with the Doctrine of Scripture, and with the other before-cited Expressions of the Liturgy. And This is absolutely necessary to be done by every one, who when he prays with his Mouth, desires to pray with his Understanding also.[86]

To achieve this aim, twenty-seven liturgical expressions are tested against the 'more antient and usual Forms' not only in the ancient (for him, ante-Nicene) Church, but also against the expressions in Scripture. In the latter case (the test of Scripture), Clarke either cited passages from specific texts or referred the reader to his own digests in Part I and in the first chapter of Part III. In the former case (the test of 'ancient and usual' expressions), Clarke cited some ancient fathers, as well as some modern theologians, supplementing these citations with comment of his own.

As will become apparent later on,[87] North recognises that the most controversial parts of the second chapter are the numbered passages concerning two creeds. The first is the so-called Athanasian creed, so-called because it was written by an unknown author, not by Athanasius.[88] There are two reasons why this creed presented a particular problem for Clarke. First, it summarises the consubstantial doctrine that Clarke sought to replace. Second, it begins with what Clarke described as a 'damnatory' clause—'Whoever will be

---

[85] This chapter was considerably revised in Clarke, *SDT* [1719].

[86] Clarke, *SDT*, Introduction, p. xix.

[87] See *infra* 5.5.2: No. 1, RN→SC 1713 (ff. 25v–28).

[88] Clarke, *STD*, Pt III, Ch. 2, pp. 418–54 (Nos 2–7).

saved'—for the words that follow this incipit state that belief in the consubstantial doctrine is necessary for salvation. In an attempt to solve this problem, Clarke focused on determining the literal meaning of certain of the doctrine's key words (e.g., one, unity, co-equal, equal),[89] which, of course, also meant interpreting them consistently with his own subordinationist doctrine.

At the conclusion of this exposition, he stated:

> Thus have I endeavoured to explain intelligibly this very difficult Creed: understanding several of the expressions therein contained.... If any One shall here object, that probably the Sense I have now given, does not express the intention of the Compiler: I answer that it is not easie to know certainly what was the Intention of an unknown Author, who lived in those dark Ages, the 7th or 8th Century: That, if it was never so certainly known, yet all sincere Christians are bound to interpret every *human* Composition according to the Rule laid down in the 6*th*, the 8*th*, the 20*th*, and the 21*st* of the *XXXIX Articles*,[90] and not according to what they may imagine was the intention of any uninspired Author....[91]

But Clarke was not yet finished with the Athanasian creed, for he went on to observe that despite attempts 'by way of *Apology* for, or *Explication* of, this Creed',

---

[89] Op. cit., pp. 441–3.

[90] Arts 6, 8, 20 and 21 treat, respectively, 'Of the sufficiency of the holy Scriptures for Salvation'; 'Of the Three Creeds' (i.e., Apostle's, Athanasian, Nicene); 'Of the Authority of the Church' and 'Of the Authority of General Councils'; see *Articles ... touching the true Religion*. Note that of the articles cited by Clarke, the very first (faith in the consubstantial doctrine of the Trinity) is missing; see *infra* 4.4.5 n.139.

[91] Clarke, *SDT*, Pt III, Ch. 2, p. 444.

...it cannot be denied to be a matter worthy of the most serious consideration of the Governors of the Church,[92] whether it would not be more advantagious to the *True Interest* of *Christian Religion* (the Thing of the greatest importance in the World,) to retain only those more indisputable Forms and Profession of Faith, which were received unanimously in the Primitive Church, and which (without affording Matter for Controversy) confessedly contain all that is explicitly necessary, to the *Baptism, Absolution,* and *Salvation* of a Christian.[93]

And he followed this observation with six recommendations concerning the creed, including 'whether it had not better be quite omitted'.[94]

The second controversial part of the chapter relates to the Nicene creed and, in particular, the phrase 'one substance with the Father'.[95] This creed dates from the revision, made at the Council of Constantinople in 381, to the earlier Nicene term *homoousios* ('*omousios*' in Clarke and North's writings).[96] While Clarke admitted that the correct translation of the word is indeed one substance with the Father, he pointed out that, as the Greek word is not found in

---

[92] Two meanings could be given to the phrase, 'Governors of the Church'. On the one hand, it could refer to the two provincial synods or assemblies of the clergy, constituted by statute and called on to deliberate on ecclesiastical matters. If this is Clarke's meaning, then he probably had in mind the more important provincial synod of Canterbury that was commonly designated as 'Convocation', although, properly speaking, the provincial synod of York was also a convocation. On the other hand, since the Church of England was established by law, the phrase could refer to Parliament, which, with the Sovereign, forms the supreme legislature of the realm consisting of the Lords Spiritual and Temporal (forming together the House of Lords) and the representatives of the counties, boroughs and universities (forming the House of Commons). See also *infra* 4.4.5 and n.160.

[93] Clarke, *SDT*, Pt III, Ch. 2, p. 446.

[94] Ibid., pp, 446–54, p. 448.

[95] Ibid., pp. 446–76 (Nos 14–17).

[96] The Nicene term affirmed the divinity of the Son, whereas the revision also affirmed the divinity of the holy Spirit; see Phan (ed.), *The Cambridge Companion*, p. 9.

Scripture, not only is it ambiguous but also 'much harder to explain intelligibly, than any of the Expressions which we meet with in Holy Writ'.[97] What is more, he claimed that different understandings of the word led not only to logical contradictions but also increasing disputes, as well as schisms, the consequence of which was a corruption from perfection (the Eusebian myth). For these reasons, perhaps, he concluded that if in antiquity those 'in Power' had condemned and forbidden the use of unscriptural expressions, then the controversy begun by Arius[98] would not have occurred. And since Clarke unfolded this supposition at some length and followed it with a catena of authorities, both ancient and modern, a careful reader might infer that those now in power should learn from the mistakes of the ancient authorities and censor all unscriptural expressions.[99]

4.4. SEMANTIC FIXITY VS. SEMANTIC CHANGE

In the hope of reinstating 'Catholic Unity' by resolving disputes and divisions within the Church of England, Clarke attempted to secure consistency of interpretation, as well as certainty for his subordinationist doctrine of the Trinity. To achieve consistency, he applied a textual method, described in the previous section, which is reducible to four principal assumptions. First, since God's intention is represented in the texts of the inspired writers, 'whatever is there delivered, is infallibly True'.[100] Second, since the texts themselves

---

[97] Clarke, *SDT*, Pt III, Ch. 2, pp. 465–6.

[98] For a brief summary of the controversy and its aftermath, see Phan (ed.), *The Cambridge Companion*, pp. 62–3.

[99] Clarke, *SDT*, Pt III, Ch. 2, pp. 470–5.

[100] Ibid., Introduction, pp. vi–vii. Note that this assumption raises the question, what was the 'original' manuscript to represent this intention? According to his gloss on Text No. 180, Clarke, ibid., Pt I, Ch. 2, asserted that 'the Alexandrian MS [is] *the antientest and best of all our Copies*, says the Learned Dr. Mills', i.e., John Mill, who, for thirty years, had worked on an edition of the Greek New Testament.

are a key to that intent, textual analysis is the only way to unlock their meaning. Third, from this kind of analysis, the literal (univocal) meaning is to be derived that, he claimed, should 'signify neither more nor less than the words of Scripture necessarily and indisputably do'.[101] Fourth,

> *As* in reading a Comment upon any Book whatsoever, he that would thence understand the true meaning of the Text, must not barely consider what the words of the Comment may of themselves possibly happen to signify; but how they may be so understood, so as to be a consistent Interpretation of the Text they are to explain: *So* in considering all Forms of Human Composition in matters of Religion, it is not of importance what the words may in themselves possibly most obviously signify ... but in what Sense they can be consistent Expositions of those Texts of Scripture....[102]

In other words, an hypothesis or unifying universal proposition is required for providing a consistent interpretation or exposition of scriptural texts; and, as indicated in the previous section, it was just such an hypothesis that enabled Clarke to unify his interpretation in a system or code of propositions.

But Clarke also attempted to secure demonstrative certainty for his unifying hypothesis (Locke's stage three[103]). To achieve such certainty, he conceived reason as a logical method of proving that his subordinationist doctrine was true and the rival consubstantial doctrine, false. For this purpose he relied on the logical principle of

---

Published as *Novum testamentum ... Accedunt loca scripturæ parallela ... & appendix ad variantes lectiones* (London, 1707), it included a long Prolegomena in which the Alexandrine Codex of the Septuagint is described; see Fox, *John Mill*, pp. 68–9. At that time, the Codex itself was in the Royal Library at St. James's Palace; and between 1707 and 1709 the biblical scholar, John Ernest Grabe, published the first and fourth volumes of his edition of the Codex (the second and third volumes were published posthumously). Today, the Codex is no longer considered as the oldest copy.

[101] Clarke, *SDT*, Introduction, p. xxix.

[102] Ibid., p. xxii.

[103] See *supra* 4.4.2.

non-contradiction, whereby no proposition can be both true and false; and he used this principle in a number of different ways. For example, he employed it when setting up his code of propositions as a rational system that is self-consistent and, hence, non-contradictory with itself. He also employed it in Part II and in the second chapter of Part III, where he argued that the consubstantial (three-in-one) doctrine is a contradiction in itself, as well as a contradiction to Scripture.[104] And when arguing that logical necessity occurs where the opposite of a proposition also implies a contradiction, he used the same principle to demonstrate that the two rival doctrines themselves are contradictory rather than to prove the ways in which both doctrines may be true.

Clarke's theological method, therefore, consists of two rational procedures, textual and logical, in both of which he focused entirely on the understanding. But North will argue that the doctrine of the Trinity cannot be rationalised by 'an army of interpretations and categorys ... led on with plenty ... of genus, species, individualls, complex, absolute, ordinate, coordinate, self existent and such (applied to the devine essence) metafisicall and chop-logick nothings'.[105] Indeed, since the nature of the Godhead was not revealed, it will 'bear no interpretation att all', much less an interpretation that reduces a fundamental doctrine to a 'political plan'[106] so as 'to make it square with the scanty model of our [human] capacity'.[107]

North thus rejects the demands of reason to represent that which cannot be represented; and he does so, because he approaches God as a worshipper, not as a philosopher. For him, therefore, the function of the consubstantial doctrine is symbolic or instrumental— to assist the mind on its journey towards God—and it offers this

---

[104] See Clarke, Pt III, Ch. 2, p. 433, *et passim*.

[105] See *infra* 5.5.2: No. 1, RN→SC 1713 (f. 25v).

[106] Ibid. (f. 4v), the 'plan' being Clarke's monarchial model of king, prince and prime minister.

[107] Ibid. (f. 12).

assistance in the form of paradox, which, of course, is a contradiction. Thus North will remark: 'All that a reasonable man can say is, that it [the paradox] is wonderfull and strang[e]' and, hence, leads to 'adoration' more than 'to contradiction'.[108]

In this statement we are reminded that all paradox, not just divine paradox, operates 'at the limits of discursive knowledge'.[109] For, as Rosalie Colie has pointed out,

> Because paradox manages to be at once a figure of speech and figure of thought, appropriate to a view of the universe profoundly metaphysical ... it served to mediate all sorts of ideas and things which, under strict categorical arrangements, do not at first glance appear to 'fit'. Because of its deliberate lack of limitation, its conscious blurring of distinction and difference, the asymptotic mode of paradox managed to bear the burden of doubleness imposed by a metaphysical world view.[110]

It is to be noted, therefore, that some of the 'abstracts' encountered in previous chapters belong to the realm of paradox—for example, space, time, eternity, infinity and nothing, as in creation *ex nihilo* or vacuum or nought (zero).[111] Moreover, in one of the fragments of his 1706 correspondence with Clarke, North himself evokes, but does not develop the paradoxical encomium. For he uses the epithet 'wonderfull' to refer to things that some would consider 'low' and without value—the little animals, observable only with a microscope, and their inferred even smaller animal spirits. For the paradoxist,

---

[108] Op. cit. (f. 20).

[109] Colie, 'Some Paradoxes', p. 109. See also Colie, *Paradoxia Epidemica*, p. 11, the entire book of which contains immense insight into the varieties and functions of paradox during the height of its flowering.

[110] Ibid., pp. 508–9.

[111] Ibid., pp. 307–8 (such paradoxes operate 'across limits, across beginnings, ends, and the boundaries man sets to his knowledge' and that the boundaries include 'the limits of his discourse') and p. 223 (nihil paradoxes aspire to imitate God's unique original act).

however, these and other such things are 'wonderfull', because they display the wisdom of the Creator.[112]

But paradox was beginning to lose its function as an instrument of transcendence in the journey to God, for the trend was towards a rational (discursive) basis for language.[113] Clarke's book is symptomatic of this trend, because in accepting the ideal of Eusebian purity, he aimed to cleanse not only a fundamental doctrine of the Church of England but also the language by which it was to be expressed.[114] And this meant finding the literal meaning of the parabolic and symbolic language of Scripture itself so as to make it internally consistent, as well as systematising the internal contradictions, ambiguities, redundancies and anomalies in and between the scriptural texts. What is more, Clarke's textual procedures, which are anchored to a doctrine of original intent,[115] imply that the language of Scripture *ought* be read univocally in terms of *timeless* and, hence, *unchanging* meanings of the words that represent God's intention. Indeed, in his gloss on the scriptural passage, 'Jesus Christ the same yesterday, and to day, and for ever' (Heb. xiii.8), Clarke stated: 'The Meaning in this Place (as appears from the Context) is, that the *Doctrine of Christ*, once taught by the Apostles, ought to be preserved unchanged'.[116]

---

[112] See *supra* 3.3.2: No. 4, RN→SC [1706] (f. 293v).

[113] See Cohen, *Sensible Words*, pp. 34–42, *et passim*.

[114] I.e., Clarke is a language purist. For some others, see Kassler, *HRN*, pp. 149–50, 157; and for an extreme example, see Lodwick, *On Language, Theology and Utopia*. For the religious (*vs.* descriptive or logical) meaning of the passage in Zephaniah iii.9 concerning the restoration of 'a pure speech', see Frye, *The Double Vision*, p. 83.

[115] Today, this type of method underpins the theory of legal interpretation called 'originalism', the aim of which is to lock in subsequent generations to the specific 'intentions' of the framers of the U.S. Constitution, 'intentions' meaning, of course, the originalist interpretation of what the framers' words mean.

[116] Clarke, *SDT*, Pt I, Ch. 2, sect. iii, Text No. 662, p. 127. See also his 'immutable' morality in Clarke, *A Discourse* [1706].

For North, however, the 'letter' of Scripture is not fixed; rather, it is responsive to semantic change. And this is so, because language, the instrument of thought, is an arbitrary invention of humans and a representation of their changing history and customary practices.[117] Thus, 'language and its modes' are created by 'fashion' and 'by like means as plants ... grow from many litle occasions, and the usage of many persons'. Even orthography, as well as style and pronunciation will never be 'at any stay, but roll from age to age with perpetuall chang[e] and variety'.[118]

The consequences of North's theory of language and meaning for scriptural interpretation may be illustrated briefly from certain passages contained in his extensive analysis of the historiography of some of his contemporaries.[119] In the passages in question, which contain the kernel of his common-law jurisprudence, he describes the Gospel's two commandments not only as sacred arguments but also as consummate wisdom. As sacred arguments, the revealed commandments establish our duty to God, ourselves and others and, hence, have the force of a fundamental law (imagined as the ancient constitution). But as consummate wisdom, they take the form of maxims[120] or rules of living, including such, for example, as 'honesty is the best policy', 'to condemn none without hearing' and 'that rule which is worth them all', do unto others as we would be done by. Accordingly, he offers two suggestions: first, that the revealed commandments revert from the written scriptural text ('letter') to custom in the form of maxims; and second, that ancient as well as

---

[117] See Kassler, *HRN*, pp. 33, 129–53, 173–9. RN's premiss indicates that he is an anomalist or one who holds that language is purely conventional or arbitrary in origin and without any natural analogy between names and things, whereas the language purists were analogists who argued for the natural origin of language.

[118] Ibid., p. 138.

[119] RN, *Examen*, which he began writing sometime prior to 1711 and completed sometime before the end of 1713; see Kassler, *HRN*, pp. 389–92, also for some of the manuscripts relating to this work and to which should now be added: RN, 'Examen of an Historical Libel mistitled a compleat History of England vol 3d by an Impartial Hand' (UK:Mch) Allen Deeds E.3.4 Folder N no. 34.

[120] E.g., adages, aphorisms, proverbs, sententia, etc.

newly coined maxims do not repeat the original words of the commandments.[121]

North, therefore, allows for the possibilities of natural growth in a living language, possibilities that, for him, include rare words or terms of art, 'antique' words or archaisms, 'newfangled' words or neologisms (including his own), sonorous words, idiomatic expressions both native and foreign[122] and figurative expressions which he understands as any device or pattern of language which renders meaning more comprehensive. Indeed, he holds that the 'more copious' a language, and repleat with variety of ... words and phrases, the richer is the treasury of those that deal in words'.[123] Consequently, it is vain to hold language close to 'any idiom or analogy', in the hopes it will lead, 'as some pretend', to improvement.[124]

These statements might seem to be the sentiments of a literary writer who follows in the footsteps of Cicero and Horace,[125] which in North's case they are. But they also represent the sentiments of a professional lawyer, who, in drawing on the *copia* of expression, relies on some of the possibilities for semantic change itemised in the foregoing paragraph. For example, terms of art help the lawyer communicate within professional ranks, whereas less technical words communicate with those to whose needs the law administers. While some precision 'in the small' is possible (e.g., by means of terms of art), the bulk of law language cannot be reduced to exact meaning, because the common law itself is built on the ambiguity that is part of life. Consequently, its basic notion is that 'things cannot be too precise, that no matter how thin a principle is sliced, particularization

---

[121] RN, *Examen*, p. 352; see also Kassler, *HRN*, pp. 44, 112–14, 156, *et passim*; see also RN, 'Of Etimology', p. 228.

[122] E.g., provincialisms or dialects, slang and pithy sayings.

[123] RN, 'Of Etimology', p. 241.

[124] Ibid., p. 253, and pp. 253–64 for a critique of some of the language purists who hoped to 'institute or reforme language by swift or larg stepps'.

[125] See Kassler, *HRN*, p. 150.

is always possible.' And this flexibility is reflected in its common language, which also shares the imperfections of language itself.[126]

4.5. CLARKE AND NORTH RESUME CONTACT

By the end of 1712, the year in which Clarke's book was published, only a few critics had responded to it.[127] But North tells us that when the book 'came first out, and made much noise', he was 'informed by letter from an incomparable lady' that Clarke 'had a mind to know what I thought of his book'. These statements occur in a letter written in 1719,[128] in which North does not identify the lady who acted as an intermediary. Nevertheless, from that letter it is possible to glean three facts about her: first, that Clarke 'used to be kindly received' in her house, which suggests that his visits had ceased before 1719; second, that during the period of his visits, she probably had a residence in London; and third, that she was still living in 1719, the year when North requested her permission to publish the letter that is the focus of the next chapter.

Although these facts do not provide sufficient clues to enable certain identification, yet North's adjective 'incomparable' is striking, for his use of the term occurs rarely and, it seems, chiefly in

---

[126] Melinkoff, *The Language of the Law*; for problems with exact meaning, see especially pp. 387–98, and for illustration of the *copia* of expression in law language, see the entire book, which also is suggestive for understanding some aspects of RN's style. Unfortunately, I was unaware of this important book until after my own (*HRN*) had been published.

[127] For the replies by Whiston and two anonymous critics, see Ferguson, *ECH*, pp. 55–6, 59–60, who seems to have been aware of another anonymous reply published just after the end of 1712; see *An Inquiry into the Manner of Assenting to the Articles and Liturgy of the Church of England Practised and Recommended by Dr. Clarke* (London, 1713); see also *The Daily Courant* (Thursday, 22 January 1713), p. 2.

[128] See *infra* 5.5.2: No. 3, RN→Bedford 1719 (f. 1v).

relation to his cousin 'Captain' Charles Hatton.[129] Perhaps, therefore, the 'incomparable' lady was Hatton's sister, Anne, second wife of Daniel Finch, 2d earl of Nottingham. In the 1690s they resided in the parish of St. James, initially in Cleveland Row, and then in St. James's Place.[130] However, by 1719, when their residence was in Bloomsbury Square, the Countess was seldom in London, because she preferred to remain with her numerous children at Burley-on-the-Hill, Rutland;[131] and this could explain why Clarke ceased to visit the 'incomparable' lady.

But if this lady was indeed Lady Nottingham, then there may be another explanation why the visits ceased. For Clarke had a friend in Cleveland Row by the name of Charlotte Clayton;[132] and her hostility to the Countess was due to the latter's belief that Clarke's writings were tainted with heresy. However, this is merely court tittle-tattle, told by a diarist who also disliked the Countess,[133] so without more concrete evidence, the question whether or not the 'incomparable' lady is the same as Lady Nottingham will remain a matter of conjecture.

---

[129] See RN, *General Preface*, p. 84 ('incomparable'), *Life/JN*, p. 117 ('incomparable') and *Life/FN*, p. 464 ('truly noble'). The 'Captain' in question was second son of Christopher Hatton, baron Hatton of Kirby, and his wife Elizabeth, sister of RN's mother.

[130] See Chancellor, *Memorials*, pp. 95, 119.

[131] See her correspondence in the British Library.

[132] In 1714 she was appointed as one of the women of the bedchamber to Caroline, then princess of Wales, who subsequently came under her influence. As a consequence, Clayton became a patronage broker not only for Clarke but also for some of his supporters, including Benjamin Hoadly, afterwards a leading Whig and low-church propagandist; see Taylor, 'Benjamin Hoadly'.

[133] I.e., Mary Cowper, who, like Clayton her close friend, was a woman of the bedchamber and a patronage broker; see Cowper, *The Diary*, pp. 18–19. Note that on p. 1 the stated aim of the diarist, to present an accurate account of events at court, must be taken with a grain of salt because of the diarist's very evident biases.

In the same 1719 letter North states that because the request from Clarke came through an intermediary whom North knew and admired, 'a letter or two past' between him and Clarke,

> ...in which I could say nothing on the subject, but that from the report of his book, I did not like it. For I then understood his designe was to goe off from the Scripture into philosofy. But I was given to understand that was a clear mistake, and that his interest was the direct contrary, viz., to hold to the Scripture against all incertein reasoning.[134]

Although this 1712 correspondence is no longer extant, North himself provides a few details about its subject matter. First, he indicated that he understood 'the scope' of the book was 'to set up reason against revelation'; and he told Clarke as much. In return, the latter replied that 'uncertein reasonings ought to be layd aside, and the revelation in Scripture (of the holy Trinity) ought to have bin kept to'.[135] These details are consistent with statements in the 1719 letter.

Second, in 'a paper' answering Clarke, North provided a brief summary of his common-law conception of reason as the trans-subjective 'experience of worldly things', for

> Wee know nothing properly but thro the means of bodily contact, and the impressions of that reflected by our memory. These enable us to lay antecedents and consequents together, and by a sort of reasoning called induction, which depends on frequency, if not on constancy, to conclude that when the antecedents, as have bin observed, arrive, the consequents are not farr off. This branched out into all the instances of experience, made up of history, observation, instruction, conformity, and the like, is what wee call reason.[136]

---

[134] See *infra* 5.5.2: No. 3, RN→Bedford 1719 (f. 1v).

[135] See *infra* 5.5.2: No. 1, RN→SC 1713 (f. 3v).

[136] Ibid. (f. 19).

Finally, to the above-mentioned 'paper' Clarke replied that he 'would give but short answers'.[137]

As a consequence of this 1712 exchange, North found himself 'under a necessity to say somewhat to' Clarke's book, so he borrowed it from 'a neighbouring minister', 'perused it' and in 1713 'dispatched' his answer 'directed to our mediatrix', the 'incomparable' lady.[138] Since this letter and material related to it occupies the next chapter, it is important to note that the Ariadne thread that knits together the various parts of North's response concerns the nature and conditions for religious truth—here, the doctrine of the Trinity—and this issue involves, but is not restricted to the theological method by which that truth is to be sought. In addition to this issue, there are some others that require comment.

First, although the two men had different approaches to the doctrine in question, the one as a theologian, the other as a worshipper, and although the latter approach is an important theme in North's main letter, yet the nature of Clarke's book requires him to focus on '*theologia*', namely, the subordinationist doctrine and its implications for the consubstantial doctrine, which, for North, is the 'rock' on which the Church of England's faith stands.[139] It should be noted, therefore, that the two different doctrines represent two

---

[137] Op. cit. (f. 30v).

[138] See *infra* 5.5.2: No. 3, RN→Bedford 1719 (f. 1v). The 1713 letter has been discussed briefly by two previous commentators; see Korsten, *Roger North*, pp. 68–73, and Stewart, 'Samuel Clarke, Newtonianism, and the Factions of Post-Revolutionary England'. For a brief critique of the methodological whiggism inherent in their assessments, see Kassler, *HRN*, pp. 4–8.

[139] I.e., it is the first article of faith; see *Articles ... touching the true Religion* (unpaginated), Art. 1 ('Of Faith in the Holy Trinity'), which reads: 'There is but one living and true God, everlasting, without body, parts, or passions; of infinite power, wisdom, and goodness, the maker, and preserver of all things both visible and invisible. And in unity of this Godhead there be three Persons of one substance, power and eternity; the Father, the Son, and the holy Ghost.'

different patristic traditions, one pre-Nicene and Eastern, the other post-Nicene and Western.[140]

In the Eastern tradition,[141] the doctrine was expounded as a functional theology that emphasised God the Father as God in the primordial sense—He was God in his own right, unbegotten and source of the Son and holy Spirit. Consequently, all processions in God are hypostatic, that is to say, personal. To reconcile this theology with monotheism, some Alexandrian fathers insisted on the metaphysical and moral subordination of the Son to the Father;[142] and it was this doctrine that was adopted by the previously-mentioned Arius, the fourth-century presbyter from Alexandria, who stressed not only the oneness of God to the detriment of the divinity of the Son, but also a hierarchical relationship, not a co-equal one. And thus began dissensions in the ancient Christian church, which lasted for nearly a century.

In the Western tradition, the doctrine of the Trinity was expounded as an ontological theology of the consubstantial union of three co-equal hypostases or persons in the Godhead.[143] Accordingly, all processions in God are from one principle—the unity of God. In the case of the second person, the consubstantial union also involves the divine and human natures in the hypostases of Christ. But, as North points out, since the Son's divinity is not revealed in Scripture

---

[140] For the chief theological factor in the rupture between the Eastern and Western Christianity, see Stormon, 'The Procession of the Spirit', pp. 81–2. Note that the Church of England is a branch of the Western church, which at the Reformation repudiated the supremacy of the Pope and asserted that of the Sovereign over all persons and in all causes ecclesiastical as well as civil in his or her dominion.

[141] For this Greek-language tradition, see Phan (ed.), *The Cambridge Companion*, pp. 49–69. For its three 'currents'—Alexandrian, Antiochene and Cappadocian—see Stormon, 'The Procession of the Spirit', pp. 88–98, *et passim*.

[142] This was the opinion of Origen, perhaps the most important of the religious Platonists of Alexandria; see ibid., p. 89. For him God the Son is less than God the Father and is the direct source of the holy Spirit, although the ultimate source remains the Father. Note, however, that the theology of the holy Spirit was underdeveloped at this time; see Phan (ed.), *The Cambridge Companion*, e.g., pp. 52–3, *et passim*.

[143] For the beginning of the Latin-language tradition, see ibid., pp. 70–84.

and, consequently, involves difficulties which human understanding is incapable of solving, we still may speak of that nature indirectly, that is, symbolically. And since the Son's Incarnation is revealed in Scripture, his humanity may be spoken of not only literally but also figuratively, because we understand it by analogy with our own experience.[144]

In his book Clarke argued that the consubstantial doctrine conceives God not as absolutely simple[145] but as complex.[146] And since the latter conception implies a plurality of gods, he claimed that it leads to the sin of idolatry, which is in breach of the first commandment.[147] To avoid this breach, he insisted that because the metaphysical 'substance' of the divine persons is nowhere revealed in Scripture, not only is it unknowable but also cannot be spoken of at all, not even in terms of the symbols used in '*Scholastick and Philosophical Inquiries* concerning the *metaphysical Nature and Substance of each of the Three Persons*'.[148] He therefore rejected the consubstantial doctrine in favour of a doctrine in which the functional hierarchy of the three persons may be spoken of, and understood in literal terms.

---

[144] See *infra* 5.5.2: No. 1, RN→SC 1713 (ff. 20–23v, *et passim*).

[145] I.e., Clarke's conception; see *supra* 4.4.1 and n.22.

[146] See *infra* 5.5.2: No. 1, RN→SC 1713 (ff. 19, 25–25v). See also Clarke, *SDT*, Pt III, Ch. 2, p. 455, who identified three notions of complexity: 'Not *One compound Being*, constituted or made up of *Three Parts:* Not *One Species*, consisting of *Three coordinate Individuals*: Not *One Person*, considered *only under Three different Denominations*. For the *first* of these Notions is repugnant to the *Simplicity and Perfection* of the Divine Nature: The *second*, entirely destroys the *Unity* of God: The *third*, either wholly takes away the very *Being* of the *Son* and *Holy Spirit*, or at least introduces ... a total *Confusion of Persons*'.

[147] See ibid., Pt III, Ch. 2, pp. 415, 425–6, 428, 435, 437, 447, 455, 465 and 479. According to Wiles, *Archetypal Heresy*, p. 80, Newton thought that a breach of the first commandment was 'a more dangerous crime' than atheism. But the basis of both men's concern with such a breach seems to have been antipathy to the Church of Rome and its worship of saints rather than polytheism or tritheism.

[148] Clarke, *SDT*, Introduction, pp. xxvi, xxviii–xxix.

The second general point to be noticed here concerns the nature of North's 1713 letter, which begins with a brief statement of his own position, namely, that 'in the general' his faith is 'with that of the Church of England ... establisht by law', that is to say, legally established as the publically recognised form of religion. This is followed by another brief statement that in responding to Clarke, he will adopt the *persona* of 'an opponent' and will 'talk alltogether as an adversary', 'which will not much vary from the usage in our [legal] trade, for how hard soever wee wrangle, it is not observed wee are the less good friends for all that'.[149] In concluding the letter, he once again resorts to the terminology of a lawyer when he states he has completed his 'harrangue'; but in future he prefers to avoid subjects that might 'grate upon' their friendship, 'which I would preserve with all the industry I have'.[150]

Despite these lawyer-like beginnings and endings, North's letter does not represent one side of an advocate's wrangle in a courtroom. Rather, it consists of a summary of the facts of the case with reference to points supposed applicable to the issue under scrutiny, that is to say, it represents a barrister's brief drawn up for the instruction of a Church of England clergyman who had previously conducted his own case not in a court of law but in a book. Leaving aside, then, the introduction and conclusion, the main part of the letter consists of nine enumerated points, the first five concerning Clarke's 'designe', and the next four, his method, to which North adds a brief conclusion. This is followed by five 'particular' matters which are not enumerated but which had been mooted earlier in the letter. They concern (1) the deists who acknowledge the existence of God upon the testimony of reason but reject revealed religion; (2) Clarke's charge that the consubstantial doctrine makes God complex; (3) and (4) the two creeds that

---

[149] See *infra* 5.5.2: No. 1, RN→SC 1713 (ff. 1, 3v).

[150] Ibid. (f. 31).

symbolise the consubstantial doctrine; and (5) the use of the church fathers.[151]

In both main and minor points, North considers Clarke's 'performance'; and it is in these considerations that we find the instructional part of his brief, some of which are substantive and others, hints only. The former includes advice concerning distinctions that Clarke failed to make, two of which are especially important. One is an epistemological distinction between faith and understanding,[152] a distinction that reflects North's fideism.[153] The other is a distinction between a literal ('just') meaning of words and phrases and a figurative interpretation of them, in which he develops some aspects of his theory of language and meaning.[154]

In addition, North reminds Clarke of two subjects that he had not addressed at all. One is the subject of the Incarnation and Passion and, thus, of Christ's role in our redemption.[155] The other is 'the true designe of our Saviour's discourses upon earth'—first, that he offered the world 'a perfect example' of how to live; and second, that he used language that tended 'to assist devotion, excite the extatique, and furnish incomparable matter for perpetuall and pious meditation, such as might hold men endeared to his goodness, shewed forth by his infinite good will and suffering for them'. In treating these subjects North touches on the affective element in religion, for as he points out: 'The ideas of the almighty power are more apt to excite fear, then love, but those that attend our Saviour's Passion are all

---

[151] This fifth matter is the subject of a book once owned by RN; see RN Books (1). Written in French by Jean Daillé, the elder, it was published in an English translation as *A Treatise concerning the Right Use of the Fathers in the Decision of the Controversies that are at this Day in Religion* (London, 1651, re-issued 1675).

[152] This distinction, avoids the problem of spiritual knowledge, a concept as much contested as the Trinity; see *supra* 4.4.1 and Dockrill, 'Spiritual Knowledge'.

[153] See *infra* 5.5.2: No. 1, RN→SC 1713 (ff. 1v, 2v, 18v).

[154] Ibid. (ff. 16–17).

[155] Ibid. (ff. 20–21).

inducements to love and charity, and therein lay the perfection of the Gospell'.[156]

As for the hints, these cover a wider range of issues. For example, one relates to the issue of toleration,[157] when (in North's words) Clarke attributes the 'concern' of the doctrine in question 'to us Protestants in especiall manner, as if all Christians whatever had not as much reason to have a right faith as wee Protestants'.[158] Other hints concern Clarke's subordinationist version of the doctrine and its implications for changes in the liturgy, which North looks upon 'under providenc[e] to be the stay of the Christian religion in England'.[159] He therefore reminds Clarke several times that such changes need to be made in 'a regular way', that is, by 'lawfull authority'.[160] And still other hints intimate that Clarke's book plays into the hands of the deists.[161]

There are grounds for this last hint, since, like the deists, Clarke sought to demystify faith by turning it into a cognitive act, namely, an act of assent to his universal proposition; and he also sought to promote Christianity as a thoroughly rational faith. In these two ways, he was carrying on (though without acknowledgment) Locke's theory of mind and language, the thrust of which was 'the elimination of all mystery and obscurity from philosophy, science, and theology, and the advancement of these subjects through

---

[156] Op. cit. (ff. 21–22). RN and Clarke conceive 'almighty power' differently, the former as an omnipotent creator or first cause and the latter as a monarch or all-governing ruler (i.e., *pantokrator*).

[157] Clarke excludes not only non-Christians but also members of the Church of Rome. But the extent of his toleration is not altogether clear in two cases: (1) the different kinds of sectaries within Protestantism, some of which he sought to bring into conformity with the Church of England; and (2) the orthodox within the Church of England who held the consubstantial doctrine.

[158] See *infra* 5.5.2: No. 1, RN→SC 1713 (f. 7).

[159] Ibid. (f. 9v).

[160] I.e., either Convocation or Parliament; see *supra* 4.4.4 n.92 and *infra* 5.5.2: No. 1, RN→SC 1713 (ff. 8v, 10, 13v, 26v, 28, 30).

[161] Ibid., e.g., (ff. 10v, 14v, 24v–25, 26–27).

reason'.[162] Indeed, in 1696 that theory was utilised by the author of *Christianity not Mysterious*,[163] whose book became a manual for deists.

And this leads to my third remark, namely, the meaning of terms such as 'deist' and 'Socinian' which North and others of his day tended to use loosely, not precisely. And therein lies the difficulty in providing a definition of such terms. According to one commentator, for example, although the term 'Socinian' had various meanings since the Reformation, yet in the last decade of the seventeenth century it was applied to those who rejected the pre-existence of Christ and the Person of the holy Spirit.[164] According to another commentator, the same term had separate, though allied meanings. On the one hand, it described English writings that propounded in whole or part the doctrines of Fausto Paolo Sozzini, including his version of the subordinationist doctrine of the Trinity. On the other hand, the term characterised a method that placed 'a greater accent on human reason than is proper in the delicate balance between the reason of the interpreter and the fullness of scriptural revelation'.[165]

Similar difficulties occur with the term 'deist'.[166] North, for example, describes deists both as the 'modish apostates' of his day and as freethinkers who refused to submit their reason to the control of authority in matters of religious belief. According to Geoffrey Cantor, by the beginning of the eighteenth century, the latter term was a designation claimed especially by the deists, who were often classed with atheists and libertines. For although the term

---

[162] Cantor, 'Berkeley's *The Analyst* Revisited', p. 671.

[163] See *infra* 5.5.2: No. 1, RN→SC 1713 (f. 12v and n.87).

[164] Colligan, *The Arian Movement*, pp. 1–4. According to Wiles, *Archetypal Heresy*, pp. 83–5, this meaning is found in Newton's unpublished writings on the philosophy of religion.

[165] Reedy, *The Bible and Reason*, pp. 119–20, p. 120. For details of Sozzini's theology, see Mortimer, *Reason and Religion*, Chapter 1.

[166] Colie, 'Spinoza and the Early English Deists', pp. 29–30.

'freethinking' was applied to characterise an attitude to theology, a dominant theme in the literature of free thought was the supremacy of reason for avoiding error and for bringing knowledge to a state of perfection.[167]

Although North mentions the Socinian controversy in passing, he is far more concerned with that generated by deism, which, notably, he opposes to ancient Arianism.[168] For example, in remarking on Clarke's use of patristic texts, he points out that during the period of the Arian controversy, the church fathers responded to 'hereticall tho otherwise sincere Christians' called Arians, whereas 'now' an Anglican clergyman requires a strategy different from the ancient fathers, because deism is anti-Christian.[169] This distinction hints at the importance of taking into consideration historical and cultural context from which controversies emerge, so it is unfortunate that North did not consider one issue to fall among the 'most material',[170] namely, the Eusebian myth 'that Christianity was perfect at first', for this is a key to Clarke's theological method.[171]

Finally, in presenting a summary of the facts of the case in question, North produced a response to Clarke's book that is quite different from the responses of other critics published from 1712 and,

---

[167] Cantor, 'Berkeley's *The Analyst*', pp. 670–1. Although there has been a proliferation of studies of deism since the nineteenth century, the classical study of this movement repays rereading; see Stephen, *History of English Thought*, vol. 1, Chapters 3–5.

[168] During RN's lifetime the term 'Arian' was as problematic as the other terms noted above, for when applied loosely it could designate any unorthodox opinion concerning the Trinity; see, e.g., Hutton, 'The Neoplatonic Roots of Arianism', p. 140; see also n.169 below.

[169] See *infra* 5.5.2: No. 1, RN→SC 1713 (ff. 29–29v). RN's distinction between Arians and deists does not follow the polemical practice of his time, when they were grouped together as a class; see Wiles, *Archetypal Heresy*, p. 199.

[170] See *infra* 5.5.2: No. 1, RN→SC 1713 (f. 30v).

[171] Clarke, *SDT*, Introduction, p. viii, differentiated the method of treating Christianity from the method of treating 'things of humane invention, experience, or disquisition', for these latter things, which are not perfect at first, 'improve generally from small beginnings, to greater and greater Certainty, and arrive at Perfection by degrees'.

in increasing numbers, from mid-1713 onwards. Those responses were written chiefly by academically-educated clergymen whose scholastic style of argument, including its reliance on the principle of non-contradiction, may have been more congenial, because more familiar to Clarke. Indeed, it was from these responses that he made selections for inclusion as appendices to subsequent editions of his book.[172] It should be noted, therefore, that, for reasons still to seek, North's response, which had been solicited by Clarke, was never included among the selections.

---

[172] See, e.g., the appendix to Clarke, *SDT* [1832] reprinted in Clarke, *The Works*, vol. 1, pp. 391–532.

1713 CORRESPONDENCE WITH CLARKE ON THEOLOGICAL MATTERS

5.1. THE CORRESPONDENCE AND RELATED MATERIAL: DESCRIPTION OF THE MANUSCRIPTS

North's letter commenting on Clarke's book is preserved in only three versions. I say 'only' because, as will become apparent, there were at least two other versions. Of the three now preserved in the British Library, version 1 and 2 are bound together in the same volume, Add MS 32550: ff. 1–31v, 32–79,[1] whereas version 3 is bound in a separate volume, Add MS 32551: ff. 2–33. The first two versions are written in the style of handwriting that most often appears in North's manuscripts, whether neat or messy. Version 1, a draft, is complete as it stands. Version 2, also a draft, is incomplete, the final folios (ff. 71–79) consisting of four discards from, and three fragments relating to the draft. From certain of its physical features, to be mentioned later, version 2 represents work in progress. Version 3 is the most 'finished' text; but it is not the only text in Add MS 32551, which contains four documents, all of which are written in North's very tiny hand. Since this style of handwriting is rare, at least in the British Library manuscripts, it appears to have been reserved for special occasions only. The four documents are as follows:

---

[1] The old foliation of version 1 (ff. 190–220v) and version 2 (ff. 221–266) has been crossed out and the new, inserted. On the new f. 78v is recorded the date when the British Library (then British Museum) acquired the manuscript (January 1886).

(1) Folio 1v[2] is a copy of North's undated letter 'To the Reverend Mr. -------' (for the recipient, see below), in which he presents 'a short account of the reall inducement I had to medle in controversie so much out of my element'.

(2) Folios 2–33 (ff. 33v and 34 are blank) are a copy of his letter to 'Reverend Sir', i.e., Samuel Clarke, with the old-style date 20 February 1712/3. This is what I have called version 3 of his letter, to which North has added (on f. 6v) a footnote citation to a book that was published in 1719 (see document 4 below).

(3) Folio 34v is a copy of a letter from George Hickes[3] to North, to which the latter has assigned a title, 'A Copy of a Letter from the Reverend Dr. Hicks' and a date, 'May 23 1719'. However, the year 1719 reflects the date when North prepared all four items in volume 32551 and not the date of the original letter, since in 1717 North sent a copy of the same letter, dated 23 May 1713, to Hilkiah Bedford, Jr.[4] which is now preserved at (UK:Ob) Eng. Hist. b.2: f. 170 (see further below).

(4) Folios 35–36v are a copy of North's letter to 'Reverend Sir' (i.e., Hickes[5]) with the date when it was first written: 'Rougham 1713. 2 June 1713'. It begins: 'I received lately a letter from your adversary', meaning Clarke, part of whose letter, as well as part of

---

[2] Folio 1 has the British Library stamp, above which and in a later hand are the words: 'North MSS 6'.

[3] I.e., bishop of the non-juring Church of England, controversialist and philologist; see further *infra* 5.5.2: No. 3, RN→Bedford 1719 (f. 1v nn.256, 261).

[4] I.e., the son and namesake of a mathematical instrument maker in London; see Brown, 'Guild Organisation and the Instrument-Making Trade', pp. 15–6, 27–8, 33–4. As a non-juror and defender of Hickes, Bedford wrote books and pamphlets on theology and religious polemics. In 1713 he was in London, when he received a letter from RN concerning his nearly finished work, *Examen* (published posthumously); see Kassler, *HRN*, pp. 389–92. In the following year Bedford was found guilty of seditious libel and imprisoned for four years, his supposed deed being writing, printing and publishing a book (the real author was the non-juror, George Harbin). In January 1719, after his release, he became chaplain to Heneage Finch, 5th earl of Winchilsea (another non-juror). In the same year he opened a boarding house at Westminster for the scholars of Westminster school.

[5] See *infra* 5.5.2. No. 2: RN→Hickes 1713 (f. 36v).

North's response is then quoted. This is the sole evidence that the correspondence between Clarke and North continued after the latter sent his legal brief to the former. In this third version of the brief, North also added two footnotes (ff. 35, 35v) relating to the book mentioned in document (2) above, namely, David Martin, *A Critical Dissertation upon the Seventh Verse of the Fifth Chapter of St. John's first Epistle* ... tr. into English [from the French], London, 1719.[6]

As the date of the book in document (2) and the erroneous date in document (3) indicate, all four documents were copied by North in 1719, probably sometime after 26 May, when a second, much revised edition of Clarke's *Scripture Doctrine of the Trinity* was published.[7] By this time Hickes was dead; and his literary executor, Bedford, was collecting material for the purpose of writing his biography. On 2 July 1717 North had written to Bedford concerning some of his own recollections of Hickes; and he also enclosed a copy of Hickes's letter of 23 May 1713 described in document (3) above. In addition, he provided some further information, including the following remarks concerning that letter, in which he describes Hickes as 'a true profet', because North's first reply to Clarke[8]

> ...did produce a short answer, and as he [Hickes had] desired, it was imparted to him [by North] in a letter, dated 2 June 1713,[9] in which also [was included] the replye made to Dr. Clarke. I presume you may find that letter among the others

---

[6] Both the translator, Samuel Jebb, and the printer, William Bowyer, Jr., were non-jurors. The latter also published a short reply to Martin's *Critical Dissertation*, the author of which was Thomas Emlyn, perhaps the first unitarian minister and a supporter of Clarke; see Maslen and Lancaster, *The Bowyer Ledgers*, pp. 510, 544.

[7] *The Post-Boy* for 23–26 May 1719, p. 2. Note that in 1719 Clarke's Boyle Lectures were also re-issued in revised versions.

[8] See document (2) above.

[9] See document (4) above.

you were pleased to mention; if not, I can procure a copy, which will be at your service, if any occasion requires it.[10]

On the basis of the above evidence, I conclude that document (1), the covering letter to the unidentified 'Reverend Mr.', was written to Bedford.

The editions in the next section are restricted to the three letters written by North himself, the copy texts of which are as follows:

No. 1 (Draft of a letter) North→Clarke 20 February 1713 (UK:Lbl) Add Ms 32550: ff. 1–31v

No. 2 (Copy of a letter) North→Hickes 2 June 1713 (UK:Lbl) Add MS 32551: ff. 35–36v

No. 3 (Copy of a letter) North→Bedford 1719 (UK:Lbl) Add MS 32551: f. 1v

Only the first manuscript requires further comment and description, because as copy text I have chosen version 1 of North's letter to Clarke. That version represents what we might call his first thoughts, whereas the case is different (for different reasons) with versions 2 and 3.

Moreover, version 1 provides a template for assessing the other two extant versions. For example, it is clear that all three versions of the letter have a four-part structure. In the first part, North sets out his own position concerning the doctrine in question and lays down the approach he will take in presenting objections to Clarke's book. This plan is included in all three versions. In the second part, he assesses what he believes to be the main issues in a series of numbered criticisms (eight in version 1, eleven in the other two versions). In the third part, he assays, without enumeration, five

---

[10] See (UK:Ob) MS Eng. Hist. b.2: RN→Bedford, 2 July 1717, ff. 171–171v; copy of RN→Hickes, 23 May 1713, f. 170; with RN's added information, f. 170v. Note that MS Eng. Hist. b.2 is a continuation (from b.4) of Bedford's collections for a biography of Hickes.

miscellaneous points (in version 2 these are included under enumerated point 12, as they are also in version 3, except that here the miscellaneous points are enumerated as five sub-points). In the fourth part, he brings his letter to conclusion in a manner that is similar in all three versions.

Finally, by starting with version 1, it is possible to work out a tentative chronology of the 1713 correspondence, including letters that no longer are extant. If we begin with the May 1713 letter from Hickes to North (document (3) above), this letter contains the following statement: 'I have now read over the whole letter to Dr. Clerk with the same great satisfaction that I told you I had read the first 29 pages'.[11] This statement suggests that North sent Hickes a copy of the first twenty-nine pages of version 1; that Hickes replied in a letter, no longer extant, offering comments on those pages; and that the incomplete draft of version 2 represents the beginning of North's revisions. That version has in both left- and right-hand margins a large number of marks, some of which are an 'X' and some, a hyphen. Both marks seem to indicate portions of the text that needed correction; and in most cases the lines of text adjacent to the marks do have corrections. Note that in document (1), the letter to Bedford, North states that he had 'communicated a copy, with some small alterations of expression in a few particulars, to the Rev. Dr. Hicks'.[12] This statement suggests that the actual letter that North sent to Clarke was a revision of version 1 and that he sent Hickes a copy of that revision, though neither the revision of version 1 sent to Clarke nor the copy of the revision sent to Hickes are now preserved. Then, in 1719, he made a copy of the revised letter, with some additions, which he sent to Bedford with the other documents detailed above.

As for the copy text itself (version 1), the style is epistolary and familiar, for North writes directly to 'you' or 'Sir' and refers to himself chiefly in the first person, occasionally using the royal 'we'.

---

[11] See RN, 'A Copy of a Letter from the Reverend Dr. Hicks', (UK:Lbl) Add MS 32551: f. 34v, described as document (3) above.

[12] See *infra* 5.5.2: No. 3, RN→Bedford 1719 (f. 1v).

Physical features. The four-part structure of the initial draft is as follows: introductory (ff. 1–3v), main issues in nine enumerated points (ff. 3v–23v), minor issues in five unnumbered points (ff. 24–30), conclusion (ff. 30–31), with an appended final folio (f. 31v), which contains three entries with directions for emendations to, or insertions in the main text (see ff. 6, 8v, and 10v). I have not followed these directions, preferring to retain the main text separate from the emendations in f. 31v.

North himself provides consecutive pagination, using arabic numerals at the bottom, centred, of each folio, recto and verso, beginning with page 1 (f. 1) and ending with page 31 (f. 16). Folio 4v has no pagination, which ceases after f. 16v. I have not included this pagination in the edition, even though it may be important as a guide to changes in the other versions of the letter.

The handwriting is that which North commonly used during his middle period, although the emendations in f. 31v are written in his smaller, neater writing, not unlike that used for all four documents in Add MS 32551. Some words or letters are obscured by inkblots (ff. 29–30) or by tight binding (f. 8v). Other folios contain a significant amount of emendations (deletions and insertions), sometimes making parts of the text difficult to decipher—in particular ff. 5, 8v, 9v, 10v, 11 (first paragraph); but from f. 11v on such emendations are much reduced.

Dating. The watermark of the paper is Arms/LA.[13] Even though the letter is dated old style 20 February 1712/13, a date retained in versions 2 and 3, it is not certain that this date represents the exact day or even the month when the letter was actually sent to Clarke. However, from the evidence in documents (3) and (4) above, we may assume that Clarke received North's letter sometime before 2 June.

---

[13] See Chan, 'Dating the Paper', pp. 54, 91, whose date for Arms/LA is the same as the dates of the letters in volumes 32550 and 32551. Note, however, that RN also used other paper in these volumes; and in any further investigation, the date of these papers would require consideration.

5.2. EDITIONS OF THE MANUSCRIPTS

NO. 1 (DRAFT OF A LETTER) NORTH→CLARKE, 20 FEBRUARY 1713
(UK:Lbl) Add MS 32550: ff. 1–31v

[f. 1]

Rougham, 20 February 1713

Reverend Sir,

I have at last procured and read over your book intituled *The Scripture Doctrine of the Trinity*, but as to the giving a judgment of it, I hope you will hold me excused, for it would be a strang[e] confidence [or] a [*]*misi pari* to censure a theologicall work of yours.

I must owne in the generall that my faith is with that of the Church of England, as establisht by law; a lawyerlike confession, you will say, which under that caracter might be thought sufficient. But that I might not seem to decline wholly answering your expectation, I will venture to be a litle more particular, and avow that as to the doctrine of the holy Trinity[,] which is the subject of your book, I understand it in the plain and obvious sence of the church creeds and articles,[14] which I conceiv carry a sufficient declaration and explication (as may be made to us) both of the words and sense of the holy Scriptures concerning that venerable mistery,[15] and such as hath bin handed downe to us from the primitive Christian times, to which, by reason of a better tradition and notice of the example and teaching of the Apostles then had[,] wee ought to yeild more deference then to any other later humane authority whatsoever. **[f. 1v]** And I say further, that neither by all or any of the texts collected in your book, nor by all the reading and hearing I can value

---

[14] I.e., the Apostle's, Nicene and Athanasian creeds and the thirty-nine articles of the Church of England; see *Articles ...touching the True Religion*, Art. 1 ('Of Faith in the Holy Trinity'), quoted *supra* 4.4.5 n.139, and Art. VIII ('Of the Three Creeds'), which states that the creeds in question 'ought throughly to be received and believed: for they may be proved by most certain warrants of holy Scripture'.

[15] I.e., a religious truth known only from divine revelation, usually a doctrine of the faith involving difficulties which human reason is incapable of solving.

my self upon, have I found any reason to controvert, or vary from the sence of those creeds as they stand in our liturgy, in any particular, and l[e]ast of all that of the holy Trinity in unity to be beleeved and worshiped. But herein I may be more explained afterwards.[16]

In the meantime I cannot but think it strang[e] that any person should interpose his litigating talent in refining and cavelling about formes and expressions concerning the devine nature, used of ancient time, in the Christian church, under pretence of modifying religious misterys to square them with our groveling conceptions, or experiences of materiality, and especially that of the holy Trinity, a subject that is extra[,] or rather supra, to all humane thought or tryall, so as no words can be adequated to it, either in analogy, paralell, shaddow, or otherwise. My sentiment therefore upon the point is, that wee have no power or posibillity of reasoning at all about that matter, but must take the doctrine upon pure faith, grounded upon the authority of sacred text, as it stands there declared, that is, revealed to us, being in all other respects incomprehensible.

As to the devinity of our blessed Saviour in [f. 2] unity and equality with the Father, the authority is most express in his owne words— The Father is in Me and I in Him.[17] What is that but a perifrasis[18] of perfect union and equality as full as words can express it[,] for what porpose significant beyond what the witt of man could have invented? Then— He that sees Me sees the Father,[19] and againe— My Father and I are one.[20] I doe not [*]deal more authoritys, becaus these are categoricall [and] sufficient, nor doe I at present medle with the interpretation you are pleased to bestow upon the words here cited.

But to goe on, the devinity of the holy Ghost is proved by the baptismall comission— Goe and baptize in the name (that is, by the

---

[16] See *infra* ff. 8–10v, *et passim*.

[17] John x.38.

[18] I.e., circumlocution.

[19] John xiv.9.

[20] John x.30.

power or authority) <u>of the Father, Son and holy Ghost</u>.[21] This includes unity and equallity of power for which I appeal to the comon use of speech and language. The Apostle's words are positive— <u>There are three that bear record in heaven, the Father, Son and Holy Ghost, and these three are one</u>.[22] I might produce other passages in the apostolik wrightings, for they run generally in that style and tenor, shewing the three persons in unity to be worshiped and adored[,] of which they themselves are in practise examples. I pass here also the latitude of interpretation you have used, as to these passages.

Here I think is enough positive in affirmance of the church's creeds and doctrine of the holy Trinity in unity. And as for those texts which have bin alledged[,] **[f. 2v]** and are in your book amassed together to impeach it as to your sence of coequality[,] I am fully satisfyed they are all to be understood, not as explications of the devine [*]essence, with analogy to humane affaires and practises in the world, that is, not in a literall but in a similitudinary manner. For who can hear the words father, son, mission, authority and the like pronounced without recurring strait to those ideas that comonly belong to, and are excited by them? Therefore such could not be intended to explain the nature of the devine essence, which is incomunicable by any ideas or words, and that brings it[,] as I sayd[,][23] to be a matter of pure faith, and not of understanding. And when wee pretend, (blind as wee are,) to discerne any glimering of the devinity, wee use other kind of words, such as omnipotent, omniscient, omnipresent and the like[,] which have nothing to do with any created being whatever. But I may prosecute this matter afterwards a litle further.[24]

Having considered that the mistery of the holy Trinity in vnity, as to the so being, is expresly revealed to us, but in no sort as to the

---

[21] Matthew xxviii.19.

[22] 1 John v.7.

[23] See *supra* f. 1v.

[24] See *infra* (point 7) ff. 16v–18v, *et passim*.

manner how, and that nothing appears to shew there was any intention it should be formally revealed, whereby it is matter of faith and not understanding, it is not to be wondered that wee have had in all times so many words made about it, for so comonly it happens, when thought doth not clear the way for speech. As when men labour by words of comon use which one understands to describe things out of all notice, and which no man doth or can understand[,] and which words belong to [**f. 3**] any thing more then that for which they are brought. This is [the source of] some unhappyness in the wrightings and discourses of very good men,[25] as also in your book, which would explaine what cannot be explained, and bring in words, and passages out of the holy Scripture, as direct revelations of an [*]economy of the divine [*]essence, which clearly belong to something els. I may hereafter shew somewhat to be sayd in their excuse,[26] of which[,] if you pleas, you may put in for your share. But thus farr I have gone without much clashing and onely to justifie my owne profession [of faith], as here it is made, which I think every Christian ought to doe in a reasonable manner. But I fear I have overdone it.

As to the designe, method, and performance of your book, I am sensible of so much incompetency to judg that it is farr from me to pretend to it.[27] But since you have required that I should give some sentiment upon the reading it, I shall not altogether stand off, but declining the part of a censor, take that of an opponent, and talk alltogether as an adversary. I shall accordingly collect what objections occurr,[28] leaving it to your self to judg if there be reason in any thing I produce or not. This being a liberty indulged in freindly debates, and academik disputes, you will have the candor to

---

[25] I.e., of the church fathers, as indicated from the cross-reference in n.26 below.

[26] See *infra* ff. 28–30.

[27] RN here and elsewhere (e.g., ff. 4v, 9, 12v, 27, 28) resorts to irony in its original Greek meaning of simulated ignorance.

[28] For the nine numbered principal objections, see *infra* ff. 3v–23v; for the five unnumbered miscellaneous objections, see *infra* ff. 24–30.

take nothing I shall write [**f. 3v**] as reflection [against you personally] but consider it as human antagonist in controversie that levells all his [*]artigliery against you, which will not much vary from the usage in our [legal] trade, for how hard soever wee wrangle, it is not observed wee are the less good freinds for all that.

1. You well observed upon my other paper,[29] that I understood the scope of your book was to set up reason against revelation; and indeed I did apprehend so, having no knowledg of it but from occasionall talk, which I did not much attend to, having then no thought of being engaged to consider the subject matter of it. You added that you thought uncertein reasonings ought to be layd aside, and the revelation in Scripture (of the holy Trinity) ought to have bin kept to. By this one would think wee were perfectly agreed, for I say litterally the same thing, and yet wee shall appear to differ [*]*toto cælo*. For that which you take to be certein reasoning and revelation as concerning that holy mistery, I take to be neither the one, nor the other.

You have indeed, by way of comon place of your owne contrivance, made a pandect[30] of all or most of the words in the New Testament that relate to the subject, which is a redundant paines, and perhaps better spared, for I think it had bin more concise and to the porpose, to have layd open the Bible, and [**f. 4**] bid us read it all over. For passages and words in holy Writt are best understood in their context and relation. And there are also many texts which are of one and the same tenor or tendency, and fall under one and the same construction, and being questionable, are resolved by one and the same note or distinction; so that any one of them may to the porpose be [*]*instar omnium*. Why [then] should so many more of them be set forth at larg? O it is specious, looks bigg, and promises much more then hath bin held forth upon the subject[,] altho wee find really no more weight then it hath borne before. But it is usefull to leav the beaten track, and take a new one, with a banderole

---

[29] I.e., his 1712 answer to Clarke *supra* 4.4.5; for other references to this paper, see *infra* ff. 19 and 30v.

[30] I.e., a treatise covering the whole of a subject.

extraordinary, to seem great to those who think superficially, [*]*et herent in cortice*.

But as to the matter of revelation, which you would be thought to advance, I fear the dealing is like that in Tully, of the Epicureans with the gods, [*]*nomine ponunt re tollunt*.[31] For it appears that you understand literall texts figurately and figurate or allegoricall texts literally [whereby] it is all one which is figurated allegoricall and which is to be understood directly. As wee say (with your leav to recurr to our trade) that if a judg will interpret in the one side and hold the other to the letter [then] which side he pleaseth will carry it[,] and so seems to me that you doe. For where a text pinches your opinion you [*]aneantize it by an interpretation, which you manage after a rationale of your owne, so that a text decisive against you is made to speak your language, as if you were inspired, and not the text. This I think[,] if it appears to be so[,] differs very litle from a setting up of your owne reason against the revelation of the sacred Trinity. For according to my weak understanding that rationale of yours lays aside the [**f. 4v**] doctrine of the Trinity in unity, coequall and coeternall and in the room of it substitutes a politicall plan of a king, prince, and [*]*primier ministre*,[32] that is, three in one, not by way of any humane reconciliation whatever but in verity and in truth according to the termes in which it is revealed to us, beyond which no modifying gloss or interpretation can reach. What else is the consequence of your subordination? But of that more afterwar[ds].[33]

In the meantime it is obvious that such a terrene representation of the cheif mistery of the Gospell concerning the devine nature[,] should it prevaile, hazzards or rather confounds Christianity it self. For, according to that notion[,] I mean subordination and not

---

[31] See the exposition and critique of Epicureanism in Cicero, *De naturâ deorum*, in which the sum of the critique, 1.119–24, is that by deifying nature, the Epicureans do away with the gods but retain them in words.

[32] MS has, at the bottom of f. 4v, the rest of the sentence squeezed in and its lines indented. To the left of the indentation are the words 'only otherside'.

[33] See *infra* ff. 5, 5v, 7, 11v, 20, 23.

equallity in the holy Trinity[,] it is indifferent[34] at best, nay, perhaps unlawfull to worship Christ, pray to or adore him, or to pay to him any other honour or observance, then would be reasonable to pay[35] to any angel, secular magistrate or inferior being, that is, as you say, with a regard onely to the post and office assigned them. For such is the relation with which you qualifie praying to Christ and the holy Spirit, and so you construe the practise of the Apostles themselves. So that if those inspired saints, when they so prayed, as wee find by their wrightings they often did[,] had not your mentall reservation, [then] they were no better than mistaken idolators; for prayers to send guiftes and graces ought to be made to the supream God onely, and not to any inferiour nature or subject whatever. So the holy martirs[,] that rather then comitt idolatry, went out of the world in torment, praying to Christ to forgive their sins, and receiv their spirits, litle thought of being in danger of idolatry in so doing. At best according to your model their excuse must be the same as the more literate heathen made for their polytheisme. For, sayd they, wee worship Vulcane and Neptune as[36] [*]enarches of fire and **[f. 5]** water with regard onely to their offices and ministrations, not hereby derogating from the great superior authority that is over them all.

But now Sir to shew how you imploy your reasoning faculty in the use of this prærogative of interpretation[,] I shall give but an instance or two. The first[37] is that in Saint John— These are three that bear record in heaven and these three are one[,] says the text,[38] that is, say you, one record or testimony and not one [*]essence.[39] If it were so, [then] it had bin sayd (not are, but) agree in one [in which] there is difference sure between the thing, and an agreement of divers

---

[34] I.e., not particularly good; poor, inferior, rather bad.

[35] MS has 'to pay' crossed out, but for sense the words should be retained.

[36] MS has 'onely as'.

[37] MS has 'it'.

[38] 1 John v.7.

[39] Clarke, *SDT*, Pt I, Ch. 4, p. 238: 'Not ... One and the same *Person*; but ... One and the same *Thing*, One and the same *Testimony*.'

in it. If before judges in a secular court three wittnesses swore one and the same [*]fact, [then] he that should say[,] there was not but one wittness and then come off with an interpretation of words [that] carried not that sense as to be understood of the fact sworne[,] would be called a prevaricator. But if we construe the word supposing record in the [scriptural] text to [refer to the three heavenly witnesses] themselves, as I take it to be there intended[,] then it is flatt,[40] that whither you take the testifyers or thing testifyed[,] it is three, and all these [are] one. But grant further that the testimony was given from heaven to us mortalls on earth by the Father, Son, and holy Ghost, which (subordinately) are three not one in heaven, that is, in truth setting aside all humane modes of thinking[,] these three are one, which in the holy mistery of the Trinity in unity, wee are required to beleev and confess.

The next is wors. One? say you, that is, in the exercise of one authority, or the authority of one,[41] as the [*]*paterfamilias* is truely the sole potentate of the family, tho the son and servants act and doe buisness by vertue of it, and so the father, sons and servants are one.[42] These are such groveling shifts, that a devine [*]*ad cleram*, should be ashamed of, [but] what may be done [*]*ad populum*, I shall observe afterwards.[43] I remember [that] Dr. Sherlock had found out,

---

[40] I.e., absolute, undeniable truth.

[41] See Clarke, *SDT*, Pt II, in Prop. 39, pp. 332–3: 'The reason why the Scripture, though it stiles the *Father* God, and also stiles the *Son* God, yet at the same time always declares there is but *One God*; is because, in the *Monarchy* of the Universe, there is but *One Authority*, original with the *Father*, derivative in the *Son*'. Note that monarchy implies a dominion or kingdom; see *infra* ff. 18v, 20.

[42] In ibid. p. 333, Clarke noted that some ancient fathers (e.g., Eusebius of Caesarea) represented the monarchy of the universe 'under very handsome Similitudes: As, that a *Father* in his own House, and his *Son and Heir* in the same House, are not *Two Masters*; because there is but *One Authority*, viz. that of *the Father*, exercised *by the Son*'. For RN's response to Clarke's use of the simile of the household or family, see *infra* ff. 18–18v, 20.

[43] See *infra* (point 4) ff. 8–10v.

[*]*idem cogitans*, as the center of unity; [44] another may say somewhat els, as fancy **[f. 5v]** suggests. There have bin divers of these adumbratory [*]conceipts. Sir John Suckling mentions the spring, fountain and stream, all one water; [45] others[,] the three demensions of one body;[46] and more appositely (if possible), the three facultys of one humane soul.[47] These [similes] have bin found out to answer in a plausible manner, tho not adequately[,] the possibility of a tri-unity without a contradiction and so farr they are different. But nothing within the orthodox pale was ever thought of so low, as a subordination of authority in the exercise of one power, to be taken for the unity in the text. A notion extracted from the frailetys and defects in the human exercise of power is not onely unworthy of the deity, but, [*]*verbo venia*, debases Christ into a near [*]enarch or subgovernour.

Where this way of interpreting into nothing the plain import of a text will stopp[,] short of the destruction of the Christian religion in the world, is hard to define. The Quakers are gone so farr away with it already, as to say, Jesus was a man that had in him a good spirit called Christ, and all good men, that is, Quakers have that spirit within them, which is Christ, whereby they are as perfect and

---

[44] I.e., William Sherlock, *A Vindication of the Holy and Ever-blessed Trinity* (London, 1690). The book was written in answer to those Socinians who, in denying the divinity of the Son, fell into the error of unitarianism. But some critics accused the author of advocating the older error of tritheism, according to which the three persons of the Trinity are three distinct Gods.

[45] For his 'adumbratory' notion, see the conclusion of his essay, *An Account of Religion by Reason* (London, 1636). Note that Suckling was a friend of RN's grandfather, Dudley, 3d baron North; see Beaurline, 'Dudley North's Criticism', pp. 303–4, 312.

[46] I.e., John Wallis, who, in *The Doctrine of the Blessed Trinity* (London, 1690), defended the consubstantial doctrine against Socianism by comparing the Trinity with a cube that has three dimensions but one geometrical shape.

[47] I.e., Augustine of Hippo; see the last eight of fifteen books of *De Trinitate* (written between 399 and 419), in which he explored the possibility of understanding 'three in one' by a series of analogies drawn from human psychology.

adorable as he was.[48] It is true, you doe allow the mistery of the holy Trinity in unity so farr as to owne this subordination and delegated power of the three persons was not in priority of time [but] from eternity, the manner of which[,] you say truely[,] is not revealed. No, nor was any manner at all in the [*]essence of the holy Trinity, [—] offices, imployments superiority **[f. 6]** or inferiority, power originall or delegated, or any modifications whatever [—] ever revealed or for ought appears designed to be revealed[,] as may be fuller made appear afterwards.[49]

2.[50] In the mean time I shall observe you doe not deal fairely with this text— these three are one.[51] For you say it is suspected, and that the Greek manuscripts want it, and it is not to be much relyed on in controversie, which your integrity would not permitt you to conceal. But you were permited to give us a true state, or crittique of that matter, if you had so thought fitt. And it was reasonable to have done it, that your reader[,] learned or unlearned[,] might have had somewhat more then your bare word to rely on in a caus of such consequence to the Christian faith and profession.

---

[48] For RN, sectaries like the Quakers confuse the natural and the supernatural in their accounts of spiritual knowledge, whereas he seeks to maintain a sharp distinction between the two, as he hints *supra* ff. 1v–3 and as he develops more fully *infra* ff. 16v–23v, *et passim*.

[49] See *infra* (points 7–9) ff. 16v–24.

[50] MS has in the left-hand margin an instruction to replace a portion of text with revision 'A', for which see *infra* f. 31v. The portion to be replaced is the paragraph that follows (which begins with the words, 'In the mean time', and ends with the words, 'Christian faith and profession').

[51] I.e., 1 John v.7. For his gloss, see Clarke, *SDT*, Pt I, Ch. 4, p. 238: 'Though it ought not indeed to be concealed, that This Passage, since it does not certainly appear to have been found in the Text of any Greek Manuscript, should not have too much stress laid upon it in any Controversy.' In the second edition, Clarke, *SDT* [1719], p. 121, added the name of John Mill to his categorical statement that 'as to the *Greek* Manuscripts', it has 'never yet been proved to be found in the Text of *ANY ONE* of them, elder than the Invention of Printing'. According to Foxe, *John Mill*, p. 72, the longest note in Mill's 1707 edition of the Greek New Testament concerned the text in question.

I have read in some English book,[52] that the Arrians[53] [*]rased
out of some manuscripts that text, and the [*]rasure is yet to be seen,
so it might well be wanted there. They had in their faction (then
called heresie) all the power imperiall and episcopall, [and] onely a
few zealous bishops continued to oppose them, whom they [the
Arians] persecuted bitterly. And it may as well be supposed that they
supprest, as that the Catholiks[54] (for so the other side was then
called) forged texts in their owne favour, unless wee should be
perswaded that all the honesty lay on the Arrian side, and none on
the other. They who [*]raged so heathenishly against the persons of
the Catholiks [**f. 6v**] may more probably be thought to divert some of
their fury upon the sacred text, till they brought it to their [*]gree[,]
which their opposers could never be brought to.

It is to be observed that this doctrine of the holy Trinity in
unity and equality is not among those [matters] as (by abuse) have
crept into Christianity to subserve any[55] ends of greatness or wealth
of the clergy, but [rather is] a matter of pure faith and devotion
without any respect to temporall consequences; and the hoary noise
of preistcraft cannot be brought against it.[56] Nor is it to be conceived
why the ancient [church] doctors should suffer as they did in the caus
against the Arrians, with a constancy that in the end carried the
ground, if there had bin fraud at the bottom by imposing a fals text
which in those days must be knowne when almost every Christian
had a Bible in his hands.[57] The fraud (if any) was more likely to be

---

[52] Numerous divines treated 1 John v.7, so without more information it is not
possible to identify this 'English book'.

[53] I.e., lit., followers of Arius; see *supra* 4.4.5. For some non-literal meanings up to
1713, see Wiles, *Archetypal Heresy.*

[54] I.e., those belonging to the ancient Christian church, as it existed undivided,
prior to its separation into Eastern and Western Christianity.

[55] MS illegible.

[56] I.e., the cry of the deists that after Adam's fall, human self-interest manifests
itself in the practice of priestcraft—the so-called 'imposture theory'; see Harrison,
'Religion' and the Religions, pp. 68, 72–85.

[57] Perhaps; but how many early Christians were literate?

on the side of power and oppression, then on the suffering side[,] against which nothing could be desired better then [that] such a charg had bin [made]. But if the Catholiks then [*]paumed a text, [then] you have set the matter right by slurring an interpretation. I wish therefore this insinuation had bin left out of your book which some may find savours of the Socinian.[58]

But wee must not let it goe so, for wee look upon the authority of this text to be finally determined and setled upon 1400 or 1500 years tryall, and now [*]*transit in rem judicatam;* nor ought it to be so much as reflected on, till those learned books they say are wrote and published in the controversie, be answered, **[f. 7]** which I doe not hear is or is likely to be done yet.[59] And therefore wee insist that wee may keep our Bibles as they are handed and delivered to us by the Christian church since the primitive times, free from aspersions[,] at least not to depreciate the doctrine of the holy Trinity, the rather becaus all the [*]batterys of the [*]theisticall and atheisticall people are now a days levelled, and swarmes of paper pellets continually flirted at it. Their management is [*]artificiall, for when a text is not to be altered what cours is left but to overturne it all at once, [*]*una litera potest*. But you have taken another cours, having left it to stand in your list of texts but with a black mark. For you had an interpretation at hand to bring it to nothing, and the reflection having prepared the pallat of the reader[,] your other must needs goe downe the better, and well was it for the poor text [that] you were so well provided of an expedient.

3. And some persons that are not your freinds may take an umbrage to caluminiate[60] your conduct as not ingenuous, or upright, but savoring a litle of art or trick, becaus you attribute the concern of

---

[58] For Clarke's sense of the term, see *supra* 4.4.3. n.68; for various other meanings, see *supra* 4.4.5.

[59] Books on 1 John v.7 continued to be published long after 1713; see, e.g., [Thomas Turton], *A Vindication of the Literary Character of the late Professor Porson, from the animadversions of the right reverend Thomas Burgess ... in various publications on 1 John v.7* (Cambridge, 1827).

[60] I.e., to charge falsely and maliciously; to slander.

this doctrine to us Protestants in especiall manner, as if all Christians whatever had not as much reason to have a right faith as wee Protestants. But by this, you seem to insinuate as if the doctrine opposite to your subordination, [th]at is, the unity and [*]ømousios, were popish inventions **[f. 7v]** or at least some pretension of argument for their impositions.[61] This is a figure [of speech] called a [*]wheadle, to catch the people as fish to an opinion with a popular bait. It is well known that this doctrine was solemnly setled[,] I think they say about the third century.[62] And then the Pope himself, if I may so say, was a Protestant; and as many centurys were past before any of the abuses took root in the Romish Church,[63] for the caus of which wee have most [*]justly and necessarily devided from them. But how? Not by running away so much further from Christianity then they have gon[e], like some pretended reformers who run they know not whither and are running on still, but by stepping back into the times primitive and antecedent to all these [Romish] abuses, and there wee find our church's doctrine of the holy Trinity in unity. And should wee now leav that holy mistery or debase it into a most worldly [*]economy, for fear the Papists should make a fallacious

---

[61] Clarke used the word 'popish' in relation to the Scholastics; see, e.g., Clarke, *SDT*, Introduction, p. xxv ('the Popish Schoolmen'). He also claimed, ibid., Pt III, Ch. 2, p. 465, that the Schoolmen's definition of '*one Individual Substance*' as one subsistence or one person only was equivalent to the modalist 'Notion' of the third-century Christian priest and theologian, Sabellius, who held that the Persons of the Trinity were not distinct but only different aspects or modes of one Being.

[62] I.e., after the accession in 379 of Theodosius I as Roman emperor of the East, who in 381 summoned the Council of Constantinople, from which the Nicene creed derives. According to Kelley, *The Human Measure*, pp. 53–4, his Code was designed 'as a complete "guidance for life" for the Christian community of the West'; and Theodosianism in general 'not only softened the rigors of ancient Roman law and showed favor to the Barbarians but also marked, along with the demise of paganism, the beginning of religious uniformity'. For a less favourable viewpoint, see Wiles, *Archetypal Heresy*, pp. 32–4, 45.

[63] I.e., ambition, avarice and profligacy during the Renaissance, the consequence of which eventually produced the Reformation, thereby robbing the papacy of spiritual authority; see Russell, *A History of Western Philosophy*, pp. 495–503, *et passim*.

use of it, our case would be wors then those who— [*]*dum vitant [stulti] vitia in contraria currunt.*

The onely countenance that I ever heard the Papists took in the controversie against us from the doctrine of the holy Trinity hath bin in the caus of transubstantiation.[64] For say they[,] you hold the Trinity which is you confess a mistery but reconcilable to humane understanding, why then should you differ from us that hold transubstantiation, becaus it is a mistery. **[f. 8]** This is a shallow reasoning at best; but the pretence hath bin exposed past shame, and in this [*]minuit[65] of difference it is dissolved. The subject of the holy Trinity is without our knowledg but by the means of revelation; but the subject of transubstantiation is the proper object of our sences by which we know all things that are immediately communicable to us. Where lys then the danger of arguing from the one to the other in any respect? And why should the glory of the holy Trinity be eclipsed by bringing sensible representations afore it, least the Papists take occasion to argue weakly? All this I have bin moved to say against that politick consideration of yours directed to us Protestants by way of prevention, as you pretend. For wee are not ignorant of what force or charme the word popery is when a turne is to be served.[66]

4. There is more matter scattered towards the end of your book in divers places,[67] which hath some affinity with the nature of this,

---

[64] I.e., the doctrine that in the sacrament of the Eucharist the bread and wine are converted into the body and the blood of Christ. According to the Church of Rome, Christ's body and blood are literally present, whereas in the Church of England, they are present only spiritually and, hence, defined in figurative terms.

[65] RN uses the same spelling in versions 2 and 3 of his letter.

[66] After the accession of James II, who was ambitious to advance the cause of Catholicism, the deep-rooted anti-Catholicism, evident in English politics between 1673 and 1678, once again came to the fore and along with it a renewed anti-scholasticism. In the latter year numerous Roman Catholics were executed as part of the so-called 'Popish Plot', which elsewhere RN treats as a libel; see RN, 'Of Etimology', p. 222 n.52, and Kassler, *HRN*, pp. 363–4.

[67] I.e., Clarke, *SDT*, Pt III, Ch. 2, pp. 415–80, which includes his comment on both Nicene and Athanasian creeds; see further *infra* ff. 25v–28.

which is popularity, and for that reason I will not stay, but speak with it alltogether now. It is where you recomend a reformation to be made in the publik liturgy in order to reduce it to the plainest [*]formes, and clear to all capacitys, as you say the ancient [ones] were. But more particularly [you recommend] expunging the Athanasian creed,[68] and I suppose [also] the [*]*omousious* of the Nicene[69] [**f. 8v**] whereof the drift seems to be, that the doctrine of the holy Trinity in unity should be setled and declared by the nationall church according to the mould of your interpretation by way of governour and subgovernourship. And why? Least by us[ing] [*]formes of speech that carry us beyond our clear understandi[ng,] Christians should not all agree, and then divers persons be[come] dissatisfyed, and sc[h]ismes perpetuated to the end of the world. [*]*Belle & beau.*[70]

I beleev no reasonable person would think mu[ch] if by lawfull authority[71] and in a regular way any fittin[g] alterations were made in the forme of the liturgie, tho it extende[d] even to drop the creed called the Athanasian, which is not essentia[l] to the service and might have bin left out at first without being imposed, as y[ou] improperly speak,[72] for the word impose belongs onely to the Apostles creed.[73] But whither it be a proper method for a reverend devine in the first place to fall out with the liturgy in print [is]

---

[68] Op. cit., p. 448, who urged 'those in Power' to consider seriously whether the Athanasian creed 'had not better be quite omitted'.

[69] Ibid., pp. 470–1, who supposed that requiring 'men to forbear the use of' unscriptural expressions would have been most effectual for suppressing ancient schisms (and by implication, modern ones).

[70] MS adds: 'Vide B', referring to the words that should be added here from *infra* f. 31v 'fol. 16. B.'.

[71] I.e., either by Convocation or by Parliament; see also *infra* ff. 10, 13v, 26v, 30 and *supra* 4.4.4 n.92.

[72] Clarke, *SDT*, Pt III, Ch. 2, p. 450.

[73] In f. 27v of version 3 of RN→SC 1713 (UK:Lbl) 32551: ff. 2–33, RN indicates that the word 'impose' refers only to a test for reception into the Church of England, namely, baptism, when the Apostle's creed is read.

without[,] and I think against law, charging it (if not construed after
your particular manner) with litle less then heresie, or it may be
wors[,] idolatrie or imposing (that is your word) a fals creed (I think
you say or quote somewhere that the Athanasian, as it is exprest[,]
cannot be [true] or [words] to that effe[ct]⁷⁴) and then setting forth
the sighs of good people, that things might be regulated into greater
plain[n]ess, with storys of attempts towards it by great and good men
which alas came to nothing.⁷⁵ I say whither this be a decent
proceeding and consonant to a desire of unity and to avoid dissention
and sc[h]ismes when wee have authoritys to applye to, I submitt to
your owne reflection, for I will lay no charg upon what declares it
self.

As to the liturgy, tho I am unworthy **[f. 9]** to be an advocate in
a caus of that nature, yet I shall venture so farr in its defence as to
distinguish the two respects in which it is considered: first[,] the
matter of faith and doctrine declared or confessed in it; and then
next[,] the forme and style of it. As for the latter I doubdt not but you
were satisfyed of its excellence even to your owne hearty wish, who
have recomended plaineness, which vertue shines in it to perfection.
I have often attended sollicitously to observe if I could find one word
ambiguous, or that had an incertein sence, indecent or capable of a
construction to any porpose foreign to the designe of the [church]
service, and I never could meet with any one such. Which is an
excellence very few books in the world have a [*]just pretence to. So
that I am sure as to plaineness in the style and manner of expression
it needs no correction.

Then it seems that somewhat is amiss in the doctrinall part
which is to be sett right by expunging the Athanasian creed. Nor is
the Nicene creed safe. And your expressions have such latitude as

---

⁷⁴ In his comment on the Athanasian creed, Clarke did not say this explicitly; but
in his comment on the Nicene creed, he did state that if the word *omousios* is
understood to signify either one substance or one essence, then in those senses,
'strictly and metaphysically taken, 'tis plain it cannot be True'; see Clarke, *SDT*,
Pt III, Ch. 2, p. 465.

⁷⁵ See ibid., p. 450; see also *infra* ff. 26–26v.

wee have reason to fear that is not all, for the same quarrell will extend to the address in the litany,[76] the *gloria patria*, the suffrages, the *kyrie eleason* and many other passages wherein devine honnour is payd to the Persons of the holy Trinity distinctly. All this, and no body can tell how much more must goe out before the reformation you propose **[f. 9v]** or recomend can come to perfection. And when that is done, it will be a miracle if there be not abundance of reformers more [who] rise up one after another, as dissenting sects are apt to engender and swarm, when ye [*]*barriere* of the church liturgy is broke[,] [but] which I look upon under providenc[e] to be the stay of the Christian religion in England. And with like priveledg as you have taken, and for ought I see with as much reason[,] I say a perpetuall succession of reformers may arise and demand away [with] all the rest and [*]grutch if they be not satisfied. For all your reconciling and complaisant overtures will hold good, so long as any persons shall pleas not to be satisfyed.

Now the liturgie may be [*]convict of error in doctrine, in some or other of its parts[,] which by its most [*]animose and pestilent enimys hath not yet bin done and I presume cannot be done. Nor has the Athanasian creed it self, upon which the present controversie bears[,] been confuted by any thing I have met with objected to it; and you yourself have found out [*]salvos for the difficult parts of it. If this be so that the liturgy doth not err in doctrine, then I say every thought of making alterations expressed so as to countenance any pretence, as if it did so err, is altogether unjustifyable, and all attempts that way are unreasonable and with defyance to be resisted. For there is a great difference between points of Christian doctrine and faith, and matters of behaviour and discipline. The latter may be ordered **[f. 10]** in any manner as being [*]indifferent in themselves, but doctrine is not so, nor is it in the power of the governours of the church or state together to dispose of

---

[76] I.e., an appointed form of public prayer, usually of a penitential nature, in which the clergy lead and the people respond, as in the *gloria patri*: (Lat.) glory be to the Father; the versicles and responses called 'suffrages'; and the *kyrie eleison*: (Gk.) Lord have mercy upon us. For his comment on the litany, see Clarke, *SDT*, Pt III, Ch. 2, pp. 454–61.

them. And if all were so much of a free thinking principle to chang[e] such doctrine by any law or canon whatsoever [then] it were [*]cass, [*]*ipso facto* and null, and no person obldged by it, but on the contrary obldged to hold the true faith, whatever all the powers of the earth say to the contrary.

And it being so as to the positive part[,] it follows with litle less force in the negative, viz., that if every part of the service of the liturgie be grounded upon any doctrine of the Christian faith, as for instance it is upon that of the holy Trinity in unity and equality[,] then to alter the forme so as to lay [it] aside [or] so much as to discountenance that doctrine, would be, in my thinking, impious and anti-Christian. And [though] admitting that the Athanasian creed be not essentiall and might have bin omitted at first,[77] yet[78] it is publikly demanded [to omit this creed] as if some fals doctrine were conteined in it, or for complaisance to those that pretend to think so, which falsity cannot be proved by any man or men living. **[f. 10v]** Nor will it I [*]hope be undertaken by any, since you have taken so much paines to clear it. I say it is absolutely unfitt, and against all reason and conscience, now to drop it. For the so doing[,] which is that you are pleased to recomend[,] would give a countenance (it is a sort of reasoning used by yourself) to those who oppugne the Christian faith in the holy Trinity as it is in some manner confessed thro out the whole liturgie; and it would be construe[d] a giving up not of Saint Athanasius but of the sacrosanct Trinity. And after th[at is] done there would remain no reason at all to retein any of the expressions that depend upon it, altho taken out almost verbatim from the wrightings of the holy Apostles themselves.

Therefore Sir, the reformation you plead for is unobvious reasoning[,] very ill timed at best[79] and very unhappily introduced

---

[77] MS has a partly illegible insertion: 'and so by due ...ing[?] be yet left out,'.

[78] MS has 'yet since'.

[79] I.e., ill-timed because of the present state of affairs described in f. 28 of version 3 of RN→SC 1713 (UK:Lbl) Add MS 32551: ff. 2–33: that Arianism 'as now it stands' is 'more treacherously drest up under other denominations, as

upon a reason that makes absolutely against you,[80] whither all powers and potentates in the universe say yea or nay to it, and no hair scruple or iota of that kind ought to be abated to pleas any of them. The Church of Rome hath bin charged for compounding with the heathen idolaters[,] and I doe not see how the Church of England can be excused in compounding with any sort of unbeleevers, or such as[,] whatever they preten[d,] have no religion at all.

5. But then it is sayd [by you],[81] why must such misterys in religion be imposed in the publik service as men ordinarily cannot reconcile to their rea[son], which makes many (as was sayd[82]) stand off from conformity with the **[f. 11]** church service, and others fall off [from the church itself]. Were it not better that such stumbling blocks were removed out of the way? That is, that the service should not containe or impose (as your word is) any doctrines, or use any expressions that are not plain and familiar to be understood, such as all might agree in, which you say might be if nothing were put forth but what was conteined in the Scripture, as some suppose the Athanasian creed, etc. is not; that matter I may consider afterwards.[83] But at present the buissness shall be with the people's understanding and unanimous agreement.

The meaning of this is that the Scripture should be so construed, as that the subject matter may be understood, and the liturgy must square with such interpretation, as nothing may be misterious to give offence to the self-conceited vulgar or to [*]opinionative doctors. It is very obvious to comon reflection what dismall consequences attend this sc[h]eme [of yours], which pretends to mould and [*]levigate the misterys of the Christian religion for

---

Socianianism, Whistonisme, theism [i.e., deism], and free thinking', all of which are 'now raging in print as well as in promiscuous discourses'.

[80] MS adds '.C.', referring to the sentence to be inserted from *infra* f. 31v ('Fol. 20 C.'). But the mark for the insertion should have been placed at the end of the present sentence.

[81] Clarke, *SDT*, Introduction, pp. xxviii–xxx.

[82] Ibid., Pt III, Ch. 2, pp. 447–8.

[83] See *infra* ff. 13v, 25v–28 and 29v.

complaisence to humours, however specious and demure they may
be. For if the mistery of the holy Trinity in unity must be interpreted
into nothing and rescued[,] as you[84] pretend[,] from implying a
contradiction as it stands expressed in the liturgie, [then] it is plaine
that the same way of reason[ing] and discours may be carryed on to
the abolishing of all religious worship whatsoever.

[f. 11v]

　　　For religion itself, and everything that belongs to it, as to our
capacity, is as misterious and hard to be understood as the holy
Trinity is. As if wee consider but the soul of man, or any spirituall
being without any regard to body, [then] what can wee make of it?
Much less how it can be united to or move a body. What then is the
devine [*]essence and its allmighty power? What think you of
creating out of nothing and annihilating? May not a freethinker with
his scandalous liberty, if he pleas, as in other blasfemous instances,
say that such a power implys a contradiction, and cannot be, and that
the comon prayers ought to be purged of all expressions that suppose
it; otherwise he cannot (as he may) say[,] away with such
impositions. And all this with as much [*]colour as you are pleased
to regard those who cannot [do] away with the holy Trinity in vnity.
If you say that other[85] is revealed in Scripture, [then] I say this is as
expresly as that, or any other doctrine is. But then, say you, wee find
reason from some termes of subordination to qualifie the latter. I
reply then, that I may as well pretend from some termes of infirmity
to contradict the other. There is no mean therefore. [For] if any one
religious mistery must be melted downe for reconciliation or
compromise with sc[h]ismaticks or [*]humorists for no **[f. 12]** better
reason then that wee cannot understand it, [then] all the rest [of the
mysteries], even Christianity it self, and the comon principles of all
religion whatever must flow away with it.

　　　It is not enough to say, No, wee will not carry it so farr. I desire
to know, how it will be stayed. [*]*Eadem ratio, eadem lex*. If wee

---

[84] MS has 'they'.

[85] I.e., the almighty power to create out of nothing.

quitt one mistery, or which is the same thing, interpret away the misterious part of any Christian doctrine [then] is there any answer to those who demand another, or for brevity[,] religion it self to be dissolved? Therefore let us not pretend to modifye that revelation which is given us of the devine [*]essence, or any other [*]essences, of which wee can have no notion or imagination, but in the very termes in which they are delivered [in Scripture]. Revelation of that sort will bear no interpretation att all, for by what measures should wee judg of it? And it is a gross error to pretend to any such interpretation, much less thereby to file and pare a mistery revealed to make it square with the scanty model of our [human] capacity. I doe not here enter upon the caus of the holy Trinity in unity, as it stands expressed in divers places of our liturgy[,] intending a place particular to it.[86]

But I shall in the meantime take the freedome to remarque that things will not lye tho men may[,] and the latter even in so doing discover their **[f. 12v]** falsity by the nature of those things they medle with. As for instance, the declared enimys of all revealed religion doe not or very seldome apply their force against religion it self directly, or say there is no revelation of any; but they fall foul upon the misterys of our faith, and endeavour to demolish them, supposing that if they can once prevail to have misterys layd aside, or which is as bad, or wors, interpreted into worldl[y] resemblances, and so reduced to nothing, that then all revealed religion must tumble. For the argument is [*]thro-stitch't, and must conclude, all or none. Witness that self-declared imposter that 'tituled his book, *Christianity not Misterious*,[87] a falsity held forth in the very front of his book, for not onely the Christian, but all religion whatever is, and must of necessity be misterious. And all the rest of the atheisticall

---

[86] See *infra* ff. 26–30.

[87] I.e., published anonymously in 1696, the author, John Toland, argued against the supposed tyranny of the priest class, who used mystery to keep the faithful in ignorance of the true Gospel, whereas he proposed that in the Gospel there is nothing contrary to reason or above it and that no Christian doctrine can properly be called a mystery. Elsewhere, RN, 'Of Etimology', pp. 284–5, refers to the book and comments on Toland's successors, the deists.

society, who make not religion in generall the subject of their
derision, but onely the misterious part and especially what concernes
Christianity, they know their work well enough. But why any
devines of the Church of England, if any such be, should at this time
of the day assist, and as it were chime in, as if they were in
compliance with such fellows, is past my understanding to resolve.

If it had not bin found by experience that most if not all
projects of late years[,] [though] advanced in sheep's cloathing, for
amendment of the liturgye upon the **[f. 13]** pretence of larger
comprehension,[88] were at the bottom but treacherous and wolvish,[89]
[and if those projects] demanded onely that matters may be ordered
any way, as ceremonys, behaviour, and [liturgical] formes, for which
reason and upon manifest discovery that the power and revenews of
the church and not the formes of the service were the beloved [and
that] there hath not bin any such chang[e] made [in the forms
themselves], [then] there were a litle more reason to hearken to these
familiarizers of Christianity.[90] Tho, as I sayd,[91] in truth, there can be
no reason, less then a conviction of error, such as induced and yet
iustifies the reformation, to alter the doctrinall expressions one
scruple [in order] to pleas any one man or number of men. Wee allow
those anti-cermonialists might have more [*]colour in demanding the
alterations pretended to be so reconciling. But at this time when the
mouths of all the atheists and anti-Christian [*]theists are wide open
bauling against the revelation and misterys of the Christian faith,

---

[88] I.e., inclusion of the sectaries.

[89] RN here, ff. 12v–13, and *infra* f. 26v, draws on an idiom of biblical origin; see
Matthew vii.15.

[90] In MS this long sentence reads: 'If it had not bin found by experience that most
if not all projects of late years advanced in sheep's cloathing, for amendment of the
liturgye upon the pretence of larger comprehension, were at the bottom but
treacherous and wolvish, demanded onely that matters may be ordered any way, as
ceremonys, behaviour, and formes, for which reason, and upon manifest discovery,
that the power and revenews of the church and not the formes of the service were
the beloved, there hath not bin any such chang[e] made, there were a litle more
reason to hearken to these familiarizers of Christianity.'

[91] See *supra* (in point 4) ff. 8–10v.

there is no manner of reason to expect a better hold of religion for complyance with them, or with such as speak their language.

On the other side the reason is strong that (untill errors are demonstrated) all men[,] who under title of the holy mistery of Baptisme wear the name of Christian, should unite in a firme resolution to defye all these familiarisers and **[f. 13v]** to yeild to them not one jota of doctrine, nor word or syllable in the liturgie that depends on any such. Much more is it the charg of our ecclesiasticall superiors to be carefull to give no countenance to such boldnesses. And if I had the liberty of making a motion to the reverend assembly of the church's representatives,[92] [then] it should be not to palliate or take away the trinitaria[n] [*]formes of devotion in the liturgie and the[ir] expressions in the creeds, particularly those in the Athanasian, but on the contrary to declare with as much zeal[,] as the author of [*]*Quicunque vult*[93] well affirms, that whoever is informed of the holy Scriptures and taught the misterys revealed in them, and especially that of the holy Trinity in unity, must so beleev of it, if he will be saved. This would carry a face of primitive zeal; and if the impudence of such as those times called heriticks could possibly justifie such positive expressions, [then] I am sure the audacity and effrontery of those wee in our times call atheists and [*]theists demand it.

So litle reason is there in my opinion to modifye the misterys of the Christian religion, in hope to satisfie nice pretenders, under which mask lurk the utter enimyes of all religion. Doe wee not see the Quakers have renounced the holy Sacraments of Baptisme and the Lord's Supper **[f. 14]** and in all their discourses, and most of their wrightings blaspheam them? Doe wee hope to gain them by

---

[92] I.e., Convocation, consisting of an Upper and Lower House in which the church's representatives deliberate on ecclesiastical matters. On 2 June 1714 the Lower House issued a complaint against Clarke's book, and thus began a series of deliberations involving both Houses, after which a resolution was issued on 7 July; see *infra* 5.5.3. For details, see Ferguson, *ECH*, pp. 83–97, and Wiles, *Archetypal Heresy*, p. 114.

[93] The first incipit of the Athanasian creed and, hence, its Latin name; see also *infra* f. 26.

introducing lay, that is, no baptisme, and renouncing the real presence,[94] because wee doe not comprehend the manner and efficacy of either? The Arrians in old time betrayed Christianity to the Turks,[95] where that heresie lives at this day;[96] and doe wee think that [*]Arrianizing the doctrine of the holy Trinity will make any to become Christians that are not so? The method of proceeding in that manner by delivering up [*]*tradendo* the misterys of our religion[,] whither directly or by soothing and humanizing interpretations, is so farr from at reconciling, that it exposes Christianity; and in consequence a fairer [*]colour is had to deride such [*]traditores then can possibly be by the worst of enimys found out to depreciate any Christian mystery or to despise such as have most rigorously asserted it.

All this Sir, I have bin moved to say against that politik tendency of your book, directed so fairely and plausibly to us Protestants by way of antidote against papall abuses, and for [*]wheadling into conformity the sect of free thinkers (for I know none, out of that caracter, who doe allow of misterious expressions). And making the publik service so plain, as you pretend[,] any one may be **[f. 14v]** satisfyed with, that is, most plainely and clearly understand every expression in it. And where the termes hard to digest are taken into the liturgie out of express Scripture, to interpret

---

[94] I.e., of the body and blood of Christ; see *supra* f. 7v n.64.

[95] According to Russell, *A History of Western Philosophy*, p. 333, regional feelings which had seemed extinct since the Roman conquest were revived by the Arian controversy, so that, e.g., Constantinople inclined to Arianism, whereas after the Arian controversy ended, new controversies 'of a more or less kindred sort' arose that impaired the unity of the Eastern empire and facilitated the Mohammedan conquest.

[96] See f. 86v in RN, 'Of the clergy of England' (UK:Lbl) Add MS 32526: ff. 79v–87v (early period, probably written shortly after he became a non-juror): 'if the story of Mahomet were not super induced [then] a Turk were a reall Arian Christian of reasonable faith in most points but that of the devinity of our Saviour'. The chief, but not the only source for RN's interest in the law and traditions of the Turks is his brother, Dudley, who spent many years in the Levant, including a period in Constantinople; see RN, 'Of Etimology', pp. 167, 200, 300–1, 313, 323, *et passim*, and RN, *Life/DN*, especially pp. 49–56.

that so as shall signifye nothing or wors [is] to debase the devine [*]essence into humane [*]formes. I say Sir, I have bin moved to alledg thus much against the general designe of your book, as la[u]nching out farther towards incouragement of [*]theisme and irreligion, then I [*]hope you are well aware of. But I shall goe on.

6. I observe the method of your book hath a spice of that wee call mathematicall, for you have a grand postulatum,[97] and then data or axiomata,[98] which wee must grant, and to them all the following reasonings or articles, after the guise of propositions, theoremes, sc[h]olia, and corrolarys, are numerously referred. I doe admitt that very sound reasoning may be cast in this mould, but withall, my fancy is that in matters controversiall or perswasive, it is a better way to subjoine the proofs immediately to the proposition; for so the reader is less intangled and troubled then to goe to and fro comparing one thing with another. It hath bin my observation that the mathematique method out of its element, which is [*]pure and pute quantum, hath proved more [*]obnoxious to fallacy, and particularly that of [*]*petitio principi* then any other. I am sure that if I would instruct a fallacious [**f. 15**] argument, [then] I would begin with data and postulata and plant the fallacy there, and conclude [*]*Quod. E. D.* Perhaps some might not be aware at first of the use intended to be made of the data, and then rather then goe back to a reexamination of the principles half forgott, pass foreward and think the proof very strong. As I remember in Westminster Hall,[99] the judges would not let a man argue the law [so as] to adjust that before he had stated the case, but [would] hold him to state the case first, and then apply the law as he might. For one cannot tell what sort of

---

[97] I.e., *sola scriptura*; see Clarke, *SDT*, Introduction, pp. iii–iv: 'in every Inquiry, Doubt, Question or Controversy concerning Religion, every man that is sollicitous to avoid erring, is obliged to have recourse ... to the original Revelation'.

[98] I.e., that which needs no proof, namely, the digest of texts in ibid., Pt I, pp. 1–240.

[99] I.e., a hall in the Palace of Westminster, built *c.*1099, in which the common law courts sat until 1882; see RN, 'Of Etimology', pp. 245, 301, 335, 336, and Kassler, *HRN*, p. 21.

scruples in the law may be to the porpose, till it is knowne to what sort of [*]fact it is to be applyed. Now Sir, how farr I may challeng your method upon these considerations is next to be considered.

First[,]^{100} the proceeding seems very [*]artificall if not ensnaring to the unwary, for some matters are not capable of demonstration, as for instance, principles themselves. For in all argumentation some matters must preceed others, els how should they^{101} prove them? And what should prove those which are first of all? Therefore where the question perpetually recoyles back upon the principles and is [*]couched wholly there, the synthesis^{102} of the argument is all trifling and comes to naught. But having the shew of a shrewd chaine of argument, [it] looks bigg; and out of pure [*]visage, gaines a careless observer, who is not apt to goe back to distinguish upon the principles, and so is caught.

You have one grand postulatum, which is that wee are not to admitt any doctrine, which is not conteined [**f. 15v**] in the holy Scriptures; and upon this all your fabrick is built. And you are very much in the right for choosing so firme a foundation, for it is a maxime that all Christians in the world agree in. But now if the question happen to be, what is conteined and what not, then wee are at difference again[,] and your postulatum is meer [*]fucus and comes to nothing, only it has a popular air, as if you brought nothing forth but what will be agreed to as readily as that. This is also a stroke of art, and such are very useful in oratory, where perswasion is more aimed at, then display of truth. And the reverend devines have had somewhat to doe, to ferret the sectarys out of that subterfuge. For, say some sectarys, shew us the surplice, infant baptisme, and divers other matters (which your self owne, and practise) in the Scripture. Therefore the advancing this generall reference to the onely judg and rule of faith that wee can recurr to, and which all

---

^{100} I.e., to begin with. (There is no 'second' or further enumeration.)

^{101} I.e., those who argue.

^{102} I.e., in logic, the action of proceeding in thought from causes (discovered) to effects or from laws or principles to compositions.

agree in, is to look bigg and perhaps sooth[e] some sectarys with nothing at the end on't. For all the difference remaines[:] what is, and what is not contained in the Scriptures.

Then follows your axiomata, the comon place of texts, from the generall view of all which, as from a majority by voting, you inferr your propositions[103] and referr back by fiftys and hundreds to the texts. Now, Sir, either you take the texts in a literall sence or [you take them] with an interpretation. For many texts, taken literally, cannot be true, otherwise then as they are that of figures of speech or **[f. 16]** as if it were sayd[,] I am like a door, and like a vine,[104] with such extent of application as suits the discours, or designe of speaking. Wee should have bin very glad, if you had pleased to have made this distinction one of the heads in your catalogue of texts concerning the devine [*]essence, so as wee might have understood how much of them was to be taken literally, and what figuratively or with similitude to humane state and condition, [for by this understanding] wee had received more satisfaction then is like to be gathered from your book.[105]

Your postulatum is generall,— viz., conteined in the Scriptures. You must allow here wants— according to a just[106] sence or [an] interpretation of them. Els I deny your postulatum; and then downe falls your fabrick. But adding that distinction all is fair and well, and to this I expect you will agree. But it is yet strang[e] to me that you should make a shew of such strikt adherence to the texts, even to preciseness, I had almost sayd, in the style of a sectary[,] and yet [*]*in petto* retein this distinction of a just sence and [an] interpretation, as if you reserved to your self alone that liberty, and no one els was to medle at all in that way. For I observe you are very buisy with interpretation where you are pinch't, but all the rest that bears a shew on your side **[f. 16v]** you lett goe without any note or

---

[103] I.e., the fifty-five 'Propositions' in Clarke, *SDT*, Pt II, pp. 241–378.

[104] See John x.9; see also John xv.1 and 5.

[105] See further *infra* ff. (point 7) 16v–18v.

[106] I.e., literal, proper.

distinction at all. So there lys the frailety of your reasoning. You call out[,] Scripture, Scripture; and when wee come to the point, instedd of Scripture, you [*]paum and slurr upon us your interpretation, as if that were Scripture, which in truth, wee must, with your leav, controvert in the main of your articles. And on the other side, if wee dispute your propositions, [then] you pretend to curb us with express Scripture, and shew us the words. Wee answer, those words are figurate, and cannot be true [*]*ad literam*, but must be understood as pure resemblances. There you dash us againe with humane inventions that are not to be imposed on the consciences of pious Christians. And this advantage you hold to your self by that single artifice of advancing all the texts by way of catalogue, without any distinction made which are to be understood directly and which figurately or with reasonable interpretation. And this is a new way of mathematicall demonstration in devinity.

[7.] But it is time to be more particular. There is a necessity that all matters spirituall should be expressed by resemblance to things corporeall, for so farr wee can understand and no farther. And accordingly the devine nature it self is represented to us under such resemblances. As for instance, how should infinite power joyned with goodness and love towards us be **[f. 17]** exprest so well as by the word father? because such (naturally) hath absolute power and withall the greatest care and tenderness that can be. Therefore wee are taught to say[,] Our Father, etc.[107] And the like is to be alledged in every instance of Scripture, wherein the devinity is expressed to us, by words which have a meer humane relation; and wee cannot from them argue to the devine nature. The doing so is [*]*grossiere* and would suit better with those ignorants that use[d] to picture[108] the holy Trinity, which is litle less then blasphemy. Therefore wee must make a [*]*differènce* where the scripture expressions have no relation to the world but are positively affirmed and where they are delivered in words of pure humane sence and relation. The former admitt no

---

[107] Presumably, Matthew vi.9.

[108] I.e., to represent in pictorial form.

sort of gloss or interpretation, but the other must be understood as figurative and referred by way of resemblance[,] how faint soever it may be, to those things which the words in comon speaking signifie, and not at all to the devine nature, as if such could be revealed to us in that manner.

To exemplifye this distinction. The texts upon which the doctrine of the holy Trinity in unity is founded are of a positive or direct import and cannot be [*]glozed, for who can pretend to be lett into [f. 17v] the secret so farr as to know more of the devinity then our blessed Saviour thought fitt to comunicate or reveal? These texts are, as I noted,[109]— My Father is in Me, and I in Him.[110] Who shall interpret this and say, how? Then— Thou shalt not tempt the Lord thy God.[111]— sayd to the divel[112] then tempting our Saviour. Who shall refine upon this text and say how, in what order, degree or manner? So also the Apostle,— These three are one.[113] Who shall justifie to say, one testimony, or one authority, neither of which are in the text?[114] Where doe wee find those [*]glozes in Scripture? Are they not humane invention?[115] What is become of your postulatum now?

But on the other side when the expressions are in termes humane, wee must consider the relation and interpret them according to that, or els wee shall have mad work. If a man should gather all the texts where the devinity is exprest as active and passive after the manner of men [then] it would make a great shew in a catalo[gue], but by no means prove that the truth was so. For it was done as a

---

[109] See *supra* f. 2, where RN 'noted' only the first and third texts cited here.

[110] John x.38.

[111] Matthew iv.7 or Luke iv.12.

[112] I.e., divul or deevil (dial. var. of devil).

[113] I.e., 1 John v.7.

[114] But Clarke said this; see *supra* f. 5 n.39.

[115] According to Clarke, *SDT*, Pt I, Ch. 4, p. 238, the phrase (these three are one) from 1 John v.7 was not found in the early manuscripts but was a later interpolation and, hence, a human invention.

means to touch the spirits of men, who are sensible of such things, but not as explications of the devine [*]essence. Who therefore from thence will argue that the deity is angry, repents, expostulates, and the like? as they must doe that will affirme such texts in a literall sence, that is, such as the words in humane use and ordinary application import.

[f. 18]

Let me give one example, and a considerable one[,] and that is of the real presence in the holy Eucharist, which is implyed in the words, This is my body, doe this etc.[,][116] but wee neither presume to say how that presence is, for wee are not capable to know, much less speak it. Nor doe wee assent to the litterall sence, as the Papists, who say it is bodily, as the words in comon use import, which they explaine by the greatest of absurditys, transubstantiation.[117] This wee utterly deny, saying the words were not to be understood litterally, but figuratively onely.

Now Sir[,] you have amassed abundance of texts in which the Persons of the holy Trinity are expressed under the termes of Father and Son and holy Spirit, and according as the relation of these termes, as the proper subject bears, you ascribe order, preeminence and dignity; and on the other side, mission, deputation, office and the like. As— My Father is greater then I,[118] — I doe the works of my Father;[119] I goe to the Father[120]— the Father hath sent Me;[121] and many others of the like expressions. You are not pleased to bestow your gloss and larg collections upon these passages, so as to allow any relation to humane affairs, as you have done upon the other texts that are hard upon you; but you argue from these as if they were literall and positive revelations of the devine [*]essence. So wee may

---

[116] Luke xxii.19.

[117] See *supra* ff. 7v–8 and n.64.

[118] John xiv.28.

[119] See John x.37.

[120] John xiv.28.

[121] John xx.21.

as well argue according to you, an union of a family or **[f. 18v]** kingdome,[122] (for all the persons of a family are to other familys as one, and so [too] a king and his subjects to other kingdomes[123]) as an unity of the devine Trinity, which you say is but father, son, and servant,[124] so low doth your fancy creep in your representation of the devine nature. And I beleev you think there is a force in the number of the texts of this kind (to which I shall give some further answer afterwards[125]) as if there were any weight in that. But surely if wee will consider things aright, [then] the distinction I have stated[126] answers all at once. I must confess your conduct in that [regard] is [*]artificiall, [for] you have put all together without distinction, and depend [that] your reader will not make any. And for your owne part, where any sit hard upon your thesis, that is, subordination in the deity, there shall not want [*]colours and representations enough, ready drest[,] to bring them to nothing, altho they are such, as will bear no distinction or [*]commentation at all.

8. That the mistery of the holy Trinity in unity stands proved in Scripture by direct and positive texts, not modyfyable by any gloss, and so is become a pure point of faith, I think, is already shewed. The objections that have bin made by those that oppose the doctrine, as divers sorts of men under various denominations have done, whereof your self, as I take it[,] bring up the [*]*arriere*, or as the best orators use to make the [*]*perroratio* to all the rest, amount to two onely, and those are, first[,] the seeming **[f. 19]** contradiction that one should be three, and three one, which you mollifie with the terme of

---

[122] Clarke's monarchy of the universe implies a kingdom; see *supra* f. 5 n.41.

[123] For the former simile, see *supra* f. 5 and n.42; for the latter, see Clarke, *SDT*, Pt II, p. 333: 'That a *King* upon the Throne, and his *Son* administring the Fathers [*sic*] Government, are not *Two Kings*'.

[124] Ibid., Pt III, Ch. 2, p. 451, the Son taking upon himself 'the Form of a Servant'.

[125] Missing in this text.

[126] I.e., between a literal sense and an interpretation.

a complex being[,]^{127} the other, the seeming inconsistency with our Saviour's Passion, as if the deity suffered, which, say they, may not be conceived. As to the former[,] they harp much upon the prerogative of reason, as the light every one must follow, and bring what authority you will, unless our reason entereins it, no effect of perswasion can follow.

This doth not directly concerne the case you make, because you advance the authority of the Scriptures as decisive in this question; and how you manage it has bin, and will be more diffusedly observed.^{128} But as to them that idolize reason, as they call it, I must here reiterate what was [*]couched in my other paper,^{129} which is in effect that what they call reason is onely experience of worldly things. Wee know nothing properly but thro the means of bodily contact, and the impressions of that reflected by our memory. These enable us to lay antecedents and consequents together, and by a sort of reasoning called induction, which depends on frequency, if not on constancy, to conclude that when the antecedents, as have bin observed, arrive, the consequents are not farr off. This branched out into all the instances of experience, made up of history, observation, instruction, conformity,^{130} and the like[,] is that which wee call reason.^{131} But what is all this to any knowledg of the devine [*]essence, or of the nature of any immateriall being whatever?

The very words which these opponents frequently use[,] such as vnion, separation, identity, number, order, **[f. 19v]** proportion, etc.[,] are but termes of comparative quantity and carry an idea of [geometrical] extension, which abstracted, all those words become insensible and improper, so that without the limits of extension, all

---

^{127} See *infra* ff. 25–25v, where RN elaborates on this issue.

^{128} See *supra* ff. 16v–19 and *infra* ff. 22v–23v.

^{129} For this paper see *supra* f. 3v n.29.

^{130} I.e., in regard to some respect or matter, correspondence to or with a pattern.

^{131} I.e., common-law reason as trans-subjectivity.

our imagination as well as language failes.[132] I know some idolizers of analitik[133] value their art much becaus it is applicable indifferently[134] to all entitys, even those that are imateriall, as angells and spirits which being distinct or nombrable may fall under computation thro all the operations. But with their peace be it sayd, they have no idea of an angell or spirit or of the distinction of one from another but onely [*]*sub specie quanti*, so that fancy is nothing to the porpose. Therefore from an impossibility that three and one should consist, when applyed to things extended and locall, to argue the like impossibility when applyed to the devine [*]essence, is a most rampant non sequitur.[135] It is a much sounder reasoning to say, there are birds, beasts, fish in America, therefore the [*]*ansæ* of Saturne cannot be without them, becaus wee know probably the *ansæ* and America are [*]*ejusdem generis* and both capable [of supporting life],[136] but wee know nothing at a[ll], distinctly or properly of the devine nature.

Now wee must consider that the revelation of the holy Trinity in unity was made for our sakes, and it was necessary that it should be put in some termes, whic[h] wee could apply a sence to, as that of three persons subsisting and yet one. For unless we had a miraculous exaltation of our nature, it may be supra angelicall, wee could not receiv a revelation of the deity it self. **[f. 20]** [Therefore] it was necessary to substitute ideas suitable to our nature; and if those are

---

[132] In f. 18 of version 3 of RN→SC 1713 (UK:Lbl) Add MS 32551: ff. 2–33, RN adds: 'I know there is a disposition in a certein Doctor [i.e., More] to hold immaterial beings extended; but ... he may as well hold body not to be extended as non-body (for so onely wee conceiv of spirit) to be extended....'

[133] I.e., the resolution of problems by reducing them to equations. As the rest of the sentence suggests, RN's reference here is to Wallis, who asserted that numbers are applicable to all things, geometry as well as angels; see Barrow, *The Usefulness*, (Lect. III), pp. 29–49, p. 40.

[134] I.e., without difference or distinction.

[135] I.e., an inference or conclusion that does not follow from the premiss.

[136] For a long version of this 'sounder reasoning', see Huygens, *Cosmotheoros*; see also *infra* f. 24 n.163.

[*]wonderfull and unintelligible by us, [then] is it not the more suitable to [apply them to] the incomprehensible subject? All that a reasonable man can say is, that it is wonderfull and strang[e]; and so I presume the Desciples, with amazement[,] thought and were silent, when our Saviour sayd, The Father is in Me, and I in Him.[137] But for them to have sayd, that cannot be, had bin impious, as it is for any one now, from the proof onely of his owne sences, to say the Trinity in unity cannot be. And there is reason to think it is layd afore us in such termes as induce [*]wonder and amazement, rather then knowledg, that wee may be excited more to adoration, then to contradiction. And really it is a sort of apostacy that men should be, as in this age is seen, so much more propence to the latter, then to the other, although it[138] is the duty of amost every minute wee live.

9. I fear you will accuse me of overdoing in this last article, becaus it may seem to accuse you, as one of those that are against the doctrine of the holy Trinity in unity. For you doe owne the Trinity in the terms of Father, Son, and holy Ghost, which you say have an eternall dependance and add that how such dependance was or is made hath [not] bin revealed. So farr of a mistery you allow, but then you stick so close to the notion of subordination and delegation of power, opposing the coequality, that you reduce the matter to the image of a kingdome or family,[139] which is very neerly, if not **[f. 20v]** altogether a subversion of the church's faith in that point, which made me insist so much on the question at larg. But [now] wee shall come neerer to you.

And that will be in considering the next difficulty, which is of the holy Incarnation and Passion[,] which say some is hard to be assented to, and I doe not remember much in your book tending to it,

---

[137] John x.38.

[138] I.e., adoration of the deity.

[139] See *supra* ff. 5, 18–18v.

which is, perhaps, becaus it is not your proper subject.[140] The reason
the others alledg is that it doth not accord with their understanding,
the frailety of which argument hath bin touched already.[141] The
question is not what wee can understand, or reconcile, but what our
evidence oblidgeth us to beleev; for if wee are not to beleev till wee
understand [then] I am sure, brutes would have as much religion as
men. But in this mistery of the holy Incarnation, one part is the
proper object of our knowledge and understanding, and that is the
humanity of our blessed Saviour who[,] having taken our nature upon
him, thought it no robbery to be equall with God. Now so much of
our blessed Saviour as concernes our redemption and the caus, with
his miracolous coming into the world, and conversing among men,
etc. is accordingly fully and plainely related, or revealed to us, in the
termes of comon use or conversation of one man with another. And
this is what concernes us cheifly, being the signification of our
[*]wonderfull redemption and instruction as to our behaviour, with
especiall regard to it. But is there any thing that shews it was **[f. 21]**
intended the people of the world should be made speculatively wise
in a kind of celestiall theology, so as to vnderstand the nature of the
devine [*]essence and the manner of those amazing operations? Or is
not the direct contrary rather declared, when wee are comanded not
to pretend to such high flights, but to be as litle children,[142] faithfull
and innocent? How comes it then, that from our Saviour's ordinary
conversation, wee must inferr modes of the devinity? It was enough
that he declared to his Disciples that the Father was in Him;[143] but to
render that consonant to their reason seems to be neither his porpose
nor their capacity.

---

[140] Christ's Incarnation and Passion (suffering) has direct bearing on Clarke's
'proper subject', because the Christian experience of redemption is illusory if
Christ, as the Redeemer, is not God in the fullest sense.

[141] See *supra* ff. 2, 2v, 6v, 11v, 18v.

[142] See Matthew xviii.3.

[143] John x.38. Note that in the following paragraph RN includes a number of other
paraphrases from Scripture, but as they are extremely short, I have left them
unidentified.

But farther, in order to shew the true designe of our Saviour's discourses upon earth, wee must reflect that his whole cours of life was framed and directed to shew forth to the world a perfect example of humility, obedience, and doing good. He ordinarily required that men should beleev in him and [should] follow the patterne he set before them. And instances are very rare if any are to be found of his aggrandizing himself. He did not owne as from himself even his good works, but styled them the works of his Father. When he was demanded if he was that profet, etc., he affirmed nothing, but bid them say the blind see, etc. and then let them judg. And altho he spake often of the Father, yet when he spoke of himself [**f. 21v**] he almost allwais styled himself the son of man, and so lowly, as to add that he had nowhere to lay his head. And in his teaching he used the [*]formes of ordinary signification among men, for their capacity would not admitt other. Such were coming, going, sent, dispatch, authority, power. Altho at certein times he declared enough to imply his devinity also, as when being warm and provoked by the Parasys[,][144] [yet] (if I may so speak) he sayd, Before Abraham was, I am.[145] Which is a flight of the highest revelation of the devine nature extant in the whole Bible and accordingly struck dumb all his treacherous and insolent opposers. And of this sort were those sayings which the Apostles layd up in their hearts and probably understood them not till after his Resurrection. And then they preached the doctrines of them to the whole world, from which preaching and testimony wee have derived to us at this day the misterys of the holy Trinity and Incarnation, which are made good by

---

[144] I.e., Pharisees (Heb., *perusim*, from *perash*, to separate) means those who have been set apart, i.e., the Jewish party of this name who strove to ensure that the state was governed in strict accordance with the traditional and written law as the will of God revealed to Moses and contained in the Old Testament.

[145] John viii.58. See Clarke, *SDT*, Pt I, Ch. 2, sect ii, p. 98, Text No. 591, according to whom the 'plain Meaning is, that he [Christ] was really *with God in the Beginning, and before the World was*'. As RN indicates above, for him the passage is inspirational; and this probably accounts for his citation of it in a number of his writings; see, e.g., f. 269v in RN, 'Inceptio', (UK:Lbl) Add MS 32546: ff. 169–173v, *continued* ff. 266–270v (undatable).

reflecting back upon those texts, which shew them, but before our Saviour's Passion were not understood.

Besides this designe of being an example or patterne of a religious life[,] there is another, to which our Saviour's expressing himself in the language of the world, in the Person of his humanity and after the modes of men's ordinary **[f. 22]** capacity[,] tended. And that is to assist devotion, excite the extatique, and furnish incomparable matter for perpetuall and pious meditation, such as might hold men endeared to his goodness, shewed forth by his infinite good will and suffering for them. The ideas of the almighty power are more apt to excite fear, then love, but those that attend our Saviour's Passion are all inducements to love and charity, and therein lay the perfection of the Gospell. And what could more effectually answer those great ends then for our Saviour to convers with men in those [*]formes of his humanity, wherein he was allyed and really an infinite benefactor to the whole state of human kind, and which they might understand and apply to themselves? It was the grand mistake of the Jews to expect the power of heaven to come downe and with lightning and flaming swords to reigne among them and drive away the Romans.[146] And our sight will be found shorter then theirs, if wee gather the humble and submissive speeches of our Saviour, which relate to the state of poverty and suffering he had, for ends plainly enough declared in the Gospel, taken upon him as explications of his devine nature, or as any way applicable to the majesty of heaven.

But whatever wee argue of that sort[,] it must be from such oracles as he gave out, directed to that very end and porpose, that is, for speculation of his devinity and not for example or instruction to men. As when he say'd, The Father is **[f. 22v]** in Me, and I in him,[147]— Before Abraham was, I am,[148] and such like, and not when

---

[146] See Bk IV, Ch. 5 of *De bello Judaico Libri VII*, written by Josephus, one of the generals of the Jews in their revolt against the Romans. But RN may have consulted the 1701 edition of his works translated by Roger Le Estrange; see RN Books (1).

[147] John x.38.

[148] John viii.58.

he sayd, Be you followers of me,[149] for who can follow the deity; so
[also] in that place where he sayd, My Father is greater then I,[150]
which to humane understanding is no more then the word father in it
self signifyes. I might alledg more of this kind, but I fear I spend too
much of your time. Therefore here I add onely, that [it is important to
maintain] this distinction of understandin[g] our Saviour's and his
Apostles' words as they relate to his devinity or to his humanity,
according to the manifest end and designe of them in the Gospell.
And the latter[,][151] as no sort of explication of the devine nature, was
formed clearly in me by the reading a litle book,[152] which fell in my
way and I think termed it in Greek [*]*kat òeconomian*, which that
author sayd was an expression the [church] fathers generally used to
signifie when our Saviour's words were to be understood with
reference to his humanity or with analogy to the state of the politicall
world.[153]

Now Sir, I may appeal to your self, if any of these texts on
which you build your politicall sc[h]eme of the devinity (which
makes the Persons [of the Son and holy Ghost] subject[154] and to act
by mission and delegation)[,] thereafter proposing your reade[r]
should think of the devine nature [in such terms] and not in the
termes of same, equall, etc. as wee have them in our church creeds[,]
[then] are not [schemes] of this sort to be understood *kat òeconomian*
of humane affaires and cannot be litterally or properly applyed to the

---

[149] 1 Corinthians iv.16.

[150] John xiv.28.

[151] I.e., the words relating to his humanity.

[152] I have been unable to identify this book, which, perhaps, is the same book
mentioned *infra* f. 27.

[153] The Greek terms derive from the ancient Stoic philosopher Epictetus, whose
Enchiridion, or manual, was compiled from the philosopher's discourses by
Arrianus, his pupil. Note that RN once owned three copies of this manual; see RN
Books (1).

[154] I.e., under the dominion of the Father.

devine [*]essence? As **[f. 23]** for instance, filiation[155] cannot be understood properly but in a manner [*]*secundum captum humanum*, which may put the words similitude, analogy, allusion or what wee will to it, as if wee would explain it, but all to no porpose. For wee doe not know one graine or scruple more thereby. And if the words father, son, etc. have no direct application to the devine nature, or such as wee can any way understand [then] I desire to know by what authority wee carry on the incidents, as inferiority, dependance, trust, mission and the like[,] which are meer humane sc[h]emes, into the devinity. Nay it is wors[,] for [in regard to] the words of authority, subordination, delegation, mission, commission, dispatch, returne, and the like, as ordinarily are used in the holy Scripture in relation of our blessed Saviour's life and teaching, if you will take them as they signifye among men, [then they] presuppose legislature and all the paltry shifts and tricks of states' politiks, which the powers of the world use to governe one another, becaus the soveraines cannot doe it immediately by themselves. And therefore [those words] are improperly, I wish it wrong to say, wickedly applyed to the pure nature of the almighty power of heaven and earth, who doth all things by pure will and is adored by Christians under the revealed tri-une [*]essence, which wee neither can, nor ought we to pretend to understand. And if you oppose to me againe the express words (delivered as I sayd [above] [*]*ad captum*, etc.) [then] I returne to you againe the signification [*]*kat óeconomian*. And moreover, by the like argument I will prove to you, back't with as many positive texts[,] that **[f. 23v]** [if you interpret] the deity as active and passive after the manner of men, then, transubstantiation it self will be too hard for you.[156] Therefore Sir, I wish you would upon the next occasion contrive another topick in your digest of texts and lett it be to distinguish such as are spoken of the Godhead directly from those that are [*]*kat economian*.

---

[155] I.e., the process of becoming a son.

[156] MS has: 'And moreover, that by the like argument I will prove to you, back't with as many positive texts that the deity is active and passive after the manner of men, nay, transubstantiation it self will be too hard for you.'

My conclusion upon the whole is that the devine [*]essence is so farr revealed to us in the holy Scripture, that wee have reason and therefore are bound to beleev the holy Trinity in unity, coequall and coeternall; but the manner and [*]economy of it is in no sort revealed, nor can there be found any images or ideas by means of which wee can have any notion at all of it. Wee are furnished with representations thro the means of the humanity of our Saviour sufficient to guide our devotion, direct our [*]oraisons, and institute our behaviour and under such termes and language as suits our weak capacity. And the fathers of the church have adapted others in conformity with them for the same end and designe as neer as their utmost invention will allow [and] accordingly as wee have the pious use of in our church liturgys. But as for marshalling and modelling the subordinate offices and imployments and contriving missions and comissions, expeditions and retreats, after the manner of men, it is without warrant from Scripture, profane in itself, and trifling as to us, and considering the reasons [you] alledged for it, of the [*]last consequence to the Christian religion it self.

[f. 24] Some particular matters remain to be taken notice of.[157] One is the dread some men have upon their spirits least, by acknowledging any thing in religion to be misterious, they should deliver up their reason. On that account, Mr. C.[158] and others of his alphabet[159] strutt about in an huff, like a popish preist in dispute. Tell

---

[157] RN notices five particular matters: (1) ff. 24–25, (2) ff. 25–25v, (3) ff. 25v–26, (4) ff. 26–28, (5) ff. 28–30.

[158] I.e., Anthony Collins, author of *A Discourse of Free-Thinking occasion'd by the Rise and Growth of a Sect call'd Free-Thinkers* (London, 1713), which was published anonymously. Of its three sections, the second concerns the 'duty' to think freely on such topics as the nature and attributes of God, the truth and authority of books esteemed sacred, and the sense and meaning of those books. A critical summary of the book appeared in *The Examiner* (29 December to 2 January 1712 [i.e., new style, 1713]), vol. 8, no. 12, pp. 319–20, where it was described as 'lately published'.

[159] E.g., William Coward, author of *The Grand Essay: or, a Vindication of Reason, and Religion against Imposters of Philosophy* (London, 1704).

them of scripturall authority, tell them of Comocandum[160] or Confucius.[161] They know no rule but their owne reason, and they shall not submitt that to any authority whatever. Very true. Wee say the very same thing, and doe not submit to the Scriptures but from a conviction of the most rigorous reason. Wittness the [*]*quatuor criteria* which the good man[162] against the Jews and [*]theists observes never mett in a falsity and shews it to be morally impossible they should so concurr. This is as to the evidence of our revelation, which not being made to ourselves but to others, wee take, upon the strength of our naturall facultys, to be truely that which it is declared to be. So that our [*]title to reason and judg is very safe, but when it hath brought us thus farr, then it leaves us. For what person ever pretended to argue beyond his information? As if wee had a revelation indisputable, that there were animalls in the [*]*ansæ* of Saturne, [then] should wee pretend to argue whither they had eyes, ears, or hornes or not?[163] So in this controversie, whoever is informed, and doubts of the truth of the matters conteined in the

---

[160] RN spells this name differently in all three versions of his letter, but it is probable the reference is to one of the Roman emperors, L. Aurelius Commodus, who persecuted the Christians because they rejected the state religion.

[161] I.e., the Chinese philosopher, whose name was first Latinised, and his system introduced to the West, by the Jesuit missionary Matteo Ricci.

[162] I.e., Charles Leslie, chancellor of Connor (1686), initially, a student and practitioner in law and afterwards ordained in 1680. But after refusing to take the required 1690 oath of allegiance to William and Mary, he lost his clerical and judicial offices in Ireland, whence he removed to England and became a member of the non-juring community. During this period he published *A Short and Easie Method with the Deists* (London, 1698, 5/1712), a book that was once in RN's library; see RN Books (1). It included four rules of evidence deemed to be incompatible with any imposture; for one analysis of them, see Stephen, *History of English Thought*, vol. 1, pp. 194–201.

[163] Cf. Huygens, *Cosmotheoros*, p. 33: 'But when we come to meddle with the Shape of these [planetary] Creatures, and consider the incredible variety that is even in those of the different parts of this Earth, and that *America* has some which are no where else to be found, I must then confess that I think it beyond the force of Imagination to arrive at any knowledge in the matter, or reach probability concerning the figures of these Planetary Animals.'

Gospel, is not animal rationale.[164] And those [doubters] are such as cannot subsist without a more then humane, that is, a devine ordinance, [**f. 24v**] upon which foot, wee find it reasonable to credit all that is affirmed to us there, according to a due sence and [an] interpretation of the languages, or modes of expression there used.

Here is the foundation of our faith that The Father is in the Son, and the Son in the Father[165] with the adjunct of the holy Ghost, who is worshiped and adored together with the Father and the Son, and these three are affirmed to be one.[166] Were any thing like this sayd of phisicall entitys, which wee convers with and properly comprehend[,] to instance that the substance shall be altered and the accidents remain, as the Papists out of senceless philosophy [*]cant[,][167] reason immediately takes place and rejects it. But if both substance and accidents (if I may so express my self) are intirely out of our notice and possibility of our comprehension in any degree, [then] what hath reason to doe here? or to examine whether any mode can or cannot be? Is it not grosser so to pretend, then to paint the [*]anseatick animalls? Mr. Hobbs had more ingenuity, who postulating himself to know and understand all things, declared there could be no being in the world but body,[168] because he could not by any of his facultys comprehend any such, and so upon a negative, that is, no argument, was a positive atheist.[169] But he did not pretend to affirme or deny any thing of a subject he could not comprehend, supposing its existence, as some pretend to doe concerning the holy

---

[164] I.e., the phrase, made famous in the third century and popular in RN's day; see Porphyry, *Isogoge*, pp. 35–6, and Crane, 'The Houyhnhnms'.

[165] John x.38.

[166] 1 John v.7.

[167] I.e., concerning transubstantiation; see *supra* ff. 7v–8 and 18.

[168] In *De corpore* (London, 1655) Hobbes conceived matter ('body') as the sum total of existence and laid out a method for its investigation; see Kassler, *Inner Music*, pp. 49–107, especially 95–9.

[169] As the following sentence indicates, RN avoids any categorical statement concerning Hobbes's supposed atheism. For Clarke's position, see *supra* 3.3.5.

Trinity in unity. I doe not say all this against your self directly, becaus you are not with these men. But yet by your setting up possibilitys and impossibilitys,[170] and judging by what you imagin can be or can not be, and **[f. 25]** introducing political systemes to plausifie and reconcile the doctrine of the holy Trinity to the humour of an age that is falling off apace from Christianity, [—] as if by evading the misterys, you thought to gaine unbeleevers as the Jesuits in China that worship idols to gaine (as they pretend) proselites, comes so vile near 'em, [—] I could not forbear so much notice as I have taken of the sort of modish apostates[,] all which you may pleas to consider with candor, if it deserves any.

Another matter to be observed is altogether your owne, and that is your being disturbed [*]least by this doctrine of the Trinity in vnity[,] as our church understands it[,] should make of the deity a complex being.[171] This I think needless, groundless, and out of the way. For how can any being be made complex or incomplex by those that declare, they neither know, or comprehend, or beleev it possible they should understand the least scruple of its being one way or other directly? But [*]onely wee have faith in our Saviour's and his Apostles' words, and that takes the place of understanding in us; and according to those words wee worship and adore that which is neither explicated in Scripture, nor is interpretable by us. Then it[172] is groundless, for the word complex is a [*]jingle, and hath no signification to this porpose at all. For in truth nothing in the world that really exists can be complex, and each thing how minute soever is in every self-respect distinct, unless you will say proximity or the influence of motion is a complexure. But it is our imperfect knowledg of things **[f. 25v]** that breeds a confusion in our understandings, whence wee say things are complex and confused,

---

[170] I.e., setting up the principle of non-contradiction.

[171] Clarke's concern about this issue, mooted *supra* f. 19, derives from his doctrine of God as 'One', that is to say, absolutely simple; see *supra* 4.4.1 and n.22, 4.4.5 and n.146.

[172] I.e., Clarke's perturbation.

that are not so really or in themselves.[173] So I cannot collect what notion it is of the holy Trinity in vnity that can ly [*]couched under the word complex. But for argument sake onely[,] be it complex, the complexure should have bin layd at the door of our Saviour himself, who in his Gospell hath sayd[,] He and his Father are one;[174] He is in the Father, and the Father in Him.[175] And then the Apostles are to answer next, and not wee Christians that make the Gospell our rule and regulate our faith of things not seen according to the express testimony of that, and [hence] are not [*]deturbed from that fortress by[176] an army of interpretations and categorys, tho led on with plenty (as your book hath) of genus, species, individualls, complex, absolute, ordinate, coordinate, self existent and such (applied to the devine [*]essence) metafisicall and chop-logick nothings.

You are not pleased with the Nicene creed,[177] because of the word [*]omousios, or one substance with the Father, and the word [*]filioque,[178] by which the long controversie concerning the procession of the holy Ghost from the Father and the Son is intimated. Since you seem to dislike *omousios* as not a proper word, I should be hugely glad to know what freind of yours hath bin lately let into the secret and could tell you so. Our Saviour sayd,— in Me

---

[173] This and the foregoing sentence present a succinct statement of RN's distinction between reality (as unobservable entities) and appearance (as a composition or confusion of such entities due to nature's limitation of our sensori-motor capacity for information processing); see *supra* 1.1.4, 3.3.5 and Kassler, *The Beginnings*, pp. 117–19.

[174] John x.30.

[175] John x.38.

[176] MS has 'with'.

[177] See *supra* ff. 8–8v.

[178] I.e., the Latin clause of the Nicene creed, interpolated in the late sixth century to assert the doctrine of the procession of the holy Ghost from the Son as well as from the Father. According to Stormon, 'The Procession of the Spirit', p. 81, this clause seems to express, 'in very shorthand fashion', the teaching in Augustine's *De Trinitate*, 'though the origins of the phrase go back somewhat further'. Note however, that the interpolation was not, and still is not admitted by the Eastern church.

and I in Him,[179] which looks really like one and the same; and it must be some great intimate that could discover the contrary. But this looks too like [*]rallery for the nature of the **[f. 26]** subject. Therefore I would but ask another thing, which is, if there was once a devision in the church concerning two words, [*]*omousious* or [*]*omoiusios* [then] which should be used in the Christian confession of faith? and thereby, as men are subject to heats and passions, divers fierce disputes and scandalous animositys arose and disturbed the Christian church.[180] Is it not a great happyness that it received a determination so long agoe as the Nicene Councell,[181] and owned (the Arrians onely opposing) ever since? And I would know what spirit it is that now in our days receives such disputes of words? One would have thought truely that in a matter to which no word can be adequate and which wee have no access to the thing by any knowledg wee can have of it to help us to contrive expressions[,] wee might have rested satisfyed. In a word the Christian church hath chosen [*omousios*], as also that of [*]*filioque*, which certeinly are not against Scripture, nor confutable by any mortall man or men in the world.

But then the Athanasian (or so called) creed is a stumbling block, and a rock of offence;[182] and why? [because] it begins with [*]*Quicunque vult*, etc. Whoever will be saved must thus think of, that is, beleev in the Trinity.[183] This is a minatory claus, which [you say] many cannot consent to, and it renders men avers to conformity, and therefore that creed ought to be put out; and some men once

---

[179] John x.38.

[180] I.e., the Arian controversy and its aftermath.

[181] I.e., the first Council of Nicaea (325), when the doctrines of Arius were condemned.

[182] See *supra* ff. 8–8v.

[183] I.e., Whoever will be saved, before all things it is necessary that he hold....; see also *infra* f. 27v.

were inclined so to doe.[184] And if they had done it in a regular way,
or made any other alterations in the liturgy not contradictory to the
Gospell, [then] all good Church of England Christians[,] as I sayd
before[,][185] had rested satisfyed with it. **[f. 26v]** As for the persons, I
know one [you] named, [who] was very good indeed.[186] And there
[lies] the unhappyness, if a good man shews but an inclination to
hearken to any overtures as are made in order, as some pretend, to
vnity and comprehension, tho the matter never comes to be
considered in a regular assembly, where perhaps he himself may be
convinc't that the overture is treacherous and undermining, and not
fit to be complyed with, yet his great name shall be cited as an
argument irrefrangable for doing the thing. And who are so apt to be
imposed upon, as good men are, by wolves in sheep's cloathing, that
demand one thing merely becaus the granting that may become a
reason unanswerable for demanding somewhat els for which there
was no reason at all before? Even the advantage of contempt and
scandall, which they[187] are prepared to cast thick as soon as the point
is gained, is joy and, to them, success enough. Thus the freethinking
colledg deal with the reverend complying Church of England
devines, where [if] any one out of a [*]*bonti* of disposition,
[*]incautelousness or inadvertence countenances objections made by
some of them, [then] they are cited in the next peice, with eulogys.
See, say they, what a reverend church-man says. And, Sir, you your
self are not exempt, for you may have the fortune, or rather the

---

[184] For this specific 'damnatory' clause and the reasons why it should be omitted
from the liturgy, see Clarke, *SDT*, Pt III, Ch. 2, pp. 418, 448.

[185] See *supra* f. 8v.

[186] I.e. William Sancroft, archbishop of Canterbury and non-juror, who, according
to Clarke, *SDT*, Pt III, Ch. 2, p. 450, 'concerted' a design of 'reviewing, inlarging,
and correcting our Liturgy'. Clarke would have had this information second hand,
whereas RN held the legal office of steward to Sancroft before and after the
prelate's deprivation; see Kassler, *HRN*, pp. 26, 50. For a long character of 'this
most reverend prelate', 'with whose service and favour, my life is adorned'; see
RN, *Notes of Me*, pp. 176–86.

[187] I.e., the wolves.

dishonnour to be appealed to by them as an inexpugnable wrighter of the church against the mistery of the holy Trinity in unity, of which I wish you had thought before you publish't your book. For I would know if any sort of persons at this time are so clamorous against that doctrine as the freethinkers are. And [f. 27] why any devine at this time should write and publish, if not on their side, yet so as to give them umbrage to triumph, is past my understanding. The Christian side against them hath this advantage, that they cannot bring against this doctrine of the Trinity the charg of ambition or self-interest.[188] For I see no advantage secular at all by it, or otherwise[,] but onely by exciting a warmer devotion and zeal in religion, then meer rationall morality can produce,[189] which may be the ground of their offence and is best left to themselves to account for.

Then as to the creed it self, no person of the congregation is required to give his assent to it more then to any orthodox doctrine preached by the minister in his pulpet. And if any auditor doth not concurr with him in the extent of his expressions, [then] must the pulpet strait be pulled downe and no more preaching there, tho all the rest of the congregation were well satisfyed? The homilys[190] are appointed to be read, and yet all people doe not allow some doctrines in them. Wittness the case of rebellion,— are the homilys to be layd aside for that? The parson may say that willfull rebells are in a state of damnation. If any such be present, [then] they are not pleased, but yet have somewhat worthy of their reflection to carry away with them. The author of the Athanasian creed had summed up the whole state of the orthodox faith, as it had bin maintained by Saint

---

[188] I.e., the charge of the deists against priestcraft; see *supra* f. 6v and n.56.

[189] For one example of 'rational morality', see Clarke, *Discourse* [1706], who based morality on the reality of self-evident relations between situations and the kind of actions they demand, and who treated mistaken moral judgment on the same logical order as a contradiction in mathematical reasoning. For details, see Capaldi, *Hume's Place in Moral Philosophy*, pp. 59–62, 328–9, *et passim*; for RN's different conception of morality, see Kassler, *HRN*, pp. 179–86 *et passim*.

[190] I.e., practical discourses with a view to spiritual edification of the congregation; but in the Church of England especially the discourses contained in the books of homilies, published in 1547 and 1563, for use in parish churches.

Athanasius, with success [*]*contra mundum*, and therein explained
that distinction of the humanity of our Saviour, so as the various
[*]sounding texts **[f. 27v]** concerning the holy Trinity in unity are
reconciled [by] [*]*kat' oeconomian*, in a manner as he and the
orthodox of the Athanasian side took to be the catholik Christian
faith. This being the case[,] what less could be sayd then as is there
[in the creed] [*]assertorily as well as minatorily expressed[,]
Whoever will be saved, etc. For then, and I suppose now, a right
Christian faith, to those who are instructed, is held necessary to
salvatio[n.]

It would make one smile to observe, that in making objections
of this nature, no man speaks for himself but for others. If you ask
the objector, doe not you think the Christian faith to be so? Yes[,]
says he, but alack! others are not so clear, and it were good not to
disgust them. You yourself Sir, have found out a way, as may make
the Athan[asian] creed goe down glibb with any body; but yet you
would have it taken away, becaus others are not so wise to find out
your distinctions. It is sayd that in the controversies with the ancient
heretiks, and the Arrians in particular, to draw the people to their side
they used such expressions as the Athanasian creed hath, that the
people being under a terror[191] might consider. I am sure if ever that
was reasonable, [then] it is so now; and nothing is so pestilent at this
time as mincing matters as if the Christian faith were indifferent,[192]
to be held or lett goe, as the weathercock stands or according to the
genius of the people, which word has bin so Christianly used by
some of the Scotch Presbiters.[193] How much better is it to deal plain
and round, as the author of this creed hath done, and let them take it

---

[191] I.e., a state of fear or dread.

[192] I.e., of a neutral quality.

[193] I.e., the elders to the full congregation of Presbyterians. By the phrase 'genius
of the people', RN refers to the confessional tradition of Presbyterianism, for
although individuals are encouraged to understand and challenge the current
institutional understanding of Scripture, theology is carried out by the community
as a whole. For a brief outline of their ecclesiology, see Cornwall, *Visible and
Apostolic*, pp. 28–9.

home with 'em; and if they ar[e] not pleased to be Christians [then] the fault is not at the church's **[f. 28]** door.

Therefore [though] admitting the creed no essentiall part of the liturgy, imposed on none (but the clergy on whom by being such[,] it is fitt and together with the [thirty-nine] articles and the rest of the service, ought to be imposed) and used onely on some extraordinary days, and such as may in a regular way be left out, yet at this time, I cannot but think, considering who it is that in publique discourses at tavernes and coffee houses, doe affectedly revile that and the [creed's supposed] author[194] [—] and wee know for what end they doe it, not out of sincerity for decency and reformation, but for treachery, and to make a rent in the selvedg of Christianity, in order to runn it the easyer thro the whole cloath [—] considering this, as I sayd[,] I cannot but think the expunging the Athanasian creed out of the liturgy, at this time, would so much encourage a sort of non-Christians called [*]theists as would amount to litle less then a solemne tradition of the Christian faith.

Another thing I think you may say, which is that you have the [church] fathers on your side, for, to back the interpretations, you cite them all along both ancient and moderne. I am sure here I must be left behind, for I must not pretend to capp[195] fathers with you, that never read [any] of them except coetantius,[196] Arnobius and Minutius.[197] But yet I shall be no niggard of what occurrs to me in the generall on this occasion. And the first is that you cite no fathers

---

[194] Whiston is one who fits RN's description, since he delivered public lectures at Button's Coffee House; and he also published anonymously *Athanasius convicted of Forgery* (London, 1712), which afterwards was included in *A Collection of Small Tracts*, published in the same year.

[195] I.e., to quote alternately in contest, so as to try who can have the last word.

[196] I.e., living at the same time.

[197] I.e., the third-century teacher, Arnobius, and his pupil, Minucius, whose works are not included among the patristic writings listed in RN Books (1). The former wrote a defence of Christianity against the heathen, whereas the latter wrote a cultured person's introduction to Christianity minus the mysteries. Both fathers (especially the first) relied on material from Cicero's *De naturâ deorum*.

to the texts[198] [and the second is that] you annex no notes to [those texts], but onely where you have thought fitt to [*]gloze the **[f. 28v]** text to make it speak your sence, and then the passages of the fathers come in roundly. I think it had bin but fair, since you listed all the texts in order, to have[199] comon placed all the fathers upon them, els wee have not the matter set in a true light in your book, but wee must goe and study your fathers and comentators to see wherein they agree and wherein they differ. For the fathers were but doctors and men, subject to infirmitys; and [though] allowing them the honour of being pillars of the Christian church in persecuting times, yet all that they wrote is not free from humane fondness and frailety. And if one is allowed to pick here and there, [then] he may find sentences to support very odd opinions. And the right use of the fathers, as of all other wrighters in professions even of humane learning, consists cheifly in a generall cognizance and view of them all and a right judgment founded thereon. I admitt that quotations are proper enough; but then the other side (in a matter controversiall, tho I can hardly admitt that word as belonging to our faith in the holy Trinity) hath a liberty to oppose and counter quote.[200] But till that is done, it is but [*]*una pars audita*, and so I leav that matter to fitter hands.

But Sir, I will allow all that you demand out of them [viz.] that they laboured and strained hard to represent the mistery of the holy Trinity in unity as familiar to humane understanding by using adumbratory similitudes, palliating the advers [*]sound of the termes **[f. 29]** and the like as their utmost invention and witt served them. What then? They had to doe with heathen men, and perhaps ignorant but honest people infected with Arrianisme, whom they would reconcile to Christianity and [*]orthodoxicity by condiscending to their low capacity. And for that reason [the fathers] discoursed of the holy Trinity in a way, (I have no better termes then) [*]*kat'*

---

[198] I.e., the texts relating to the liturgy in Clarke, *SDT*, Pt III, Ch. 2.

[199] MS has 'you ought to have'.

[200] I.e., 'in a matter controversial' it is necessary to hear both sides. For this legal maxim, see *supra* 2.2.5 and Kassler, *HRN*, pp. 119–20 and n.173.

economian. Wee are not to conclude from thence, that they took upon themselves to know, ([*]*verbo venia*) [*]*intus et in cute*, the devine [*]essence or to thwart our blessed Saviour's express words. And if they and all the writers since had so done, [then] with their peace be it spoken, it ought to make us consider well, but not leav ye express text for their coment. And the like may be observed of the modernes, who seem to be mistaken in having mumbled the point of contradiction[,] which some have charged[,][201] as if there were any difficulty or as if[202] they would find evasions and loop holes to escape at, which injures the doctrine. For who can, from the nature of body and its consequents[,] with the language or termes invented to signify one and other of them, argue a contradiction in a nature absolutely divers from it, as immateriall beings are from materiall, by applying the termes of the former to the latter, the rather since the former [i.e., body] wee doe know and the latter [i.e., spirit] wee cannot know. [*]*Quid volet argumentum?*

But now the reason also is different, for wee have not to doe with heathen, who were disposed enough to be religious, or with hereticall tho otherwise sincere Christians, **[f. 29v]** who were mistaken but [de]sired to beleev rightly, and[,] in confidence of their prepose[ss]ion, beleeved they did so. With these such a [figura]tive style was necessary, and much of it is found in the wrightings of the Apostles, who were no [*]compounders. And it were well if our church were as steddy in her language to falterers in the faith, as the primitive fathers were. Whither it were popular or politik, or not, I will not determine, but I beleev the insolence of atheists and non-Christians had not bin so great as it is [now]. And that sort of people[,] rampant at this time[,] lead[203] the ignorant callers out for reformation upon tenderness forsooth of an expression of a primitive father that spoke plainely and perhaps too true for some of them. For which reason they cannot bear it; altho their conscience be hard, yet

---

[201] 'Some' includes Clarke, *SDT*, e.g., Pt III, Ch. 2, pp. 447–8, 465–6.

[202] MS has 'that'.

[203] MS has 'who lead'.

it hurts them, and why should they be put to such pain? And who are
they that groan so [mu]ch under this imposition of hearing the
Athanasian creed read, or perhaps the reading of it, or to goe a step
higher, signing a consent to it, but such as doe falter in the faith of
the holy Trinity in vnity, and are otherwise conformable, whither
sincere or not is best knowne to themselves.

As to the sectarys wee find they are concerned onely to cavil,
as they doe[,] at every thing in the church service but afford no reall
symptome of their disposition to union if this scruple were out of the
way. But on the contrary they have demonstrated by their behaviour
that more and more still would be their crye till they have got downe
episcopacy, for as to the liturgie they defye both **[f. 30]** all and some.
Therefore admitting there had bin reason for the primitive father to
have softened his expression, and instead of denouncing,
recomended[,] which, as I sayd,[204] was not thought so fitt in those
days.

Now whither on other occasions our church men separately
think fitt to use such emolients or not, yet for the church's body in
assembly to take away a confession[,] such as this is, doth[,] as I
said[,][205] give up the caus of the faith and worship of the holy
Trinity. And [if this should be done, then] it would be quoted as an
act of self-conviction by all the scribling atheists in their nex[t] free
thinking addresses to the publik. And as to the merit of that ancient
creed (cavilled at becaus the history of it is not well cleared, which
may reasonably be charged upon the Arrians, that suppressed what
they could that was advers to them) I am so farr convinc't that it is
right in all its parts, that if ever I am at a loss to reconcile all the
scriptural passages touching the holy Trinity in unity, [then] I shall
take up that creed (the cited termes of Saint Athanasias) whatever
becomes of your book.

---

[204] Perhaps implied but not stated explicitly *supra* f. 29.

[205] See *supra* f. 28.

I might touch upon divers other matters in your book[,] as that Christianity was perfect at first,[206] and that you annex no epethete of honnour to the holy Trinity.[207] I am sure it is not in your title page, where it ought to have bin, for so hath bin the usage among pious devines. But such matters as these would look affected and [*]piqueish, which I avoid[,] desiring to touch onely what is most materiall. And as for that **[f. 30v]** I hope you are prepared with a stock of humanity as well as Christian charity [that will] dispos[e] you to indulg so much of incurious, illogicall, and wors[,] theologicall expressions as you will find here. And as for all but the sincerity of the undertaking, I know you will have occasion for all those vertues, for I can readily imagin, as if I were at your elbow, how you would slash here and there upon every line and period; and then I must cry, hold, Reverend Sir, enough, [*]*habes confitentissimum reúm.*

I have now Sir done my harrangue on the one side I undertook, and regularly, as the academick way is. I should take the other [side], but that being yours, I fear you would think I should injure it by mismanagement. Therefore it is left to your self to dispose of this as you pleas. As to the former paper,[208] you sayd you would give but short answers. As for this [paper] I desire you will give none at all, for I declare, I shall make no returne[,] having no pleasure in such kind of debates. But as for the harmeless phisicall flights, a feild allowed us to play in, and exercise the agility of our thinking[,] if I thought you could be at leisure enough from your more weighty charges and undertakings, [then] I would be at your service upon the subject of your new[209] philosofy, touching which[,] I think I may

---

[206] I.e., the myth of Eusebius of Caesarea; see Clarke, *SDT*, Introduction, pp. viii–ix; see also *supra* 4.4.2.

[207] I.e., the epithet 'holy'.

[208] For this paper, see *supra* f. 3v and n.29.

[209] I.e., Newtonian.

with good warrant say with the master of eloquence—[210] [*]*Quod primi omnium [estis qui] veram phisiologiam compistis.*

And in those offenceless **[f. 31]** [physical] speculations[,] wee may clash to porpose, without hazzard from any difference of opinion, to grate upon good correspondence and freindship, which I would preserve with all the industry I have, esteeming the countenance you have given to such a [*]scribblaster as my self, by your taking notice of the [previous] papers[,][211] as they have come to your hand, as a greater honnour then I have a lawfull title to. And as I would by all means preserve, so I foresee an occasion I shall have to make use of your freindship. But I must not anticipate what is not yet ripe for projection, but with all sincerity and esteem[,] assure you that I am

Sir your most faithful humble servant,
Roger North[212]

[*The folio that follows contains North's alterations to the foregoing letter, as well as his instructions where they should be inserted.*]

[f. 31v]
Fol. 11. Attend .A.[213]

> I cannot but observe that you doe not deal fairely with this text, viz. These three are one,[214] for you say it is suspected the Greek manuscripts want it and it is not to be much relyed on in controversie. All which your integrity would not permitt you to conceal. Whereby you discover that the Christian church is much bound to you for your inclination to have uncovered such a desperate flaw. And so the reflection comes out but

---

[210] I.e., Cicero. But RN is relying on memory, because the supposed quotation consists of a patchwork of phrases stitched together from *De divinatione*, 2.16, where a distinction is made between real science and pseudo-science.

[211] See *supra* 3.3.2 and 4.4.5.

[212] MS, which has 'RN.', concludes with a downward flourish of the pen.

[213] For the place to be inserted (and existing text removed), see *supra* f. 6 n.50.

[214] I.e., 1 John v.7.

what hindred your giving a crittique of that matter as you might have done, and so perhaps saved yourself a great deal of trouble in distinguishing it out of the way. If you had bin of opinion against the legitimacy of that text, [then] it had bin better to have spoke so plainely, and then I could have sent you to Dr. Mills[215] and the Man of Connor[216] to be converted. But if you allow the passage, as you seem to doe by affording it a place among your scripturall texts (and indeed the popular air of using the very letter, when it is for your porpose required) why then a reflection without any reason given? But instead of that you muster the text and come off with an interpretation of your owne. If you had given in sum the crittique of that controversie with the Socinians and your judgment clear and plain, then the learned and unlearned reader had somewhat more then a [*]side-blow of your bare word to rely on in a matter of consequence.

Fol. 16. B.[217]

And so farr you are certeinly in the right, that if all people were of one opinion, or of the opinion of one man, even of yourself as well as of any other, then wee should all agree and there would be no sc[h]isme.

Fol. 20. C.[218]

that is[,] giving countenance to dissenters and sc[h]ismatiks, for I think even that better then to countenance those that are not Christians, if they have any religion at all. But of this more afterwards.[219] I stand to it at present that however all [*]indifferents are in the church's disposal, yet doctrine is not so, but ought to be maintained not onely in confession but in reputation.

---

[215] I.e., John Mill.

[216] I.e., Charles Leslie.

[217] For the place to be inserted, see *supra* f. 8v n.70.

[218] For the place to be inserted, see *supra* f. 10v n.80.

[219] See *supra* ff. 26–28.

No. 2 (Copy of a letter) North→Hickes, 2 June 1713
(UK:Lbl) Add MS 32551: ff. 35–36v

[f. 35]

Rougham, 2 June 1713[220]

Reverend Sir,

I received lately a letter from your adversary,[221] thus.

—'I shall at present onely observe that your whole superstructure being built upon the foundation of two texts must needs fall, when that foundation is removed. Now one of those texts is expresly explained by our Saviour himself, two or three times over, as it were on purpose to prevent mistake (in the 17 of Saint John[222]) to a sence quite different from that wherein you understand it. And the other text [1 John v.7][223] (besides that the gender of the Greek word makes it impossible to signifie what you would have it) the whole of it is a manifest interpolation (by accident I beleev of a coment creeping into the text rather then by any designe) it having never bin seen in one Greek copy in the world, nor ever bin quoted by any wrighter whatever either before or at the Arrian controversie. Notwithstanding those thin shaddows (very thin shaddows indeed) which my honest and good freind Dr. Mills (with whom I was well acquainted before his death) himself,[224] not without great diffidence catcht at in its

---

[220] MS has 'Rougham 1713. 2 June 1713'.

[221] I.e., Clarke.

[222] I.e., John xvii.21. In version 1 of his letter, RN did not cite this text; rather he cited John x.30; see *supra* 5.5.2: No. 1, RN→SC 1713 (ff. 2, 25v).

[223] MS has an asterisk (*) that refers to a footnote squeezed in at the bottom of f. 35 concerning the 'criticall dissertation', in folio, that 'utterly confutes all this', i.e., Martin's dissertation on 1 John v.7, for which see *supra* 5.5.1.

[224] I.e., John Mill.

defence. If you would not be deceived, you must see with your owne, and not [with] other men's eyes[225]—'

So he. To which I returned this answer.

—[']I received yours, and returne many thanks for the favour, as seeming to inferr that you thought my letter not altogether unworthy of your notice. And since you are pleased to afford a remark or two upon it, I think I ought to make you these replys.

1. You would turne the defensive upon me, which is not fair. Nor can I admitt the exchange. I stand upon old ground,[226] and from thence attaqued your book. Defend that as you may, and lett the [*]indifferent judge of the whole.

2. The two texts[227] are not my onely foundation. I did (for brevity) referr [in general] to all the rest,[228] and noted [in particular] some, as that of— the fullness of the Godhead bodyly, etc.[229] Therefore if you could shift off the two texts as you cannot, my structure stands firme.

3. Your alledging, that our blessed Saviour explained his words in a sence quite different from ours, in two or three places, as it were to prevent mistake, is first irreverent or wors. For could not our blessed Saviour foresee all the consequences of his words at the time when they were spoken, but must afterwards upon recollection, as of

---

[225] See the quote from Tillotson in Clarke, *SDT*, Introduction, pp. x–xi, which includes the words: 'I see plainly and with mine eyes'.

[226] I.e., the consubstantialist ground.

[227] See *supra* f. 35 and n.222.

[228] I.e., in Clarke's digest of texts.

[229] Colossians ii.9. Note that RN does not cite this text in version 1 of his letter. But see *supra* 5.5.2: No. 1, RN→SC 1713 (f. 16v), the paragraph beginning, 'To exemplifye this distinction', for this is the context in which RN adds that text in f. 16 of version 3 of RN→SC 1713 (UK:Lbl) Add MS 32551: ff. 2–33. It is likely, therefore, that the text was included in the no longer extant version of the letter sent to Clarke.

a defect, mend his expression? And then according to you, as will appear [below], it is no better then [*]*obscurum per obscurius.*
[f. 35v]

4. [You say that] no one text in the Bible is against [y]our sence[,] which is no answ[er] [to] [*]*kat oiconomian.*[230]

5. The [passage in] 17. John from v. 21 expresly condemnes you. For there the word ONE is applyed to a state of humanity and also to the devinity, which two cannot be taken in the same sence, but the former must be understood politically and the other[,] [*]essentially. Will any one say that the holy Trinity is one in the same manner as all good Christians are one church, or that the sociable union of Christians in any respect, resemble the unity of the devine nature because of the words— They ... one,[231] as wee are one. The reference is manifestly distributed to the severall natures: Christians in faith and charity, as they are capable, and the Persons of the holy Trinity[,] as must be of necessity in absolute perfection, one. Therefore I cannot spye out the secret of any qualifying explanation here but on the contrary, discerne an unqualifyed union of the Godhead.

6. If wee were prest to use the text [from 1 John v.7]— which three are one, as wee are not, having enough besides, [then] you cannot shuffle it off, no more then the Socinians before you have bin able to doe. And since you accused Dr. Mills of shaddow-catching and diffidence (which is your owne particular discovery, by means of your personall acquaintance and conversation with him, for nothing

---

[230] MS has 'kat' crossed out; but it is retained as 'the expression used ... to signifie when our Saviour's words were to be understood with reference to his humanity or with analogy to the state of the political world'; see *supra* 5.5.2: No. 1, RN→SC 1713 (f. 23v).

[231] MS, not the editor's, elisions.

of either appears in his book[232]), I ran strait to Dr. Hamond[233] and to Dr. Bull[234] (whose works are lately come into our parochiall library[235]) and I find them substantially, and (doubdtless) [*]*ex animo* against you. So then our foundation is [*]abutted again.[236]

7.[237] You tell me if I will not be deceived, [then] I must see with my owne and not with other men's eyes.[238] That's true, as to all objects of vision that may be come at, and look't upon. But if I should pretend to see thro a millstone, unless aided by some body's eyes on the other side, [then] I might, or rather must, be deceived. Therefore in many cases, and particularly that of the Gospell, if wee doe not see with other men's eyes, [then] wee are certeinly deceivable. The true distinction, I think, is sufficiently explained in

---

[232] I.e., Mill's edition of the Greek New Testament, published in 1707; see *supra* 4.4.4. n.100.

[233] I.e., the Church of England theologian and divine, Henry Hammond, author of *A Paraphrase and Annotations on All the Books of the New Testament* (London, 1653) and other writings.

[234] I.e., George Bull, whose *Defensio fidei Nicenæ* (1680), written to defend himself against the charge of Socinianism, became the standard argument for Nicene orthodoxy. See f. 31 in version 3 of RN→SC 1713 (UK:Lbl) Add MS 32551: ff. 2–33, where RN wonders 'how Dr. Bull comes to be a wittness in your caus, that most apertly [obs. = publicly] set up subordination of power in the holy Trinity (that is, counter to the coequality or sameness)', for Bull's writings not only are 'all to the contrary' but also show 'how arbitrary it is to extract peices and scrapps of wrighters and apply them to the avers side of the question'.

[235] I.e., the library established by RN, the catalogue of which includes a 1681 edition of Hammond's *Paraphrase* (see n.233 above) and the 1703 folio volume of Bull's Latin works, edited by John Ernest Grabe; see RN Books (1).

[236] MS has a 'Note' in the right-hand margin, as well as squeezed in between the last line of the paragraph and the end of the first line of the paragraph following, which, once again, refers to the 'criticall dissertation' by Martin, 'edit. in English Lond. 1719', in which the text of 1 John v.7 is 'vindicated to be authentick against all pretences that ever were raised against it'.

[237] In most of §7, RN recapitulates in a familiar way his common-law conception of reason as trans-subjectivity.

[238] See *supra* f. 35 and n.225.

my letter, viz., between the thing and the evidence.[239] The Apostles saw with their owne eyes and were wittnesses of the Gospell they preached, but the world received it upon their testimony, and consequently saw with their eyes. Wee are in **[f. 36]** a degree further removed; for wee have onely the tradition of their testimony, and that in language, few understood. I mean the sacred canon of the New Testament, which admitted, wee want eyes of our owne to see into it, whereby wee are forced to use those of interpreters and preachers. History, records, the holy Scripture it self, and preaching of the Gospell, are all but evidences of things unseen, supplying, as it were, eyes to the blind. Should you say to a [*]*payson*, or one, like me, dark as to Hebrew and Greek, trust no interpreter, but construe for your self, or (more grosly) of the holy canon, trust none, but hunt it out your self, you would be reckoned [to be] in jest, or madd. As [for instance,] if one should ask a countryman (under no temptation to deceiv) which is the shortest way? and he preparing to shew it, you call out, hold freind, I must find it out with my owne eyes, and not with yours. It comes to the same as to particular texts of Scripture.

Wee of the laity are comonly uncapable to judg crittically even of the evidences of our tradition, but must trust some persons. And who safer then those called [to] the church, that is, learned, good, pious, and suffering men in all times, who could have no interest or inducement to deceiv? And shall wee hearken to any fact of objectors against what thro ages is so authentically handed downe to us? And

---

[239] For an allusion to this distinction, see *supra* 5.5.2: No. 1, RN→SC 1713 (ff. 24–24v); for clarification, see ff. 58–58v in version 2 of RN→SC 1713 (UK:Lbl) 32550: ff. 32–79, and f. 19v in version 3 of RN→SC 1713 (UK:Lbl) Add MS 32551: ff. 2–33, in the latter of which he writes: 'But I must in generall add that the question is not what wee can, or cannot understand but what the evidence of our faith demands of us, as reasonable creatures, that we should beleev. ... For crediting and understanding are two acts of the mind, which are not essentiall to each other. The former belongs to the being of a thing, and the other to the manner how it is.' Note that Hickes had previously written to RN that Clarke had been 'early perverted' by Locke's notions and, in particular, that 'wee are not to beleev any thing of which wee have not clear idea; which you have boldly refuted by distinguishing crediting from understanding'; see RN, 'A Copy of a Letter from the Reverend Dr. Hicks', 23 May 1713 (UK:Lbl) Add MS 32551: f. 34v.

your own proceeding is counter to your owne rule.[240] For you pronounce positively, and then bid me to see with my owne eyes. If I had not those (others) that I have more reason to trust, [then] it is like I might use yours, and declare on your side; and then I beleev you would allow me to be very sharp sighted. I doe porpose, as farr as I am able, to use my owne eyes; but then it must be to discerne what guides or interpreters to follow. And after that I must take the subject matter (into which I cannot look my self) upon the credit of their evidence without pretending to see into it with my owne eyes. There is use enough of our judgment in determining of our guides; the rest is but an [*]ergo, like the conclusion to the premisses, and in this sence, I seeing with my owne eyes, owne the canon of the New Testament as our church receives and interprets it, and hearken to no under working crittiques whatever.

Sir, I have dwelt a little longer upon this last matter[,] because I think it a rule (in the generallity you express) very much **[f. 36v]** abused, and tending more to [*]deceipt then any other principle whatsoever that can be instilled into men. Therefore I cannot but wish, whenever you recomend it, you will not forgett to distinguish [between the thing and the evidence[241]], without which it is madness rather then discretion. The trite method of seducers is, first to insinuate to people that they are wise and can judg, for then they are prepared to swallow unthinkingly any poyson [that] is offered to them. As juglars[242] who seeming to give the choice of a card, insensibly put that into your hand which they intend you should take; to say nothing of the ancient heresiarchs and their sects. I have observed that all the simple people of our fanatick congregations[243] are fix't with high [*]conceipt of their owne sense; [for] nothing els could keep them under such a senceless separation [from our

---

[240] I.e., *sola scriptura*; see *supra* 4.4.2.

[241] See *supra* f. 35v and n.239.

[242] I.e., jugglers: tricksters.

[243] I.e., the sectaries or so-called enthusiasts, concerning which see *supra* 4.4.2 and Dockrill, 'Spiritual Knowledge'.

church]. And the teachers, knowing their pride of mind and aptness to think themselves holy and wise, doe but thro[w] out that insinuation, and then bamboozle them into a perswasion of the most extravagant nonsence. In a word such a rule (generally advised) is an hook baited with flattery and seldome fails to catch scaly fish,[244] and hath its transcendent in the Quaker's light within them. So much concerne hath pride in the prejudice of men's judgments.'

    This, Sir, is my answer [to Clarke], a great deal more then it needed; but I have not the good intent of being short, [so] I launched out a little more in the last article for the sake of the lady.[245] It is strang[e] to see how all imposters work the same way, and that people should be made fools, that instant they are made to think themselves wise. And what an open [*]fourbery it is, coming from a grave devine, to say to the blind, see with your owne eyes.

    As to the letter which you have had the trouble to peruse,[246] you must beleev I am not without the greatest satisfaction, that you have not layd it aside[,] as [that is] what I should expect, having so litle foundation for such a dispute. I thought it fittest for me to take the imposing part of it cheifly to task;[247] and as to the main, to build upon that rock[248] which he endeavours to shake or undermine. You mentioned alterations of some small matters.[249] How could you

---

[244] I.e., the figure of speech called 'a wheadle'; see *supra* 5.5.2: No. 1, RN→SC 1713 (f. 7v) and *infra* Glossary.

[245] I.e., the 'incomparable' lady; see *supra* 4.4.5 and *infra* 5.5.2: No. 3, RN→Bedford 1719 (f. 1v).

[246] I.e., the letter *supra* 5.5.2: No. 1, RN→SC 1713; see also *infra* 5.5.2: No. 3, RN→Bedford 1719 (f. 1v).

[247] Clarke, *SDT*, e.g., Introduction, p. vii and Pt III, Ch. 2, p. 450, used the word 'impose' to imply or declare outright that the consubstantial doctrine had been a later imposition to, and corruption of the fundamental doctrines of the primitive Church, as he defined it *supra* 4.4.3.

[248] I.e., the consubstantial doctrine.

[249] See RN, 'A Copy of a Letter from the Reverend Dr. Hicks' 23 May 1713 (UK:Lbl) Add MS 32551: f. 34v, where Hickes writes: 'I desire to keep your letter

oversee to [those], when the whole must want forme, justice, and that gravity of expression, as the subject demands? I wonder how I came to venture so farr. But whither publick or private, I beg you will make those alterations you think fitt. I am sure the more the better. At present the privacy of the thing, and respect to the lady, makes it difficult to send it out. If any reasonable occasion (or indemnity) emerg, [then] I shall submitt it to your [*]dispose,[250] but under a veil as to

Your most faithful
humble servant
R. North

No. 3 (COPY OF A LETTER) NORTH TO BEDFORD 1719 (UK:Lbl) 32551: f. 1v

[f. 1v]

To the Reverend Mr -------

It is reasonable, together with these papers[,][251] to send you a short account of the reall inducement I had to medle in controversie so much out of my element; what? controversie[,] say you? Yes *theologia*: and which may serve in quality of a preface, in case you send them forth past your recall.[252]

---

a litle longer, which if you would consent to have it made publik, I would consent to prepare it with a few inconsiderable alterations for the press. I hope something he [Clarke] will say to it privately or publequely will make it needfull to be published, for it is great pitty it should be confined to private hands.'

[250] Hickes was dead by the time a 'reasonable occasion' emerged, namely, publication of the second revised version of Clarke, *SDT* [1719].

[251] I.e., the 'papers' described *supra* 5.5.1.

[252] MS has the lines of this first paragraph somewhat squeezed in above the space of the second paragraph.

When Dr. ------'s book of the Trinity came first out, and made much nois, I was informed by letter from an incomparable lady,[253] at whose house the Doctor used to be kindly received, that he had a mind to know what I thought of his book. For I had once the favour of a short correspondence with him, about some phisiologicall matters;[254] and on this new occasion a letter or two pas't, in which I could say nothing on the subject, but that from the report of his book, I did not like it. For I then understood his designe was to goe off from the Scripture into philosofy. But I was given to understand that was a clear mistake, and that his interest was the direct contrary, viz., to hold to the Scripture against all incertein reasoning. In fine I found my self under a necessity to say somewhat to the book. And having borrow'd it of a neighbouring minister,[255] I perused it, which produced this answer.[256] I was conscious of great disability in such a caus, and after I had dispatched my answer, directed to our mediatrix the incomparable lady, I comunicated a copy, with some small alterations of expression in a few particulars, to the Reverend Dr. Hicks,[257] whom I had familiarly known for divers years before;[258] and that produced his letter, which [also] is subjoyned.[259] And as he foretold, the Doctor sent me a short answer, and I could as little

---

[253] See *supra* 4.4.5 and 5.5.2: No. 2, RN→Hickes 1713 (f. 36v).

[254] I.e., in 1706; see *supra* 3.3.1 and 3.3.2.

[255] Possibly, Ambrose Pimlowe, vicar of Rougham.

[256] I.e., the subjoined copy; see *supra* 5.5.1 and version 3 of RN→SC 1713 (UK:Lbl) Add MS 32551: ff. 2–33.

[257] I.e., one of the versions of RN→SC 1713 presumed lost; see *supra* 5.5.1.

[258] RN, through his brother Francis, first met Hickes about 1679, when the latter was domestic chaplain to John Maitland, then 1st duke of Lauderdale (1672). After 1690, when Hickes became a non-juror, RN acted as a legal advisor and subsequently retained a correspondence on matters relating to law and its history; see Kassler, *HRN*, pp. 47, 133–4, 170, 200, 285 n.224, 321 n.320, 329 n.360, 335–6 n.392, 378–9, 391, *et passim*; *NP 6*, pp. 43–152; and RN, *Life/FN*, pp. 54, 199–200.

[259] See *supra* 5.5.1.

forbear a reply, which letter was comunicated to Dr. Hicks[260] as he [had previously] desired of me.[261] And these letters are added to the former.[262] Dr. Hicks, either for complement, or opinion that my letter, made publick, might doe some good, (and I beleev the latter, for he was a person whose civility, when he did not approve, was silence) urged the publishing, and offered his paines to fit it for the press,[263] which I wish he had done. But cruell infirmitys and troubles fell upon him,[264] and at length procured his eternall rest from all worldly affairs. And ever since till lately the controversie [regarding the Trinity] hath layn more quiet; and it was not thought reasonable to revive it, by adding one more to the numerous answers which that book hath had.[265] But now I am called upon by some that I have shewn it to,[266] to bring forth above board my answer, becaus the dispute rages more then ever. And having leav from the lady so to doe, and I think without which, nothing could have brought me to consent, I have submitted to such as I have reason, both as to a right judgment of the controversie as of this letter to rein [it] in, to dispose of it one way or other as they think most [fit] for the good of mankind.

---

[260] See *supra* 5.5.2: No. 2, RN→Hickes 1713 (ff. 35–36v).

[261] See RN, 'A Copy of a Letter from the Reverend Dr. Hicks', 23 May 1713 (UK:Lbl) Add MS 32551: f. 34v: 'I am sorry you have desired him [Clarke] to give you no answer, but I beleev he will not be able to forbear; and if he returne you any answer, I hope you'll doe me the favour to impart it to me.'

[262] They are added only as quoted extracts in *supra* 5.5.2: No. 2, RN→Hickes 1713 (ff. 35–36).

[263] See *supra* 5.5.2: No. 2, RN→Hickes 1713 (f. 36v n.248).

[264] In 1691, as a protest against his deprivation, Hickes, then dean of Worcester Cathedral, affixed to the entrance gate to the choir a claim of right against all intruders. As a consequence of this strong legal protest, Daniel Finch, 2d earl of Nottingham, then secretary of state, issued a Privy Council warrant for Hickes's arrest; and on 11 August 1691 he was outlawed. And thus began the 'cruell infirmitys and troubles' referred to by RN.

[265] For some of 'the numerous answers' up to 1719, see Ferguson, *ECH*, pp. 47–82, 98–118.

[266] The identity of the persons in question remains to be discovered.

Whither this [letter when published] may produce any reply, or not, I doe not foresee. Perhaps contempt may serve the turne, tho neither that, nor any humane artifice of thinking or words, can answer our blessed Saviour's express declarations of himself, and the subsequent doctrine and practis of the holy Apostles [is] intirely conformable therewith, after which being understood, all other considerations [are of] litle concern.[267]

Sir, Yours etc.

---

[267] MS has this last paragraph in the left-hand margin, written perpendicular to the main text, where just below the last word of the paragraph, in tiny handwriting, are the words: 'S^r, Y^{rs} &c.'

EPILOGUE

Although North's final copies of the correspondence and related material were made in 1719,[1] the correspondence itself came to a close in 1713. With the exception of the 1719 copies and one fragment of the 1706 correspondence,[2] only two other of his manuscripts, both preserved in the British Library, mention Clarke by name.[3] This fact is puzzling, since in manuscripts written up to the time of his death, North frequently cites Newton's name, as well as his books on rational mechanics and optics. Of course, it is possible that Clarke may be named in some of North's manuscripts that have been lost[4] or that are still in private hands. It also is possible that the absence of Clarke's name is due to the main issue of the 1713 correspondence, in which North raises questions concerning the character or sufficiency of evidence for Clarke's particular hypothesis. And that the latter did not hold his hypothesis tentatively is suggested by two subsequent events. First, in May or June of 1713, Clarke omitted the communion service on Trinity Sunday; and royal displeasure led to his dismissal as one of Queen Anne's chaplains-in-

---

[1] See *supra* 5.5.1.

[2] See *supra* 3.3.2: No. 9, RN→SC, 27 November 1706 (f. 280v).

[3] See RN, *Life/JN*, p. 96, where RN indicates that the marginal headings are to be regarded as titles of chapters 'and then the series may be continual, after the model of the Dutch LeClerc in his *Parrhasiana*, whom (no more than the English Dr Clarke) I should not follow in anything but philology'. See also f. 36 in the manuscript described *supra* 3.3.3 n.266, RN, 'Physica' (UK:Lbl) Add MS 32544: ff. 1–274v, where the reference in question appears as 'Dr. Clerck in trans.', i.e., 'A Letter from the Rev. Dr. Samuel Clarke to Mr. Benjamin Hoadly, F.R.S. occasion'd by the present Controversy among Mathematicians, concerning the Proportion of Velocity and Force in Bodies in Motion', *Philosophical Transactions* (1727–8), 35: 381–8. For context, see Hankins, 'Eighteenth-Century Attempts'.

[4] I.e., during their dispersal; see *supra* 1.1.1 n.3.

ordinary.[5] Second, in 1714 a complaint concerning Clarke's *Scripture Doctrine of the Trinity* was sent by the Lower House of Convocation to the Archbishop of Canterbury,[6] as well as to the bishops of that province; and thus began a series of events that led to Clarke's decision not to accept any further position in the Church of England that involved subscription to the thirty-nine articles.[7]

But dispensing with possibilities, it is more likely that Clarke is represented in North's manuscripts under such terms as 'followers of the *Principia*', 'Newtonians', 'adversarys' or 'attractionists'. Indeed, for reasons of discretion North often (but not always) avoided naming persons who were still living, especially if he was writing about them critically; and this would explain his recourse to a more covert style of reference, as previously exemplified when considering whether North had read Clarke's Boyle Lectures.[8] If this is the case, then perhaps after 1713 he continued the practice, as in the following instance, where he writes:

> ...in England ... freedome of philosofizing, if any where, reignes here. Yet one may perceive a yearning of a sort of grave men against it, as if the new philosofy, or what is called Cartesian, corrupted men's manners, if not their religion, inclining them to atheisme; and I beleev most of them [i.e., the grave men] are in earnest. However I ... ascribe atheisme to the pravity[9] of Christian witnesses, then to any philosofy. For religion is founded on testimony more then argument ... and the Gospel and faith [concern] ... things seen and heard as well as unseen and beleeved. Wherefore the Apostles ... and fathers of the church confirmed their faith with suffering and death

---

[5] See Whiston, *Historical Memoirs*, pp. 69–70. Note that Trinity Sunday, which falls sometimes in May and sometimes in June each year, is the first Sunday after Pentecost. Unfortunately, a search of newspapers for this period did not turn up any public announcement of Clarke's dismissal.

[6] I.e., William Wake, who succeeded Tenison in 1714.

[7] See Ferguson, *ECH*, pp. 83–97, and Wiles, *Archetypal Heresy*, pp. 113–15.

[8] See *supra* 3.3.5.

[9] (OED 1, now rare or obs.) moral perversion or corruption; wickedness.

> and not ... with pre-eminences and preferments.... For who
> would beleev men that preached for their owne grandure and
> authority? ... Therefore I consider it is a corruption of
> authority, which, perhaps, cares not [that] a freedome of
> thinking should prevaile, and not this or that philosofy, while
> the people in general are neither capable of nor concerne
> themselves with any that hurts religion.[10]

Certainly, Clarke had 'preferments', although, as already pointed out,
they ceased in 1714 when Convocation met to adjudicate on his
book.

Leaving these and other problems for future investigators, two
other events that occurred during the same period deserve mention.
First, in 1713 Newton published a second edition of his book on
rational mechanics. Second, in 1714 John Edwards,[11] a Church of
England divine, published a pamphlet, the bulk of which contained
an analysis of Clarke's answers to three critics of his 1712 book in
order to show that (as the title-page has it) 'the Doctor is as deficient
in the *Critic Arts*, as he is in *Theology*'.[12] But the importance of this
pamphlet resides chiefly in its 'Postscript',[13] where Edwards
expressed his conviction that in the newly published edition of
Principia mathematica, Newton enunciated the same conception of
God that Clarke had pronounced in a reply to one of his critics,
namely, that in Scripture the name God is '*a Relative Word, and a
Word of Dominion and Power*'.[14] As evidence, Edwards provided the

---

[10] See f. 23v in RN, 'Physicks' (UK:Lbl) Add MS 32545: ff. 23–25v (middle
period?), a fragment chiefly concerned with the enemies of 'that which is called
new philosofy, a knowledge brought into the world by a set of cotemporary
hero[e]s'.

[11] During the period when Francis North was a student at St. John's College,
Cambridge (1653–5), Edwards was admitted as a sizar (1654), graduating B.A.
(1658), M.A. (1661) and B.D. (1668). Between 1695 and 1697 he published a
number of critical responses to Locke's *The Reasonableness of Christianity*.

[12] For an analysis of Clarke's answers to his critics, see Edwards, *Some Brief
Critical Remarks*, pp. 29–36.

[13] Ibid., pp. 36–40.

[14] Ibid., p. 36.

reader with the original Latin of a passage in 'Scholium generale' that had been added to, and concluded the second edition of Newton's book. And he followed the Latin with his own English translation as follows:

> *God* is a *Relative* Word, and hath Reference to *Servants*: And the *Deity* is the *Dominion* of God, not on his own Body, but on his Servants. The *Supreme God* is an Eternal, Infinite, and absolutely Perfect Being; but a Being tho' never so Perfect, without Dominion, is not *Lord God*— It is the *Dominion* of a Spiritual Being that makes *a God*. If this Dominion be true, it makes a True God; if Supreme, it makes a Supreme God, if False and Counterfeit, it makes a False God.[15]

To analyse this passage, Edwards considered the meaning of some of its words, beginning with God's 'own Body'. About these words he asked: Is God 'a *Corporeal Substance*', somewhat like the 'Conceit' of Spinoza? Or is God 'the *Infinite Extension*', like the 'Apprehension' of More? Or is this extension related to 'the *Infinitum Sensorium*, which Sir *Isaac* saith belongs to God, *and in which he moves all Bodies as he pleases, as the Soul that is in Man moves the Members of his Body*'?[16] If this is the tenor of Newton's meaning, then it is 'certain and unquestionable, that he holds God *to have a Body*, which is Unphilosophical enough, as well as Untheological'.

Next follow the words '*Supreme God*' which, Edwards pointed out, is the epithet that all Arians and Socinians use to distinguish the Father from the Son, whom they hold to be inferior. But if the superior God is '*Lord God*, because of his *Dominion*' which makes him a God, then why is '*this Attribute* chosen out before all the rest, to constitute a *Deity*'? Indeed, he cautioned Newton to take heed in

---

[15] Op. cit. ('Postscript'), p. 37. The em-dash seems to be used in place of '...'. Cf. the original passage in Newton, *Principia mathematica* [1713], Bk III, 'Scholium generale', p. 482.

[16] See ibid., p. 483: 'In ipso continentur & moventur universa' (in him are all things contained and moved), and Newton, *Optice*, Qu. 20, p. 315, and Qu. 23, p. 346.

urging this notion of dominion as absolutely necessary to constitute a deity, 'lest he deny the *Eternity* of God, for God cannot properly be said to have had *Dominion*, when there was nothing to possess'.[17] And the pamphlet writer continues in this vein at some length.[18]

I have found no evidence that North read Edwards' 'Postscript', perhaps, the first public criticism of Newton's conception of God. But, as part of his continuing search for truth, he did consult the 'noble undertaking' of the second 1713 edition of Newton's *Principia mathematica*, which 'hath polish't away some distinguishable *maculæ*'[19] in the first edition and, hence, is 'accepted now as a standard of all natural knowledge'.[20] As is well known, this second edition included a preface by the editor, Roger Cotes, who attacked the philosophy of Descartes and his followers.[21] And North may have been prompted by this preface to reconsider a subject first mooted in his draft history on authorities examined at the end of Chapter 2.[22] For in a cursorily written, but incomplete essay preserved in a notebook, he offers some further critical reflections on the philosophies of Descartes and Newton, though 'with the liberty I profess[23] and without any thought of deminishing the author

---

[17] Edwards, *Some Brief Critical Remarks* ('Postscript'), p. 39.

[18] In 1712 and 1713 Clarke had answered Edwards' two previous criticisms; but in the above case he kept silent; see Ferguson, *ECH*, pp. 60–3 and 77–8.

[19] I.e., spots.

[20] See f. 39v in RN, 'The philosofy of Cartes and Newton' (UK:Lbl) Add MS 32548: in Notebook 2, ff. 32v–50 (middle period). The final, incomplete sentence on f. 50 reads: 'I shall conclude this essay, with one note against this combined power, viz., the perseverance [i.e., inertial tendency] and centripetall [i.e., attractive force], which act against each other; that it'. But the space left at the bottom of f. 50 is insufficient to complete the 'note'; and on f. 50v a new essay begins.

[21] I.e., orthodox Cartesians, a class into which RN does not fit. For an interesting attempt to deal with this problem of classification in relation to French natural philosophers, see Clarke, *Occult Powers and Hypotheses*.

[22] See *supra* 2.2.5.

[23] According to RN (f. 39v): 'Every man's thinking freely of naturall things seems clearly his right'.

[Newton], (whom I admire for his stupendious reach of thought, especially in geometry, which science he hath advanced, I may say, to a *ne plus*)'.[24] As two statements in this text make clear, North continues to focus his attention on abstract terms. For, on the one hand, he claims (f. 42), that because Newton had 'overturned the foundations of our Royal Society' (as declared in their motto, *Nullius in verba*[25]), lesser philosophers now harp upon certain words 'to which he hath given a credit instead of any clear notion of things'; and, on the other hand, he admits (f. 45v) that 'our controversie will at length melt into words'.[26]

But this essay does not represent North's last attempt to assess critically Newton's challenge to the various versions of new philosophy. For after numerous assays of the opinions of Newton and his followers in order to 'unfold' them in the form of critical reflections, he finally included a more or less finished 'dissertation' in a manuscript, the overall title of which is (f. 1): 'The Life of the Honourable John North, S.T.P. Late professor of the Greek tongue and master of Trinity College in Cambridge, and one of the prebends of Westminster. With a dissertation of the new and moderne (new) philosofye inserted. By a Freind 1728'.[27] As I have indicated

---

[24] In this essay two Newtonians are named: William Derham and David Gregory. The context suggests that RN is referring to Derham's *Astro-Theology* (1714) and Gregory's *Elementa astronomiæ physicæ et geometricæ* (1702 or its English translation 1715). In the preface of the latter work, the author set out Newton's ideas on the antiquity of the theory of gravitation but without naming him; for details, see Cassini, 'Newton: The Classical Scholia'.

[25] See *infra* Glossary.

[26] For another essay that resulted from *Principia mathematica* [1713], see RN, 'Of Sounds' (UK:Lbl) Add MS 32537: ff. 133–148v (middle period); see also *NP 6*, pp. 3–9 (digest of the manuscript), 11–16 (editorial introduction, now requiring some emendation), 17–39 (edition) and *supra* 1.1.1 n.29.

[27] See (UK:Lbl) Add MS 32514: ff. 1–166v, after which follows: 'Notes of Dr. North', ff. 167–227v. At the beginning of the insertion is the statement (f. 62): 'Here follows a dissertation of the new, and moderne new philosofye, which may be perused or let pass to fol. 227 [i.e., RN's, not the BL's foliation] according as the knowledg of late authors may have given a tast[e] or not'. For a possible draft

elsewhere, the inserted 'dissertation', which occupies ff. 62–125v, fulfils North's obligation to his brother John, who had hoped to write a history of philosophy. But its title is somewhat misleading, for the 'dissertation' does not represent a systematic comparison of the two philosophies prominent in North's day. Rather, it represents North's final assay of the 'moderne' or Newtonian philosophy, an assay that is not restricted to the issues he first raised in his notes on reading Newton's two books.

To date, there has been no close study of North's 'dissertation' or, indeed, of his unpublished writings that relate to it. But to understand the former, it is necessary to identify the latter, because to produce the 'dissertation', North relied on the technology of common placing. In this particular instance, this meant abridging parts of some of his previous writings and then structuring his abridgments into seventy-six rationally ordered sections devised for the purpose at hand.[28] For example, an issue raised in some earlier manuscripts, discussed previously, reappears in the penultimate section of the 'dissertation'. Entitled 'Proofs of a Deity and Providence', this section begins:

> It hath bin ... pretended [by Newton] that the analiticall process, from phænomena to principles, must at last come up to a demonstration of the deity, and providence, more effectually then by the synthetick way. I wish the author had undertaken it and hope wee are not to stay, till his method produceth it (f. 123).[29]

---

outline for the 'dissertation', see the fragment in (UK:Lbl) Add MS 32546: f. 308 [f. 308v is blank].

[28] Regarding this technology, see Kassler, *HRN*, pp. 63–6. For the sequence of leaves, RN has used single and double letters of the alphabet, so that a study of these marks, some of which seem to be out of sequence, may reveal something about the composition of the 'dissertation'. But the main difficulty will be to discover the specific manuscripts that may be the sources for the commonplace notes in the 'dissertation'.

[29] Perhaps a reference to Newton, *Principia mathematica* [1713], 'Scholium generale', p. 483: 'Et hæc de Deo; de quo utque ex Phænomenis differere, ad *Philosophiam Experimentalem* pertinent'. (And thus much concerning God; to

From what was said in Chapter 3, we know that Clarke could avoid using 'the analiticall process' by incorporating some of Newton's discoveries into his physico-theological version of the teleological argument.[30] But he did not originate this version of the argument from design, for similar arguments were used by some of the Newtonians who read Boyle Lectures both before and after Clarke read his.[31]

As a consequence, perhaps, North, in a very long sentence (ff. 123–124), re-affirms the foundation of his own synthetic argument from 'pure extension and its modes, joyned with animall sensation'. For he maintains that these two principles produce 'those sublime ideas of the world' and go 'as farr in demonstration of a devine power and wisdome as the whole state and composition of nature and our reflections and judgments can reach'. But he does not include his version of the argument, because he recognises that it would 'run out into' too much length. Nevertheless, he has 'but touch't upon it, in persuance of what was hinted before, and as the observations reflected upon hath given occasion'.[32]

He therefore concludes the section, as well as the long sentence by 'wishing that the setting up so many occasionall powers, as of late have bin brought forth in the philosofick scene, may not tend to politheisme, for the heathen theology was litle other then of certein powers presiding over the various provinces of nature'. To find a plausible explanation for North's rather obscure conclusion, it will be important to remember some points made previously. First, in his 1712 book Clarke promoted a political model of the Trinity in which

---

discourse of whom from the [external] appearances of things, does certainly belong to Experimental Philosophy.)

[30] See *supra* 3.3.5.

[31] For the first lecturer, see *supra* 3.3.4 n.291. For one who read the lectures in 1711 and 1712, see William Derham, *Physico-Theology* (1713), who, in the third, corrected edition (1714), p. 21, mentions being acquainted with Clarke.

[32] See especially ff. 77v, 92, 112–113v, 121v, which relate to 'devine power and wisdome', in RN, '...a dissertation of the new and moderne (new) philosofye....'.

Christ was reduced to a ministering cause.[33] Second, in his 1713 letter to Clarke, North observed that the heathen theologians had conceived natural powers as 'enarches of fire and water with regard onely to their offices and ministrations, not hereby derogating from the great superior authority that is over all' and that Clarke's political model of 'subordination of authority in the exercise of one power debases Christ into a near enarch or subgovernor'.[34] Third, North's letter to Clarke was sent sometime before 2 June,[35] the month when the second edition of Newton's *Principia mathematica* was published.[36] We might ask, therefore: Could North's reading of and reflecting upon the contents of the new edition have led him to believe that Newton conceived his non-mechanical 'powers' as subordinate to the authority of the one almighty monarch whose name is God, because he rules a kingdom or dominion?

---

[33] See *supra* 4.4.3.

[34] See *supra* 5.5.2: No. 1, RN→SC 1713 (ff. 4v, 5v).

[35] See *supra* 5.5.1.

[36] See *supra* 2.2.1.

GLOSSARY

The terms and phrases glossed below are from the editions of North's writings in Chapters 1, 2, 3 and 5. His spelling appears in the initial term or phrase. But, when required, an equal sign (=) is followed by the modern spelling; corrections are indicated by inserting missing letters or words in square brackets; and, in a few cases, his spelling is followed by a 'see' reference to the correct spelling.

ab eterno
>   (Lat.) from eternity [3.3.2: No. 1 (f. 85)], p. 136

abutted
>   (Northism; from abutment, OED 4, ex. RN, *Examen* II.v. §81, 365) firmly supported [5.5.2: No. 2 (f. 35v)], p. 307

ad captum, i.e., *argumentum ad captum*
>   (Lat.) lit., appeal to mental capacity [5.5.2: No. 1 (f. 23)], p. 287
>   see also *secundum captum humanum*

ad cleram, i.e., *argumentum ad cleram*
>   (Lat.) lit., appeal to the clergy [5.5.2: No. 1 (f. 5)], p. 256

ad infinitum
>   (Lat.) to infinity [3.3.2: No. 1 (ff. 83, 86); 3.3.2: No. 2 (ff. 293, 293v)], pp. 132, 137, 140

ad infinitum vitum
>   (Lat.) lit., to an infinite rim; but here, to the edge of a die [3.3.2: No. 1 (f. 16v)], p. 141

ad literam, i.e., *argumentum ad literam*
>   (Lat.) lit., appeal to the literal [5.5.2: No. 1 (f. 16v)], p. 276

ad populum, i.e., *argumentum ad populum*
>   (Lat.) lit., appeal to the people [5.5.2: No. 1 (f. 5)], p. 256

aneantize
>   (OED *v.* obs.) reduce to nothing [5.5.2: No. 1 (f. 4)], p. 254

animalculi = *animalcula*
>   (Lat.) little animals [3.3.2: No. 2 (f. 293v)], p. 142

animose
>(OED *a.* obs. = animous) spirited, courageous; also, as here, hot tempered [5.5.2: No. 1 (f. 9v)], p. 265

ansæ
>(Lat.) lit., rings; here, appendages of Saturn's ring [3.3.2: No. 7 (f. 287v); 5.5.2: No. 1 (ff. 19v, 24)], pp. 158, 281, 289

anseatick animals
>(Northism) animals living on the appendages of Saturn's ring [5.5.2: No. 1 (f. 24v)], p. 290

Arrianizing = Arianizing
>(OED *ppl. a.*; 1760) professing or teaching Arianism [5.5.2: No. 1 (f. 14)], p. 272

arriere = *arrière*
>(Fr., see also OED) rear [5.5.2: No. 1 (f. 18v)], p. 279
>
>see also 'rere

artificiall = artificial
>(OED *a.* II.9, obs.) cunning, deceitful [5.5.2: No. 1 (ff. 7, 15, 18v)], pp. 260, 274, 279

artigliery
>(OED *sb.* 1, obs.) ammunition in the wide sense [5.5.2: No. 1 (f. 3v)], p. 253

assertorily
>(OED *adv.* ?obs.; 1679) by way of assertion, assertively [5.5.2: No. 1 (f. 27v)], p. 296

barriere = *barrière*
>(Fr.) fig., obstacle [5.5.2: No. 1 (f. 9v)], p. 265

batterys = batteries
>(OED 2, obs.; 1591–1625) beating of drums (to signal an assault) [5.5.2: No. 1 (f. 7)], p. 260

Belle & beau = *Bel et bon*
>(Fr.) all very fine [5.5.2: No. 1 (f. 8v)], p. 263

bonti = *bonté*
>(Fr.) goodness; benevolence [5.5.2: No. 1 (f. 26v)], p. 294

c[a]elum
>(Lat.) heavens [3.3.2: No. 7 (f. 280)], p. 157

cælum immutabile
>(Lat.) unchangeable heavens [3.3.2: No. 7 (f. 287v)], p. 159

cant

(OED *v.*[3] 3. *intr.*, obs.; 1625–98) use the special phraseology or jargon of a particular class or subject; here, of scholastic philosophy [5.5.2: No. 1 (f. 24v)], p. 290

cass

(OED *a.* a, obs.) dismissed, done away with [5.5.2: No. 1 (f. 10)], p. 266

cession (OED 1a, obs.; 1626–93) the action of giving way or yielding to physical force or pressure [3.3.2: No. 1 (f. 85v)], p. 136

colour, colours

(OED *sb.* III.12b, obs.) allegeable ground/s or reason/s; excuse/s [3.3.2: No. 2 (f. 293v); 5.5.2: No. 1 (ff. 11v, 13, 14, 18v)], pp. 141–2, 268, 270, 272, 279

commentation

(OED 1a, obs.) expository note, comment, gloss [5.5.2: No. 1 (f. 18v)], p. 279

comparatio motuum

(Lat.) comparison of movements [3.3.2: No. 1 (f. 86v)], p. 138

compounders

(OED 2, obs.; 1539–1734) compromisers [5.5.2: No. 1 (f. 29v)], p. 299

conceipt = conceit

(OED *sb.* II.4b, obs.) opinion [3.3.2: No. 1 (f. 83); 5.5.2: No. 2 (f. 36v)], pp. 133, 309

conceipt = conceit

(OED I.1c, obs.; 1588–1668) concept [2.2.2: No. 1 (f. 174); 3.3.2: No. 1 (f. 83v)], pp. 66, 133

conceipted = conceited

(OED *v.* 1. *trans.*, obs.) conceived [3.3.2: No. 1 (f. 83v)], p. 133

conceipts = conceits

(OED *sb.* III.7) fanciful notions, whims [1.1.2 (f. 6); 5.5.2: No. 1 (f. 5v)], pp. 32, 257

contracting

(OED *vbl. sb.* from contract *v.* III.9, obs.: to shorten) abridging, abbreviating [1.1.2 (f. 5v)], p. 32

contra mundum

(Lat.) against the world [5.5.2: No. 1 (f. 27)], p. 296

convict

(OED *pa. ppl.* A, obs.) proved, pronounced guilty [5.5.2: No. 1 (f. 9v)], p. 265

couched
> (OED *v.*¹ III.13, obs.) concealed, hidden [5.5.2: No. 1 (ff. 15, 25v)], pp. 274, 292

couched
> (OED *v.*¹ III.14, obs.) included [5.5.2: No. 1 (f. 19)], p. 280

cuilibat judere esto = cuilbet iudicare esto
> (Lat.) let it be left to the individual to decide [3.3.2: No. 2 (f. 293v)], p. 142 and n.109

deal
> (OED *v.* I.6, obs.; 1250–*c.*1400) deliver, give forth, render, bestow [5.5.2: No. 1 (f. 2)], p. 250

deceipt = deceit
> (OED 1b *in Law*) false dealing; concealment of truth in order to mislead [5.5.2: No. 2 (f. 36v)], p. 309

delated
> (OED 2, obs.) tendered or offered for acceptance or adoption [3.3.2: No. 1 (f. 85)], p. 135

deturbed
> (OED *v.* obs.; 1609–57) thrust out [5.5.2: No. 1 (f. 25v)], p. 292

d[e]uterai phrontides
> (Gk.) second thoughts [1.1.2 (f. 4)], p. 30

differènce
> (Fr.) difference [5.5.2: No. 1 (f. 17)], p. 276

directum
> (Lat.) a straight line [3.3.2: No. 9 (f. 287)], p. 164
> see also *in directum*

dispose
> (OED *sb.* 2, obs.; 1611–51) direction, management [5.5.2: No. 2 (f. 36v)], p. 311

dum vitant [stulti v]itia, in contraria currunt
> (Lat., Horace, *Satires* 1.2.24) while they are avoiding one fault, fools rush into others [5.5.2: No. 1 (f. 7v)], p. 262

Eadem ratio, eadem lex
> (Lat.) Same reason, same rule [5.5.2: No. 1 (f. 12)], p. 268

economy
> (OED IV.8a, obs.; 1592–1720) of the Trinity: the organisation, internal constitution, apportionment of functions [5.5.2: No. 1 (ff. 3, 7v, 23v], pp. 252, 261, 288

ejusdem generis
>   (Lat.) of the same kind [5.5.2: No. 1 (f. 19v)], p. 281

enarches, enarch
>   (Northism?; from enarchos, Gk., in authority) subgovernors,
>   subgovernor [5.5.2: No. 1 (ff. 4v, 5v)], pp. 255, 257

equabiliter fluit, et mutar nequit
>   (Lat.) flows equably, and unable to change [2.2.2: No. 1 (f. 174v)],
>   p. 68

equilibre
>   (OED 1, obs.) equilibrium [3.3.2: No. 9 (f. 287)], p. 165

ergo
>   (OED *sb.* 2, *nonce word*, obs.; 1597) logic chopper [5.5.2: No. 2
>   (f. 36)], p. 309

escume
>   (OED obs.; 1527) froth [5.5.2: No. 2 (f. 36)], p. 309

essence, essences
>   (OED *sb.* 1, obs.; 1576–1688) existence, viewed as a fact or as a
>   property possessed by something [3.3.2: No. 1 (f. 82v); 3.3.2: No. 2
>   (f. 293)], pp. 68, 131, 141, 166

essence, essences
>   (OED *sb.* 3, obs.; *c.*1532–1664) specific being, manner of existing;
>   nature, character [2.2.2: No. 1 (f. 178); 3.3.2: No. 9 (f. 383); 5.5.2:
>   No. 1 (ff. 2v, 3, 5, 5v, 11v, 12, 14v, 16, 18, 19v, 21, 22v, 23, 23v,
>   25v, 29)], pp. 76, 251–2, 255, 258, 268–9, 273, 275, 278, 280–81,
>   283, 287–8, 292, 299

essentially
>   (OED *adv.* 1b, obs., 1593–1624) on the ground of (one's) actual
>   nature; here, of Christ's divinity [5.5.2: No. 2 (f. 35v)], p. 306

et habes confitentem reum
>   (Lat.) and you have a confession of guilt [3.3.2: No. 9 (f. 383v)],
>   p. 167
>   see also *habes confitentissimum reum*

et herent in cortice
>   (Lat.) and cling to the surface [5.5.2: No. 2 (f. 35v)], p. 254

ex animo
>   (Lat.) sincerely; from the bottom of the heart [5.5.2: No. 2 (f. 35v)],
>   p. 307

facile est inventis addere

> (Lat.) it is easy to improve on discoveries already made [1.1.2
> (f. 5v)], p. 32

fact

> (OED 1c, now obs.) evil deed, crime [5.5.2: No. 1 (ff. 5, 15)],
> pp. 256, 274

faction

> (OED *v.* 1. *intrans.*, obs.; 1656) to form into factions [1.1.2
> (f. 4v)], p. 30

filioque

> (Lat.; OED 1st Engl. use 1876) and from the son [5.5.2: No. 1
> (ff. 25v, 26)], pp. 292–3

flatters

> (OED *sb.*¹ obs.; 1340–1559) flatterers [1.1.2 (f. 4)], p. 30

fond

> (OED 4, obs.) trifling, valued only by fools [1.1.2 (f. 5)], p. 31

fondness

> (OED 1, obs. or dial.) foolishness, folly [1.1.2 (f. 4)], p. 29

formes = forms

> (OED *sb.* I.7, obs.; 1382–1690) models, patterns, examples [5.5.2:
> No. 1 (ff. 8, 13v, 14v)], pp. 263, 271, 273

formes = forms

> (OED *sb.* I.10, obs.; 1297–1641) manner of doing anything; here,
> of using language [5.5.2: No. 1 (ff. 8v, 21v, 22)], pp. 263, 284–5

fourbery

> (OED obs.) piece of deception; trick, fraud, imposture [5.5.2: No. 2
> (f. 36v)], p. 310

fucus

> (OED 1b, *fig.*, obs.) cosmetic; artifice [5.5.2: No. 1 (f. 15v)], p. 274

gloze

> (OED *v.* 1. *trans.*, obs.) expound, interpret [5.5.2: No. 1 (f. 28)],
> p. 298

glozed

> (OED *ppl. a.*, obs.; 1393–1630) provided with glosses, commented
> on; speciously adorned [5.5.2: No. 1 (f. 17)], p. 277

glozes

> (OED *sb.* 1, arch.) comments or marginal notes; expositions [5.5.2:
> No. 1 (f. 17)], p. 277

gree
> (OED *sb.*² 1, obs.; 1300–1590) favour [5.5.2: No. 1 (f. 6v)], p. 259

grossier = grossière
> (Fr.) gross, rude, unpolished, uncouth [5.5.2: No. 1 (f. 17)], p. 276

grutch
> (OED *v.* 1. *intr.*, obs. dial. or arch.; 1225–1679) complain [5.5.2: No. 1 (f. 9v)], p. 265

habes confitentissimum reúm = habes confitentissimum reum
> (Lat.) you have a full confession [5.5.2: No. 1 (f. 30v)], p. 301
> see also *et habes confitentem reum*

hope
> (OED *v.* 4, obs.; *c.*1330–1632) suspect, suppose [5.5.2: No. 1 (ff. 10v, 14v)], pp. 266, 273

humorists
> (OED 1, obs.) persons subject to 'humours' or fancies; faddists [5.5.2: No. 1 (f. 11v)], p. 268

hypothecarians
> (Northism) those who hypothesise [3.3.2: No. 5 (f. 290v)], p. 153

iczen = isen
> (OED obs. variant) iron [3.3.2: No. 9 (f. 384)], p. 168

idem cogitans
> (Lat.) thinking alike [5.5.2: No. 1 (f. 5)], p. 257

illuded
> (OED *v.* 1. *trans.*, obs.; 1516–1704) deride, ridicule [3.3.2: No. 9 (f. 383v)], p. 167

incautelousness
> (OED *adv.* obs.; 1610–1734, RN ex., *Examen*, II.iv. §110, 288) incautiousness, heedlessness [5.5.2: No. 1 (f. 26v)], p. 294

inconvenience
> (OED *sb.* 3, obs.; *c.*1400–1695) misfortune [2.2.2: No. 1 (f. 177v)], p. 73

inconveniences
> (OED *sb.* 1, obs.; *c.*1400–1706) inconsistencies with reason or rule; absurdities [3.3.2: No. 9 (f. 280v)], p. 163

indifferent
> (OED *a.*¹ II.9, obs.) of two things: not indifferent; equal, same [5.5.2: No. 1 (f. 10)], p. 265

indifferent
> (OED *sb.* B.1, obs.; *c.*1570–1602) one who is impartial or
> disinterested [5.5.2: No. 2 (f. 35)], p. 305

indifferents
> (OED *sb.* B.3, *pl.*, rare) things indifferent, non-essentials [5.5.2:
> No. 1 (f.31v)], p. 303

in directum
> (Lat.) in a straight line [3.3.2: No. 9 (f. 287)], p. 164
> see also *directum*

in equilibr[i]o
> (Lat.) in equilibrium [3.3.2: No. 9 (f. 384v)], p. 169

in hypothesi
> (Lat.) by supposition [3.3.2: No. 1 (f. 87)], p. 139

in petto
> (It.) lit., in one's breast, i.e., in secret, not made known to others
> [5.5.2: No. 1 (f. 16)], p. 275

in pleno
> (Lat.) in a plenum [2.2.2: No. 1 (f. 177)], p. 72

in rerum natura
> (Lat.) in the nature of things [3.3.2: No. 2 (f. 293)], p. 140

instar omnium
> (Lat.) the equal of all [5.5.2: No. 1 (f. 4)], p. 253

intus et in cute
> (Lat.) within and on the surface [5.5.2: No. 1 (f. 29)], p. 299

in vacuo
> (Lat.) in a vacuum [2.2.2: No. 1 (f. 177); 2.2.2: No. 3 (f. 184)],
> pp. 72, 80

in vacuo infinito
> (Lat.) in an infinite vacuum [3.3.2: No. 1 (f. 86v)], p. 138
> see also *vacuum infinito*

ipso facto
> (Lat.) by the very fact or act itself; by the very nature of the case
> [5.5.2: No. 1 (f. 10)], p. 266

jingle
> (OED *sb.* 1b; 1827–65) clinking and ringing applied depreciatively
> to other sounds; here, to the word 'complex' [5.5.2: No. 1 (f. 25)],
> p. 291

jugulum causæ

> (Lat.) the main point of an argument in favour of the case [3.3.2:
> No. 5 (f. 290v)], p. 153

just

> (OED *a.* 10, obs.; 1551–1725) equal [5.5.2: No. 1 (f. 9)], p. 264

justly

> (OED *adv.* 4, obs.) properly rightly, correctly [5.5.2: No. 1 (f. 7v)],
> p. 261

kat economian see *kat' eoikonomian*
kat' economian see *kat' eoikonomian*
kat' eoconomian see *kat' eoikonomian*
kat' eoikonomian

> (Gk., Epictetus, Discourses as collected by Arrian, 3.14.7) as a
> matter of good management (of a household or family and, by
> extension, of government) [5.5.2: No. 1 (ff. 22, 23, 23v, 27v, 29;
> 5.5.2: No. 2 (f. 35v)], pp. 286–7, 296, 298–9, 306

kat oeconomian see *kat' eoikonomian*
kat òeconomian see *kat' eoikonomian*
kat óeconomian see *kat' eoikonomian*
kat oiconomian see *kat' eoikonomian*

Labor actus in orbem?

> (Lat.) lit., The ongoing cycle of work, i.e., a logical circle?
> [2.2.2: No. 1 (f. 177v)], p. 75

last

> (OED *adj.* A.II.9g, obs.; 1633) utmost [5.5.2: No. 1 (f. 23v)],
> p. 288

least

> (OED obs. = lest 2.) lest [5.5.2: No. 1 (f. 25)], p. 291

levigate

> (OED *v.* 1, *intr.*, obs.) make smooth, polish [5.5.2: No. 1 (f. 13)],
> p. 267

merum nihil

> (Lat.) mere nothing [3.3.2: No. 1 (ff. 84v, 85)], pp. 135–6

minisimuses

> (Northism; from minimus, Lat., smallest, extremely minute)
> smaller than the smallest, more minute than the extremely minute
> [3.3.2: No. 2 (f. 293v)], p. 142

minuit = minuity

    (OED obs. rare; 1612) a trifle [5.5.2: No. 1 (f. 8)], p. 262 and n.65

misi pari = *mise pari*

    (Fr.) sum (of money) staked [5.5.2: No. 1 (f. 1)], p. 249

misprizing

    (OED *vbl sb.* = misprision[1] 2, arch.) mistaking one word for
    another; here, natural history for occult philosophy [3.3.2: No. 9
    (f. 384v)], p. 170

motus relativus

    (Lat.) relative motion [3.3.2: No. 1 (f. 86); 3.3.2: No. 5 (f. 290)],
    pp. 137, 152

motus verus

    (Lat.) true motion [3.3.2: No. 1 (f. 86); 3.3.2: No. 5 (f. 290)],
    pp. 137–8, 152

motus verus & relativus

    (Lat.) true and relative motion [3.3.2: No. 1 (f. 85v); 3.3.2: No. 9
    (f. 382v)], pp. 137, 165

movant see movent

movent

    (OED *adj.*, ex. RN, *Life/FN* 1742, 292) that moves or is moved;
    moving [2.2.2: No. 1 (f. 177); 2.2.2: No. 3 (f. 184)], pp. 72, 80

Negatur

    (Lat.) It is denied [2.2.2: No. 1 (f. 176)], p. 70

nomine ponunt re tollunt

    (Lat.) grant them in name but deny them in reality [5.5.2: No. 1
    (f. 4)], p. 265

non causa, pro causa

    (Lat.) a type of fallacious argument: non-cause, for cause [2.2.2:
    No. 1 (f. 176v)], p. 71

nostrums

    (OED 1b, rare) recipes [1.1.2 (f. 5v)], p. 32

Nullius in verba

    (Lat.) lit., On the word of no one [2.2.2: No. 3 (f. 187)], p. 84

obnoxious

    (OED *a.* 1b, obs.; 1628–71) with *to*: exposed to the (physical)
    action or influence of; liable to be affected by, open to [3.3.2: No. 7
    (f. 280); 3.3.2: No. 9 (f. 383); 5.5.2: No. 1 (f. 14v)], pp. 157, 166,
    273

obscurum per obscurius
> (Lat.) the obscure through the more obscure [5.5.2: No. 2 (f. 35)], p. 306

occurs = occurse
> (OED obs.; 1621–92) meeting [2.2.2: No. 1 (ff. 174v, 175)], pp. 67, 69

omoiusios = *omoiousios*
> (Gk.) the like substance, i.e., the express image [5.5.2: No. 1 (f. 26)], p. 293

ømousios see *omousios*

omousios
> (Gk.) the same substance, i.e., consubstantial [5.5.2: No. 1 (ff. 7, 8, 25v, 26)], pp. 261, 263, 292–3

omousious see *omousios*

onely = only
> (OED *adv.* A.3, obs.; *c.*1000–1611) singularly, uniquely, specially, pre-eminently [5.5.2: No. 1 (f. 25)], p. 291

opinionative
> (OED *a.*, now rare = opinative 1, obs.) adhering to one's own opinion; here, opinionated [5.5.2: No. 1 (f. 11)], p. 267

oraisons
> (OED obs. form of orisons) prayers [5.5.2: No. 1 (f. 23v)], p. 288

orthodoxicity
> (Northism) belief in, or agreement with, what is or is currently held to be right, *esp.* in religious matters [5.5.2: No. 1 (f. 29)], p. 298

parties
> (OED *sb.* III.11b, obs.; *c.*1500–72) opponents, antagonists [2.2.2: No. 1 (f. 176v)], p. 72

paterfamilias
> (Lat.) head of the family or household [5.5.2: No. 1 (f. 5)], p. 256

paum
> (OED obs. form of palm *v.* 4. *trans.*) pass off by trickery [5.5.2: No. 1 (f. 16v)], p. 276

paumed
> (OED obs. form of palm *v.* 3. *intr.*, obs.; 1686–1724) imposed [5.5.2: No. 1 (f. 6v)], p. 260

payson
> (Fr.) lit. peasant; here, rustic [5.5.2: No. 2 (f. 36)], p. 308

penetralia

(Lat. pl.) inner chambers, interior [3.3.2: No. 9 (f. 287)], p. 163

perroratio = peroratio

(Gk.) the concluding part of an oration, speech or written discourse [5.5.2: No. 1 (f. 18v)], p. 279

petitio see *petitio principi*

petitio principi

(Lat.) begging the question; to presuppose an unproven proposition as the basis of proof [3.3.2: No. 1 (f. 84); 5.5.2: No. 1 (f. 14v)], pp. 134, 273

piqueish

(Northism; from pique, OED *sb.*[1] B.2) offensive [5.5.2: No. 1 (f. 30)], p. 301

politick = politic

(OED *sb.* B.2, obs.; 1588–1715) policy [1.1.2: No. 3 (f. 187)], p. 84

politicks = politics

(OED 3c, obs.) political practices [1.1.2 (f. 4v)], p. 30

primier ministre

(Fr.) prime minister [5.5.2: No. 1 (f. 4v)], p. 254

propriety

(OED 1, obs.) property [1.1.2 (f. 5v)], p. 32

proved

(OED *v.* B.I.1, *trans.*, arch.) put to the test [1.1.2 (f. 3v)], p. 29

pure and pute quantum

(from *purus et putas*, Lat.) purely and solely quantity [5.5.2: No. 1 (f. 14v)], p. 273

Q.E.D., *Quod E.D.* see *Quod erat demonstrandum*

quære

(OED *v.*, obs. rare) query [3.3.2: No. 7 (f. 287v)], p. 158 n.182

quatuor criteria

(Lat.) four tests [5.5.2: No. 1 (f. 24)], p. 289

quere dress

(Northism; from obs. form of quære + dress, OED, *sb.* 2: external adornment, garb) style of a query [3.3.2: No. 5 (f. 290v)], p. 153

Quicunque vult = Quicumque vult

(Lat.) Whoever will be saved [5.5.2: No. 1 (ff. 13v, 26)], pp. 271, 293

Quid volet argumentum?

(Lat.) What will the argument decide? Where will the argument lead? [5.5.2: No. 1 (f. 29)], p. 299

Quod erat demonstrandum
>   (Lat.) Which was to be demonstrated [2.2.2: No. 1 (f. 176v); 5.5.2:
>   No. 1 (f. 15)], pp. 71, 273

Quod primi omnium [estis qui] veram physiologiam compistis = Quod
>   *primi omnium [estis qui] veram physiologiam corrupistis*
>   (Lat.) Because you (people) are the first of all to have corrupted
>   true natural science [5.5.2: No. 1 (f. 30v)], p. 302

raged
>   (OED *v. intr.* 6b, obs.; *c.*1540–1603) exercised their rage [5.5.2:
>   No. 1 (f. 6)], p. 259

rallery
>   (OED obs. variant of raillery) banter [5.5.2: No. 1 (f. 25v)], p. 293

rased
>   (OED *v.*¹ 4b, obs.; 1429–1703) to alter (a writing) by erasure
>   [5.5.2: No. 1 (f. 6)], p. 259

rasure
>   (OED now rare) erasure, scraping out something written [5.5.2:
>   No. 1 (f. 6)], p. 259

Rectissime
>   (Lat.) Most correct [2.2.2: No. 1 (f. 174)], p. 65

relativus see *motus relativus*

'rere
>   (OED obs. form of rear *sb.*³ 4a) in the rear: in the hindmost part;
>   hence, at or from the back, behind [2.2.2: No. 3 (f. 187)], p. 84
>   see also *arriere*

resolve
>   (OED *sb.* 4, obs.; *c.*1625–70) solution, answer [3.3.2: No. 1
>   (f. 87v)], p. 139

reverence
>   (OED *sb.* 1, now rare or obs.) deep or due respect felt or shown
>   towards a person on account of his position or relationship;
>   deference [2.2.2: No. 3 (f. 186)], p. 83

salvos
>   (OED *sb.*¹ 3, obs.) solutions or explanations (of a difficulty) [5.5.2:
>   No. 1 (f. 9v)], p. 265

scribblaster
>   (Northism; from scribble, OED *v.*[1] 1. *trans.*: to write carelessly +
>   blaster 1: one who emits blasts) one who writes in a scribbling
>   manner, i.e., hurriedly so that what is written is faulty in style or
>   worthless in substance [5.5.2: No. 1 (f. 31)], p. 302

secundum captum humanum
>   (Lat.) according to human mental capacity [5.5.2: No. 1 (f. 23)],
>   p. 287
>   see also *ad captum*

secundum majus & minus
>   (Lat.) according to more and less [2.2.2: No. 1 (f. 177v)], p. 74

side-blow
>   (Northism) fig., blow from the opposing side [5.5.2: No. 1
>   (f. 31v)], p. 303

side-boxing
>   (Northism as *ppl. a.*) occupying a box or enclosed seat at the
>   theatre [1.1.2 (f. 4v)], p. 30

Soliditas quid?
>   (Lat.) Solidity what? [3.3.2: No. 1 (f. 82v)], p. 131

sound
>   (OED *sb.* 4d, obs.; 1614–1719) import, sense, significance [5.5.2:
>   No. 1 (f. 28v)], p. 298

sounding
>   (Northism as *ppl.a.*; from OED *v.*[1] 11. *trans.*, obs.; 1391–1671) of
>   words: signifying or meaning; implying [5.5.2: No. 1 (f. 27)],
>   p. 296

starr-kite = star-kite
>   (Northism; comb. not in OED) a species of fireworks that
>   sparkles, i.e., a rocket [1.1.2 (f. 2)], p. 25

sub specie quanti
>   (Lat.) under the viewpoint of quantity [5.5.2: No. 1 (f. 19v)], p. 281

supposititious
>   (OED *a.* 2, obs.) feigned; imaginary [2.2.2: No. 1 (f. 177v)], p. 75

suppositum
>   (Lat.) substitution [3.3.2: No. 1 (f. 84)], p. 134

tempus absolutum
>   (Lat.) absolute time [2.2.2: No. 1 (f. 174); 3.3.2: No. 9 (f. 382)],
>   pp. 66, 164

tempus absolutum & relativum

(Lat.) absolute and relative time [3.3.2: No. 9 (f. 382v)], p. 165

theisme = theism

(OED *sb.* c) belief in the existence of God with denial of revelation, i.e., the doctrine of deism [5.5.2: No. 1 (f. 14v)], p. 273

theisticall = theistical

(OED *a.*) deistical [5.5.2: No. 1 (f. 7)], p. 260

theists

(OED; ex. RN *Examen* iii. viii. § 11, 590) those who hold the doctrine of deism [5.5.2: No. 1 (ff. 13, 13v, 24, 28)], pp. 270–1, 289, 294, 297

thro-stitch't = thorough-stitched

(OED *a.* C, ?obs.) thoroughgoing [5.5.2: No. 1 (f. 12v)], p. 269

title

(OED *sb.* 7c, obs.; 1534–1701) assertion of a right; claim [5.5.2: No. 1 (f. 24)], p. 289

toto cælo

(Lat.) lit., by the whole expanse of the heavens, i.e., totally [5.5.2: No. 1 (f. 3v)], p. 253

tradendo

(Lat.) transmitting, handing down [5.5.2: No. 1 (f. 14)], p. 272

traditores = traditors

(OED 3, obs. rare; 1638) those who hand down a tradition [5.5.2: No. 1 (f. 14)], p. 272

transit in rem judicatam

(Lat.) a decided matter taken as a precedent for other cases [5.5.2: No. 1 (f. 6v)], p. 260

tromperie

(Fr.) delusion; imposture [2.2.2: No. 1 (f. 177)], p. 72

una litera potest

(Lat.) even a single letter has great force, i.e., can change the whole meaning of a passage [5.5.2: No. 1 (f. 7)], p. 260

una pars audita

(Lat.) one party heard [5.5.2: No. 1 (f. 28v), p. 298

vacuum infinitum

(Lat.) infinite vacuum [3.3.2: No. 1 (f. 84v)], p. 135

see also *in vacuo infinito*

vera quies

(Lat.) true rest [3.3.2: No. 5 (f. 290)], p. 152

verbo venia
> (Lat.) excuse the word [5.5.2: No. 1 (ff. 5v, 29)], pp. 257, 299

verus see *motus verus*

vires
> (Lat., pl. of *vis*) forces, powers [3.3.2: No. 5 (f. 290v); 3.3.2: No. 7
> (f. 280); 3.3.2: No. 9 (ff. 280v, 287, 383], pp. 153, 157, 163–4, 166

vires appetendi [et] fugiendi
> (Lat.) the powers of desiring and shunning [3.3.2: No. 1 (f. 83)],
> p. 132

vires sit liber index
> (Lat.) let powers be an independent proof [3.3.2: No. 7 (f. 280)],
> p. 157

visage
> (OED *sb.* 8, obs.; 1390–1684) fig., assumed appearance; pretence
> [5.5.2: No. 1 (f. 174)], p. 274

vis impressa
> (Lat.) impressed force [3.3.2: No. 9 (ff. 287, 383v)], pp. 164, 168

vis inertia
> (Lat.) inert force [2.2.2: No. 1 (f. 174)], p. 65

wheadle = wheedle
> (OED *sb.* now rare) an act of insinuating flattery [5.5.2: No. 1
> (f. 7v)], p. 261

wheadling = wheedling
> (OED *v.* 1. *trans.*) to entice or persuade by flattering words; here
> with relation to coaxing into conformity the deists [5.5.2: No. 1
> (f. 14)], p. 272

wonder
> (OED *sb.* II, an instance of, obs.) astonishment mingled with
> bewildered curiosity [1.1.2 (f. 3)], p. 28

wonder
> (OED *v.* 4, obs.) to affect or strike with wonder; to cause to
> marvel [5.5.2: No. 1 (f. 20)], p. 282

wonderfull = wonderful
> (OED *a.* 2, obs. rare; *c.*1380–1583) filled with wonder or
> admiration [1.1.2 (f. 3v); 2.2.2: No. 2 (f. 180); 3.3.2: No. 2
> (f. 293v); 5.5.2: No. 1 (ff. 20, 20v)], pp. 29, 78, 142, 282–3

REFERENCES

1. MANUSCRIPT RELATING TO NORTH

RN Books (1) = (UK:Nro) DN/MSC2/29. Dated 1714, the letters 'DN' stand for the Diocese of Norwich. The manuscript itself is bound as a small, oblong quarto notebook and consists of a short-title catalogue of approximately 1,150 books, the entries in which represent, with a few exceptions, the books that North transferred from his personal library to the parochial library that he established in St. Mary's Church, Rougham, where about 1709 or a little later he had a special room built adjacent to the north isle of the church for use of future incumbents, as well as for his own successors at Rougham Hall.[1] Most of the entries in the catalogue are in the hands of others, although North has occasionally made additions or corrections.

In addition to the books from his own library, there were a number of donations and bequests, including from North's niece, Dudleya North, as well as from the non-juring bishop, George Hickes, and the Norfolk lay non-juror, Sir Christopher Calthorpe.[2] Upon the death in 1693 of Archbishop Sancroft, the nephews of the prelate were 'ordered' to present North with a memorial ring, although the same nephews indicated that he could take the equivalent in money if he thought that 'more conducing' for the memorial. He therefore accepted the money, since 'by that time I had almost finisht my library at Rougham. And thought a memoriall of

---

[1] See Korsten, *Roger North*, pp. 22–3 and 267 n.231.

[2] Calthorpe, whose manor was in East Barsham, was a cousin of Sir Nicholas Le Strange; see Cherry, 'Sir Nicholas L'Estrange', pp. 314–5, *et passim*, and Yould, 'Two Nonjurors', pp. 374–81, especially p. 378 for the non-juror interpretation of the oaths of fealty to the monarch. Regarding RN's not swearing the oaths, see Kassler, *HRN*, pp. 34–47.

him there would be more lasting of him then a ring'. So he used the money to buy 'a sett of law books, had 'em bound after his manner, and wrote [an inscription] in them'.[3] These are probably the five volumes of statutes (four of Charles II and one of William and Mary), which are listed in the catalogue under the heading 'Anglice Miscellaneorum'.

The Parochial Libraries Act (7 Anne 1708 cap. 14) had required that a catalogue be produced and deposited by 29 September 1709 in cases where parochial libraries were already in existence. But since the parochial library at St. Mary's was not completed until 1712, copies of the draft regulations dated 1713, as well as the final version dated 1714 are still preserved,[4] along with some notes related to the library.[5] Unfortunately, however, in 1771 the parish library itself was destroyed and its contents, dispersed.[6]

2. PUBLISHED WORKS

Anon

Articles Agreed upon by the Archbishops and Bishops of both Provinces, and the whole Clergy, In the Convocation holden at London in the Year 1562. For the avoiding of Diversities of Opinions, and for the establishing of Consent touching the True Religion. Reprinted by His Majesties Commandment, with his Royal Declaration prefixed thereunto. (London, 1702)

The Clerical Guide, or Ecclesiastical Directory.... (2d edn, London, 1822)

---

[3] RN, *Notes of Me*, p. 186 (where the inscription is reproduced in Latin).

[4] (UK:Ccc) Misc. Doc. 165, and (UK:Nro) DN/MSC1/40.

[5] (UK:Ob) MS North b.17, no. 94 and MS Eng. Misc. c.360, ff. 259–261.

[6] See North, *Recollections of a Happy Life*, vol. 1, p. 2, and Colvin and Newman (eds), *Of Building*, p. xv.

Aarsleff, Hans
'John Wilkins', *Dictionary of Scientific Biography* ed. by C. C. Gillispie
(16 vols, New York, 1970–80), vol. 14, pp. 361–81

Alexander, Arthur Francis O'Donel
The Planet Saturn: A History of Observation, Theory and Discovery
(London, 1962)

Alexander, H. G. (ed.)
The Leibniz-Clarke Correspondence [1717] (Manchester, 1965)

Anstey, Peter
The Philosophy of Robert Boyle (London, 2000)

Ariotti, Piero E.
'Aspects of the Conception and Development of the Pendulum in the
17th Century', *Archive for the History of Exact Sciences* (1971–2), 8:
329–410

[Armstrong, John M.]
History and Antiquities of the County of Norfolk (10 vols, London,
1781–4)

Armstrong, Robert L.
Metaphysics and British Empiricism (Lincoln, Nebraska, 1970)

Ashbee, Andrew
The Harmonious Musick of John Jenkins (vol. 1, [Surbiton], 1992)

Ayres, Lewis
*Nicaea and its Legacy: An Approach to Fourth Century Trinitarian
Theology* (Oxford, 2004)

Barlow, Thomas
The Genuine Remains of that Learned Prelate Dr. Thomas Barlow ed.
by Peter Pett (London, 1693)

Barrow, Isaac
The Usefulness of Mathematical Learning explained and demonstrated:
*being Mathematical Lectures read in the Publick Schools at the
University of Cambridge* ... tr. by John Kirkby (London, 1734)

Beaurline, I. A.
'Dudley North's Criticism of Metaphysical Poetry', *Huntington Library
Quarterly* (1962), 25: 299–313

Bell, A. E.
Christian Huygens and the Development of Science in the Seventeenth Century (London, 1950)

Bertoloni Meli, Domenico
Thinking with Objects: The Transformation of Mechanics in the Seventeenth Century (Baltimore, Maryland, 2006)

'Inherent and Centrifugal Forces in Newton', *Archives for the History of Exact Sciences* (2006), 60: 319–35

Boyer, Carl B.
The History of the Calculus and Its Conceptual Development (New York, 1959)

The Rainbow: From Myth to Mathematics (New York, 1959)

Boyle, Robert
The Works of the Honourable Robert Boyle ed. by T. Birch (6 vols, London, R/1965–6)

Brown, Joyce
'Guild Organisation and the Instrument-Making Trade, 1550–1830: The Grocers' and Clockmakers' Companies', *Annals of Science* (1979), 36: 1–34

Browne, Thomas
Pseudodoxia epidemica: or, Enquiries into very many received Tenents, and commonly presumed Truths (4th edn, London, 1658).

Buchwald, Jed Z.
'Descartes's Experimental Journey Past the Prism and Through the Invisible World to the Rainbow', *Annals of Science* (2008), 65: 1–46

Burtt, Edwin Arthur
The Metaphysical Foundations of Modern Science, (revised edn, Garden City, New York, 1954)

Cajori, Florian
Sir Isaac Newton's Mathematical Principles of Natural Philosophy.... [1729] tr. revised ... with an historical and explanatory Appendix (Berkeley, California, 1946)

Cantor, Geoffrey
'Berkeley's *The Analyst* Revisited', *Isis* (1984), 75: 668–83

Capaldi, Nicholas
Hume's Place in Moral Philosophy (New York, 1992)

Cassini, Paolo
'Newton: The Classical Scholia', *History of Science* (1984), 22: 1–59

Chalmers, Gordon K.
'The Lodestone and the Understanding of Matter in Seventeenth Century England', *Philosophy of Science* (1937), 4: 75–95

Chan, Mary
'Dating the Paper in Roger North's Manuscripts', *Roger North: Materials for a Chronology of his Writings* [Checklist No. 1] by M. Chan and J. C. Kassler (Kensington, N.S.W, 1989)

Chancellor, E. Beresford
Memorials of St James's Street together with The Annals of Almack's (New York, 1922)

Chenette, Louis F.
Music Theory in the British Isles during the Enlightenment (doctoral dissertation, Ann Arbor, Michigan, 1968)

Cherry, David
'Sir Nicholas L'Estrange, Non-Juror: His Politics, Fortune and Family', *Norfolk Archaeology* (1968), 34: 314–29

Clarke, Desmond M.
Occult Powers and Hypotheses: Cartesian Natural Philosophy under Louis XIV (Oxford, 1989)

Clarke, Samuel
A Demonstration [1705] = *A Demonstration of the Being and Attributes of God....* (London, 1705)

A Demonstration [1738] = *A Demonstration of the Being and Attributes of God....* (8th edn, London, 1738)

A Discourse [1706] = *A Discourse Concerning the Unchangeable Obligations of Natural Religion and the Truth and Certainty of the Christian Revelation....* (London, 1706)

A Discourse [1708] = *A Discourse Concerning the Unchangeable Obligations of Natural Religion, and the Truth and Certainty of the Christian Revelation....* (2d edn, London, 1708)

SDT = *The Scripture Doctrine of the Trinity. In Three Parts. Wherein All the Texts in the New Testament relating to that Doctrine, and the principal Passages in the Liturgy of the Church of England, are collected, compared, and explained* (London, 1712)

SDT [1719] = *The Scripture Doctrine of the Trinity....* (2d edn, London, 1719)

SDT [1832] = 3d edn, reprinted in vol. 1 of *The Works of Samuel Clarke, D.D.* ed. by John Clarke (4 vols, London, 1738)

Cohen, I. Bernard
'"Quantum in Se Est": Newton's Concept of Inertia in Relation to Descartes and Lucretius', *Notes and Records of the Royal Society of London*, (1964), 19: 131–55

Introduction to Newton's 'Principia' (Cambridge, Massachusetts, 1978)

Cohen, I. Bernard (ed.)
Isaac Newton's Papers & Letters on Natural Philosophy and Related Documents.... (Cambridge, Massachusetts, 1958)

Cohen, Murray
Sensible Words: Linguistic Practice in England 1640–1785 (Baltimore, Maryland, 1977)

Colie, Rosalie L.
'Spinoza and the Early English Deists', *Journal of the History of Ideas* (1959), 20: 23–46

'Some Paradoxes in the Language of Things', *Reason and Imagination: Studies in the History of Ideas 1600–1800* ed. by J. A. Mazzeo (New York, 1962), pp. 93–128

Paradoxia Epidemica: The Renaissance Tradition of Paradox (Princeton, New Jersey, 1966)

Colligan, J. Hay
The Arian Movement in England (Manchester, 1913)

Collins, James
God in Modern Philosophy (Chicago, 1967)

Colvin, Howard and John Newman (eds)
Of Building: Roger North's Writings on Architecture (Oxford, 1981)

Cornwall, R. D.
Visible and Apostolic: The Constitution of the Church in High Church Anglican and Non-Juror Thought (Newark, New Jersey, 1993)

Cowper, Mary
Diary of Mary, Countess Cowper, Lady of the Bedchamber to the Princess of Wales, 1714–20 ed. by S. Cowper (London, 1864)

Crane, R. S.
'The Houyhnhnms, the Yahoos, and the History of Ideas', *Reason and the Imagination* ed. by J. A. Mazzeo (New York, 1962), pp. 243–53

Crombie, A. C., M. S. Mahoney and T. M. Brown
'René du Perron Descartes', *Dictionary of Scientific Biography* ed. by C. C. Gillespie (14 vols, New York, 1970–6), vol. 4, pp. 51–65

Dear, Peter
'What Is the History of Science the History *Of*?', *Isis* (2005), 96: 390–406

Derham, William
Physico-Theology: Or, A Demonstration of the Being and Attributes of God, from his Works of Creation. ... (3d edn, corrected, London, 1714)

Descartes, René
Principia philosophiæ = Principles of Philosophy tr. with explanatory notes by V. R. Miller and R. P. Miller (Dordrecht, 1983–4)

Discours de la méthode = Discourse on Method, Optics, Geometry, and Meteorology tr., with an Introduction, by P. J. Olscamp (Indianapolis, Indiana, 1965)

Dick, Steven J.
Plurality of Worlds: The Origins of the Extraterrestrial Life Debate from Democritus to Kant (Cambridge, 1984)

Dijksterhuis, E. J.
The Mechanization of the World Picture tr. by C. Dikshoorn (Oxford, 1969)

Dockrill, D. W.
'Spiritual Knowledge and the Problem of Enthusiasm in Seventeenth Century England', *Prudentia* (1985), supplementary number: 147–71

Edwards, John
Some Brief Critical Remarks on Dr. Clarke's Last Papers ... Shewing that the Doctor is as deficient in the Critic Art, as he is in Theology (London, 1714)

Eastwood, Brian S.
'Descartes on Refraction: Scientific versus Rhetorical Method', *Isis* (1984), 75: 481–502

Elders, Leo
The Philosophical Theology of St. Thomas Aquinas (Leiden, 1990)

Evelyn, John
The Diary ... ed. by E. S. de Beer (6 vols, Oxford, 1955)

Feingold, Mordechai
'Isaac Barrow: Divine, Scholar, Mathematician', *Before Newton: The Life and Times of Isaac Barrow* ed. by M. Feingold (Cambridge, 1990), pp. 1–104

'Mathematicians and Naturalists: Sir Isaac Newton and the Royal Society', *Isaac Newton's Natural Philosophy* ed. by J. A. Buchwald and I. B. Cohen (Cambridge, Massachusetts, 2001), pp. 77–102

Ferguson, James P.
ECH = An Eighteenth Century Heretic: Dr. Samuel Clarke (Kineton, Warwick, 1976)

The Philosophy of Dr. Samuel Clarke and its Critics (New York, 1974)

Fox, A.
John Mill and Richard Bentley: A Study of the Textual Criticism of the New Testament, 1675–1729 (Oxford, 1954)

Frye, Northrop
The Double Vision: Language and Meaning in Religion (Toronto, 1991)

Frye, Roland Mushat
'Reason and Grace: Christian Epistemology in Dante, Langland, and Milton', *Action and Conviction in Early Modern Europe ...* ed. by T. K. Rabb and J. E. Siegel (Princeton, New Jersey, 1969), pp. 404–22

Gabbey, Alan
'Henry More and the Limits of Mechanism', *Henry More (1614–1687): Tercentenary Studies* ed. by S. Hutton.... (Dordrecht, 1990), pp. 19–35

Galilei, Galileo
Dialogues concerning Two New Sciences tr. by H. Crew and A. de Salvio (New York, 1954)

Gascoigne, John
'Politics, Patronage and Newtonianism: The Cambridge Example', *The Historical Journal* (1984), 27: 1–24

'Isaac Barrow's Academic Milieu: Interregnum and Restoration Cambridge', *Before Newton: The Life and Times of Isaac Barrow* ed. by M. Feingold (Cambridge, 1990), pp. 250–90

'Samuel Clarke', *Oxford Dictionary of National Biography* ed. by H. C. G. Matthew and Brian Harrison (54 vols, Oxford, 2004), vol. 11, pp. 912–17

Gaukroger, Stephen
Descartes: An Intellectual Biography (Oxford, 1995)

Gayer, Arthur Edward
Memoirs of the Family of Gayer compiled from Authentic Sources (Westminster, 1870)

Grassby, Richard
The English Gentleman in Trade: The Life and Works of Sir Dudley North 1641–1691 (Oxford, 1994)

Grayling, A. C.
Truth, Meaning and Realism (London, 2007)

Guerlac, Henry
'Where the Statue Stood: Divergent Loyalties to Newton in the Eighteenth Century', *Aspects of the Eighteenth Century* ed. by E. R. Wasserman (Baltimore, Maryland, 1965), pp. 317–34

Guicciardini, Niccolò
Isaac Newton on Mathematical Certainty and Method (Cambridge, Massachusetts, 2011)

Hall, A. Rupert
'Mechanics and the Royal Society, 1668–70', *The British Journal for the History of Science* (1966), 3: 24–38

Hankins, Thomas L.
'Eighteenth-Century Attempts to Resolve the *Vis viva* Controversy', *Isis* (1965), 56: 281–97

Hanson, N. R.
'Hypotheses Fingo', *The Methodological Heritage of Newton* ed. by R. E. Butts and J. W. Davis (Oxford, 1970), pp. 14–33

Harman, P. M.
Metaphysics and Natural Philosophy: The Problem of Substance in Classical Physics (Hassock, Sussex, 1982)

Harrison, John
The Library of Isaac Newton (Cambridge, 1978)

Harrison, Peter
'Religion' and the Religions in the English Enlightenment (Cambridge, 1990)

Hervey, S. H. A.
A Biographical List of Boys educated at King Edward VI Free Grammar School.... (Bury St. Edmunds, 1908)

Hesse, Mary
Forces and Fields: A Study of Action at a Distance in the History of Physics (London, 1961)

Hill, Katherine
'Neither Ancient nor Modern: Wallis and Barrow on the Composition of Continua' [Part I on Mathematical Styles and the Composition of Continua], *Notes and Records of the Royal Society of London* (1996), 50: 165–78

Hoadly, Benjamin
'The Preface giving some Account of the Life, Writings, and Character of the Author', [reprinted in] *The Works of Samuel Clarke, D.D.* ed. by John Clarke (4 vols, London, 1738), vol. 1, pp. i–xiv

Holden, Thomas
The Architecture of Matter: Galileo to Kant (Oxford, 2004)

Holdsworth, William S.
A History of English Law (vols 1–13, London, 1903–38; vols 14–17, London, 1964–72)

Hooke, Robert
Micrographia or Some Physiological Descriptions of Minute Bodies Made by Magnifying Glasses with observations and Inquiries thereupon [1665] (New York, 1961)

Lectiones Cutlerianœ, or A Collection of Lectures ... made before the Royal Society on several Occasions at Gresham Colledge. To which are added divers Miscellaneous Discourses (London, 1679)

Hoskin, Michael A.
'"Mining all Within": Clarke's Notes to Rohault's *Traité de Physique*', *The Thomist* (1961), 24: 353–63

Hutchison, Keith
'What Happened to Occult Qualities in the Scientific Revolution?', *Isis* (1982), 73: 233–53

Hutton, Sarah
'The Neoplatonic Roots of Arianism: Ralph Cudworth and Theophilus Gale', *Socinianism and its Role in the Culture of XVIth to XVIIIth Centuries* ed. by Lech Szczucki *et al* (Warsaw, 1983), pp. 139–45

Huygens, Christiaan
Treatise on Light. In which are explained the Causes of that which occurs in Reflexion, & in Refraction. And particularly in the strange Refraction of Iceland Crystal [1690] tr. by S. P. Thompson (New York, 1912, R/1962)

Cosmotheoros = The Celestial Worlds Discover'd: or, Conjectures concerning the Inhabitants, Plants and Productions of the Worlds in the Planets (London, 1698)

Jones, Howard
The Epicurean Tradition (London, 1992)

Kassler, Jamie C.
The Science of Music in Britain, 1714–1830: A Catalogue of Writings, Inventions and Lectures (2 vols, New York, 1979)

Inner Music: Hobbes, Hooke and North on Internal Character (London, 1995)

'Roger North', *New Grove Dictionary of Music and Musicians* ed. by S. Sadie and J. Tyrrell (2d edn, 29 vols, London, 2000), vol. 18, pp. 53–6

Music, Science, Philosophy: Models in the Universe of Thought (Aldershot, 2001)

The Beginnings of the Modern Philosophy of Music in England.... (Aldershot, 2004)

HRN = The Honourable Roger North (1651–1734) on Life, Morality, Law and Tradition (Farnham, Surrey, 2009)

Kelley, Donald R.
The Human Measure: Social Thought in the Western Legal Tradition (Cambridge, Massachusetts, 1990)

Korsten, F. J. M.
Roger North (1651–1734) Virtuoso and Essayist: A Study of his Life and Ideas.... (Amsterdam, 1981)

'Roger North (1651–1734) and his Writings on Science', *LIAS* (1981), 8: 203–24

Koyré, Alexandre
'Les Queries de l'Optique', *Archives Internationales d'Histoire des Sciences* (1960), 13: 15–29

Lambert, John
An Illustrated Guide to St James's Church Piccadilly (London, 1958)

Leeuwen, H. G. van
The Problem of Certainty in English Thought 1630–1690 (2d edn, The Hague, 1970)

Le Grand, Homer E.
'Galileo's Matter Theory', *New Perspectives on Galileo....* ed. by R. E. Butts and J. C. Pitt (Dordrecht, 1978), pp. 197–208

Leyden, W. von
Seventeenth-Century Metaphysics: An Examination of Some Main Concepts and Theories (New York, 1968)

Locke, John
An Essay concerning Human Understanding [4/1700] ed. by P. H. Nidditch (Oxford, 1985)

The Reasonableness of Christianity, as delivered in the Scriptures ... [1695] (6th edn, London, 1748)

Lodwick Francis
On Language, Theology and Utopia ed. by F. Henderson and W. Poole (Oxford, 2011)

Lovejoy, Arthur O.
The Great Chain of Being: A Study in the History of an Idea (Cambridge, Massachusetts, 1964)

Mahoney, Michael S.
'Barrow's Mathematics: Between Ancients and Moderns', *Before Newton: The Life and Times of Isaac Barrow* ed. by Mordechai Feingold (Cambridge, 1990), pp. 179–249

Malebranche, Nicolas
The Search after Truth tr. by T. M. Lennon and P. J. Olscamp ... (Columbus, Ohio, 1980)

Malet, Antoni,
'Barrow, Wallis, and the Remaking of Seventeenth Century Indivisibles', *Centaurus* (1997), 39: 67–92

Mandelbrote, Scott
'Isaac Newton and Thomas Burnet: Biblical Criticism and the Crisis of Late Seventeenth-Century England', *Books of Nature and Scripture ...* ed. by J. E. Force and R. H. Popkin (Dordrecht, 1994), pp. 149–78

Manuel, Frank E.
The Religion of Isaac Newton (Oxford, 1974)

Margenau, Henry
The Nature of Physical Reality (New York, 1950)

Maslen, Keith and John Lancaster (eds)
The Bowyer Ledgers: The Printing Accounts of Bowyer Father and Son.... (London, 1991)

Mattern, Ruth
'Moral Science and the Concept of Persons in Locke' [1980], *Locke* ed. by V. Chappell (Oxford, 1998), pp. 261–78

McAdoo, Henry R.
The Spirit of Anglicanism: A Survey of Anglican Theological Method in the Seventeenth Century (New York, 1965)

McLeod, Ronald F.
Massingham Parva Past and Present (London, 1882)

Meadows, Peter
'John Moore', *Oxford Dictionary of National Biography* ed. by H. C. G. Matthews and Brian Harrison (54 vols, Oxford, 2004), vol. 38, pp. 967–9

Melinkoff, David
The Language of the Law (Boston, 1963)

More, Henry
Philosophical Writings ed. by F. I. Mackinnon (New York, 1969)

Morris, Charles W.
Six Theories of Mind (Chicago, 1932)

Mortimer, Sarah
Reason and Religion in the English Revolution: The Challenge of Socinianism (Cambridge, 2010)

Newton, Isaac
Opticks: or, A Treatise of the Reflexions, Refractions, Inflexions and Colours of Light.... (London, 1704)

Optice: sive de Reflexionibus, Refractionibus, Inflexionibus & Coloribus Lucis Libri Tres. Latine reddidit Samuel Clarke.... (London, 1706)

Principia mathematica = Philosophiæ naturalis principia mathematica (London, 1687)

Principia mathematica [1713] = *Philosophiæ naturalis principia mathematica* ... Editio Secunda Auctior et Emendatior (Cambridge, 1713)

North, Marianne

Recollections of a Happy Life being the Autobiography of Marianne North ed. by her sister Mrs. John Addington Symonds (2 vols, London, 1892)

North, Roger

Examen: or, An Enquiry into the Credit and Veracity of A Pretended Complete History; shewing The perverse and wicked Design of it, and the Many Falsities and Abuses of Truth contained in it ... [ed. by M. North] (London, 1740)

Life/DN = The Life of the Honourable Sir Dudley North, Kn^t ... [ed. by M. North] (London, 1744)

'Letters of the Honourable Roger North', *The Autobiography ...* ed. by A. Jessopp (London, 1887), pp. 221–81

Life/JN = General Preface & Life of Dr. John North ed. by P. Millard (Toronto, 1984)

'Cursory Notes of Building occasioned by the Repair, or rather Metamorfosis, of an old house in The Country, Reserved for private Reflection, if not Instruction, to such as succeed in it', *Of Building: Roger North's Writings on Architecture* ed. by H. Colvin and J. Newman (Oxford, 1981), pp. 1–103

Cursory Notes of Musicke (c.1698–c.1703) ... ed. by M. Chan and J. C. Kassler (Kensington, NSW, 1986)

The Musicall Grammarian 1728 ed. by M. Chan and J. C. Kassler (Cambridge, 1990)

Life/FN = The Life of the Lord Keeper North ed. by M. Chan (Lewiston, Wales, 1995)

Notes of Me: The Autobiography of Roger North ed. by P. Millard (Toronto, 2000)

'Some Notes upon An Essay of Musick, printed, 1677, by way of Comment and Amendment' [*c.*1693–98] ed. in J. C. Kassler, *The Beginnings of the Modern Philosophy of Music....* (Aldershot, 2004), pp. 179–200

'Of Etimology' [*c*.1706–15] ed. in J. C. Kassler, *The Honourable Roger North 1651–1734: On Life, Morality, Law and Tradition* (Farnham, 2009), pp. 193–354

General Preface see above *Life/JN*

North Papers
   NP 2 = Roger North's Writings on Music to c.1703: A Set of Analytical Indexes prepared by J. D. Hine *with Digests of the Manuscripts* by M. Chan and J. C. Kassler (Kensington, N.S.W, 1986)

   NP3 = Roger North's The Musicall Grammarian and Theory of Sounds: Digests of the Manuscripts by M. Chan and J. C. Kassler *with an Analytical Index of 1726 and 1728 Theory of Sounds* by J. D. Hine (Kensington, N.S.W, 1988)

   NP 5 = Roger North's Writings on Music, 1704–c.1709: Digests of the Manuscripts by M. Chan and J. C. Kassler *with Analytical Indexes* by J. D. Hine (Kensington, N.S.W, 1999)

   NP 6 = Roger North's 'Of Sounds' and Prendcourt Tracts c.1710–c.1716: Digests and Editions by M. Chan and J. C. Kassler *with an Analytical Index* by J. D. Hine (Kensington, N.S.W, 2000)

Oelsner, W.
   'Tradition as a Theological Problem', *'And Other Pastors of thy Flock': A German Tribute to the Bishop of Chichester* ed. by Franz Hildebrandt (Cambridge, 1942), pp. 147–64

Passmore, John
   Ralph Cudworth: An Interpretation (Bristol, 1990)

Phan, Peter C. (ed.)
   The Cambridge Companion to the Trinity (Cambridge, 2011)

Phillips, Hugh
   Mid-Georgian London: A Topographical and Social Survey of Central and Western London.... (London, 1964)

Popkin, Richard H.
   The History of Scepticism from Savonarola to Bayle (rev., expanded edn, Oxford, 2003)

Porphyry the Phoenician
   Isogoge tr. ... by E. W. Warren (Toronto, 1975)

.

Priestley, F. E. L.
'The Clarke-Leibniz Controversy', *The Methodological Heritage of Newton* ed. by R. E. Butts and J. W. Davis, (Oxford, 1970), pp. 34–56

Randall, Dale B. J.
Gentle Flame: The Life and Verse of Dudley, Fourth Lord North (1602–1677) (Durham, North Carolina, 1983)

Reedy, Gerard
The Bible and Reason: Anglicans and Scripture in Late Seventeenth-Century England (Philadelphia, 1985)

Rossi, Paolo
'Nobility of Man and Plurality of Worlds' tr. by A. Brickmann, *Science, Medicine and Society in the Renaissance* ed. by A. G. Debus (2 vols, London 1972), vol. 2, pp. 131–62

Russell, Bertrand
A History of Western Philosophy (New York, 1967)

Sabra, A. I.
Theories of Light from Descartes to Newton (Cambridge, 1981)

Schneewind, J. B.
The Invention of Autonomy: A History of Modern Moral Philosophy (Cambridge, 1998)

Shapiro, Alan E.
'Kinematic Optics: A Study of the Wave Theory of Light in the Seventeenth Century', *Archive for History of Exact Sciences* (1973), 11: 143–72

'Light, Pressure, and Rectilinear Propagation: Descartes' Celestial Optics and Newton's Hydrostatics', *Studies in the History and Philosophy of Science* (1974), 5: 239–96

'The Spectre of Newton's "Spectrum"', *From Ancient Omens to Statistical Mechanics* ed. by J. L. Berggren and B. R. Goldstein (Copenhagen, 1987), pp. 183–92

'Huygens' "Traité de la Lumière" and Newton's "Opticks": Pursuing and Eschewing Hypotheses', *Notes and Records of the Royal Society of London* (1989), 43: 223–47

'Beyond the Dating Game: Watermark Clusters and the Composition of Newton's *Opticks*', *An Investigation of Difficult Things: Essays on Newton and the History of the Exact Sciences* ed. by P. M. Harman and A. E. Shapiro (Cambridge, 1992), pp. 181–227

Fits, Passions, and Paroxysms: Physics, Method and Chemistry and Newton's Theories of Colored Bodies and Fits of Easy Reflections (Cambridge, 1993)

'The Gradual Acceptance of Newton's Theory of Light and Color, 1672–1727', *Perspectives on Science* (1996), 4: 59–140

'Newton's "Experimental Philosophy"', *Early Science and Medicine* (2004), 9: 185–217

Shapiro, Alan E. (ed.)
The Optical Papers of Isaac Newton, Vol. 1: *The Optical Lectures, 1670–72* (Cambridge, 1984)

Shapiro, Barbara
Probability and Certainty in Seventeenth-Century England: A Study of the Relationships between Natural Science, Religion, History, Law, and Literature (Princeton, New Jersey, 1983)

Stead, Christopher
Doctrine and Philosophy in Early Christianity: Arius, Athanasius, Augustine (Aldershot, 2000)

Stephen, Leslie
History of English Thought in the Eighteenth Century (2d edn, 2 vols, London, 1881)

Stewart, Larry
'Samuel Clarke, Newtonianism, and the Factions of Post-Revolutionary England', *Journal of the History of Ideas* (1980), 42: 53–72

Stillingfleet, Edward
Origines sacrae: or A Rational Account of the Grounds of Natural and Reveal'd Religion. To which is now added Part of another Book upon the same Subject written A.D. MDCXCVII. Publish'd from the Author's own Manuscript (7th edn, Cambridge, 1702)

Stormon, E. J.
'The Procession of the Spirit: Towards a Transcendence of the Filioque Controversy', *Prudentia* (1985), supplementary number: 81–113

Strong, Edward W.
Procedures and Metaphysics: A Study in the Philosophy of Mathematical-Physical Science in the Sixteenth and Seventeenth Centuries (Berkeley, California, 1936)

Taylor, Stephen
'Benjamin Hoadly', *Oxford Dictionary of National Biography* ed. by H. C. G. Matthew and Brian Harrison (54 vols, Oxford 2004), vol. 27, pp. 340–8

Thorndike, Lynn
The History of Magic and Experimental Science (8 vols, New York, 1923–58)

Voss, Stephen
'Scientific and Practical Certainty in Descartes', *American Catholic Philosophical Quarterly* (1993), 67: 569–84

Westfall, Richard
Never at Rest: A Biography of Isaac Newton (Cambridge, 2011)

Whiston, William
Historical Memoirs of the Life of Dr. Samuel Clarke. Being a Supplement to Dr. Sykes's and Bishop Hoadley's Accounts. Including certain Memoirs of Several of Dr. Clarke's Friends (London, 1730)

White, Michael J.
The Continuous and the Discrete: Ancient Physical Theories from a Contemporary Perspective (Oxford, 1992)

Wiles, Maurice
Archetypal Heresy: Arianism through the Centuries (Oxford, 2004)

Williamson, Tom
'Roger North at Rougham: A Lost House and its Landscape', *Counties and Communities: Essays in East Anglian History* ed. by C. Rawcliffe, R. Virgo and R. Wilson (Norwich, 1966), pp. 275–90

Wilson, Charles
England's Apprenticeship 1603–1763 (London, 1965)

Yolton, John W.
  Thinking Matter: Materialism in Eighteenth-Century Britain (Oxford, 1984)

Yould, G. M.
  'Two Nonjurors', *Norfolk Archaeology* (1970–73), 35: 364–81

Young, B. W.
  Religion and Enlightenment in Eighteenth-Century England (Oxford, 1998)

INDEX OF BIBLICAL CITATIONS

The citations below begin with the Old and proceed to the New Testament. Under each of these are listed in alphabetical order the books in which a cited text appears. For each book the chapter and verse is given, followed by the full verse from the King James Bible.

OLD TESTAMENT

Psalm 148 v.6 (He hath also established them [the heavens and the waters above them] for ever and ever: he hath made a decree which shall not pass.), p. 107

Zephania iii.9 (For then will I turn to the people a pure language that they may all call upon the name of the Lord, to serve him with one consent.), p. 228

NEW TESTAMENT

Colossians ii.9 (For in him dwelleth all the fulness of the Godhead bodily.), p. 305

1 Corinthians iv.16 (Wherefore I beseech you, be ye followers of me.), p. 286

Hebrews xiii.8 (Jesus Christ the same yesterday, and to day, and for ever.), p. 228

John viii.58 (Jesus said unto them, Verily, verily, I say unto you, Before Abraham was, I am.), pp. 100, 192, 284–5

John xx.21 (Then said Jesus to them again, Peace *be* unto you: as *my* Father hath sent me, even so send I you.), p. 278

1 John v.7 (For there are three that bear record in heaven, the Father, the Word and the Holy Ghost: and these three are one.), pp. 251, 255, 258–60, 277, 290, 302, 304, 306–7

Luke iv.12 (And Jesus answering said unto him, It is said, Thou shalt not tempt the Lord thy God.) *see below* Matthew iv.7

Luke xxii.19 (And he took bread, and gave thanks, and brake it, and gave unto them, saying, This is my body which is given for you: this do in remembrance of me.), p. 278

Matthew iv.7 (Jesus said unto him, It is written again, Thou shalt not tempt the Lord thy God.), p. 277; *see also above* Luke iv.12

Matthew vi.9 (After this manner therefore pray ye: Our Father which art in heaven, Hallowed be thy name.), p. 276

Matthew vii.15 (Beware of false prophets, which come to you in sheep's clothing, but inwardly they are ravening wolves.), pp. 270, 294

Matthew xviii.3 (And [Jesus] said, Verily I say unto you, Except ye be converted, and become as little children, ye shall not enter into the kingdom of heaven.), p. 283

Matthew xxviii.19 (Go ye therefore, and teach all nations, baptizing them in the name of the Father, and of the Son, and of the Holy Ghost:), pp. 250–1

INDEX OF NAMES

Together with God and a few names from mythology, the following index lists persons (single and collective) mentioned in this book who were alive or—in the case of bodies corporate such as schools, universities and societies—were active before 1734. Numerals in italics refer to the editions in Chapters 1, 2, 3 and 5.